Cases for Mathematics Teacher Educators

A volume in
The Association of Mathematics Teacher Educators (AMTE) Professional Book Series
Christine Browning, *Series Editor*

Cases for Mathematics Teacher Educators

Facilitating Conversations about Inequities in Mathematics Classrooms

edited by

Dorothy Y. White
University of Georgia

Sandra Crespo
Michigan State University

Marta Civil
The University of Arizona

INFORMATION AGE PUBLISHING, INC.
Charlotte, NC • www.infoagepub.com

Library of Congress Cataloging-in-Publication Data

A CIP record for this book is available from the Library of Congress
http://www.loc.gov

ISBN: 978-1-68123-625-4 (Paperback)
       978-1-68123-626-1 (Hardcover)
       978-1-68123-627-8 (ebook)

Printed in the United States of America

This book is dedicated to the memory of Beatriz D'Ambrosio. Bia's commitment to critical education was an inspiration to all of us. We were honored by her generosity in sharing her educational experiences across the U.S. and Brazil for the piece she contributed for this book. Her work and her story will live on and inspire us to continue to make a difference in mathematics education.

CONTENTS

Foreword xvii
Fran Arbaugh and Christine D. Thomas

Foreword xxi
Christine A Browning

Preface..xxiii

1   Facilitating Conversations about Inequities in Mathematics
    Classrooms.. 1
    Dorothy Y. White, Sandra Crespo, and Marta Civil

PART I
Conversations About Inequities in Mathematics
Methods Courses 7
Imani Goffney and Sandra Crespo

2   Addressing Deficit Language in Math Methods: Providing
    Critical Feedback to Preservice Teachers 13
    Julia Maria Aguirre

    COMMENTARY 1   Providing Respectful and Ability-Oriented
    Feedback to Parents: A Commentary on Aguirre's Case 23
    Monica Gonzalez

COMMENTARY 2   Missing in the Numbers: Examination of
Teacher and Racial Identities: A Commentary on Aguirre's Case....... 27
Ebony O. McGee

COMMENTARY 3   Critical Dialogues to Promote Transformative
Learning: A Commentary on Aguirre's Case...................................... 33
Beatriz Quintos

3   Understanding White Privilege: When a Good Task Is Not
Enough 39
Kristen Bieda

COMMENTARY 1   Acknowledging Personal Perspectives to
Build Mathematical Understandings: A Commentary on Bieda's
Case.. 47
Cynthia Oropesa Anhalt

COMMENTARY 2   The Importance of Context and Nuance in
Designing Learning Experiences for Teachers: A Commentary on
Bieda's Case 53
Erica N. Walker

COMMENTARY 3   Supporting Novice Mathematics Teachers'
Racial Consciousness: A Commentary on Bieda's Case..................... 57
Craig Willey

4   Why Are You Asking For These Impossible Math Lessons? 63
Sandra Crespo

COMMENTARY 1   Identifying and Supporting the Next Small
Step Together: A Commentary on Crespo's Case 71
Megan Franke

COMMENTARY 2   Equitable Mathematics Teaching for *All*
Students: A Commentary on Crespo's Case 75
Christa Jackson

COMMENTARY 3   Turning Disappointing Student Emails into
Teachable Moments: A Commentary on Crespo's Case.................... 79
David W. Stinson

5   Problematizing Gender: Trepidation and Uncertainty.................. 85
Carlos Nicolas Gomez and Eric Siy

7/17-21

COMMENTARY 1 Using Media to Problematize Gender Stereotypes in the Mathematics Classroom: A Commentary on Gomez and Siy's Case 95
Katrina Piatek-Jimenez

COMMENTARY 2 Problematizing Gender: Learning to Embrace Uncertainty: A Commentary on Gomez and Siy's Case 101
Kai Rands

COMMENTARY 3 Gender ≠ Sex ≠ Sexual Orientation: A Commentary on Gomez and Siy's Case 107
Marcy B. Wood

6 Challenging and Disrupting Deficit Notions in Our Work with ECE and Elementary Teachers 113
Courtney Koestler

COMMENTARY 1 Weakening Deficit Perspectives with Collective Agency: A Commentary on Koestler's Case 121
Higinio Dominguez

COMMENTARY 2 Building Partnerships to Challenge and Disrupt Deficit Views of Students and Communities: A Commentary on Koestler's Case 125
Elham Kazemi

COMMENTARY 3 Creating Invitations to Disrupt Deficit Discourses: A Commentary on Koestler's Case............................... 131
Amy Noelle Parks

7 Case X: Opportunities for America's Youth 135
Kimberly Melgar and Dan Battey

COMMENTARY 1 The Delicate Balance of a Three-Legged Stool: A Commentary on Melgar and Battey's Case 143
Erika C. Bullock

COMMENTARY 2 Validating and Contextualizing Preservice Teachers' Resistance to Social Justice Pedagogy in Mathematics: A Commentary on Melgar and Battey's Case.................................. 149
Niral Shah

COMMENTARY 3 Conceptions of Equity and Their Impact on Students' Opportunities to Learn Mathematics: A Commentary on Melgar and Battey's Case 155
Marilyn Strutchens

8 Hearing Mathematical Competence Expressed in Emergent
Language 161
Judit Moschkovich

COMMENTARY 1 Teaching Preservice Teachers to Successfully
Position English Learners: A Commentary on Moschkovich's Case ...171
Kathryn B. Chval and Rachel J. Pinnow

COMMENTARY 2 Preparing Our New Teachers (and Ourselves)
to "Hear Mathematical Competence": A Commentary on
Moschkovich's Case 177
Crystal Kalinec-Craig

COMMENTARY 3 Positioning, Status, and Power: Framing
the Participation of EL Students in Mathematics Discussions for
Prospective Teachers: A Commentary on Moschkovich's Case 183
Maria del Rosario Zavala

9 Tracking in a Local Middle School: Do You See What I See? 189
Dorothy Y. White

COMMENTARY 1 Unpacking Expectations and Lenses in
Mathematics Classroom Observations: A Commentary on
White's Case 197
Lynette DeAun Guzman

COMMENTARY 2 Seeing Isn't Always Believing: Recognizing
Race Dysconciousness in the Preservice Teacher Context: A
Commentary on White's Case 203
Danny Bernard Martin

COMMENTARY 3 Identity, Context, and Conversations About
Racism: A Commentary on White's Case 209
Joy Oslund

PART II

Conversations About Inequities in Mathematics Content Courses . . 215
Mathew D. Felton-Koestler and Marta Civil

10 "This Is Nice But They Need to Learn to Do Things the U.S.
Way": Reactions to Different Algorithms.. 219
Marta Civil

COMMENTARY 1 When the "U.S. Way" Is Not the Standard! A
Commentary on Civil's Case 227
Beatriz D'Ambrosio

COMMENTARY 2 Noticing Student Thinking: A Commentary
on Civil's Case 233
Eileen Murray

COMMENTARY 3 Valorization of Knowledge as a Component
of Understanding and Building Upon Students' Thinking: A
Commentary on Civil's Case 237
Randolph A. Philipp

11 Using Mathematics to Investigate Social and Political Issues:
The Case of "Illegal Immigration" 243
Mathew D. Felton-Koestler

COMMENTARY 1 Tensions and Opportunities When
Implementing Social Justice Mathematics Tasks: A Commentary
on Felton-Koestler's Case 251
Kyndall Brown

COMMENTARY 2 The Need to Be Intentional in the
Integration of Social Justice in Mathematics Content Courses: A
Commentary on Felton-Koestler's Case 255
Sylvia Celedón-Pattichis

COMMENTARY 3 "Strategic Intrusion": A Commentary on
Felton-Koestler's Case 261
La Mont Terry

12 Searching for Cohesion in a Mathematics Course
for Social Analysis 267
Jean M. Mistele and Laura J. Jacobsen

COMMENTARY 1 Embracing Tensions: A Commentary
on Mistele and Jacobsen's Case 275
Jessica Pierson Bishop

COMMENTARY 2 Less is More: A Commentary on Mistele and
Jacobsen's Case 281
Anthony Fernandes

COMMENTARY 3 Responding to Students' Needs:
A Commentary on Mistele and Jacobsen's Case 285
William Zahner

13 Not Called to Action (or Called Upon to Act): Can Social
Justice Contexts Have a Lasting Impact on Preservice Teachers? 289
Ksenija Simic-Muller

COMMENTARY 1 Becoming Political in Mathematics Education
Class: A Commentary on Simic-Muller's Case 297
Eric (Rico) Gutstein

COMMENTARY 2 Teaching Mathematics for Social Justice
as Engaging in Joint Action with Students: A Commentary on
Simic-Muller's Case 301
Arthur B. Powell

COMMENTARY 3 Mathematics and Activism: A Commentary
on Simic-Muller's Case 307
Judith Quander

14 Who Counts as a Mathematician? 311
Sharon Strickland

COMMENTARY 1 Buttons and Mathematicians: A Commentary
on Strickland's Case 319
Zandra de Araujo

COMMENTARY 2 Broadening Perspectives Through Purposeful
Reflection: A Commentary on Strickland's Case 325
Jennifer A. Eli

COMMENTARY 3 Doing Mathematics and Being a
Mathematician, These May Be Different: A Commentary
on Strickland's Case 329
Tod Shockey

PART III

**Conversations About Inequities in Graduate and Professional-
Development Contexts** 333
Joi A. Spencer and Dorothy Y. White

15 Are These Two Sides of the Same Coin? Teachers' Commitment
to Culturally Relevant Teaching While Holding Deficit Views
of Poor Communities 339
Tonya Bartell, Lateefah Id-Deen, Frieda Parker, and Jodie Novak

COMMENTARY 1 Responding to Mathematics Teachers' Deficit Perspectives About Economically Disadvantaged Students and Their Families: A Commentary on Bartell et al.'s Case..................... 347
Richard Kitchen

COMMENTARY 2 Teaching Privilege About Equity: A Commentary on Bartell et al.'s Case.. 353
Brian R. Lawler

COMMENTARY 3 What Are We Doing When Understanding Culture Is Not Enough? A Commentary on Bartell et al.'s Case....... 359
Crystal H. Morton

16 How Do I Learn to Like This Child So I Can Teach Him Mathematics: The Case of Rebecca 365
Mary Q. Foote

COMMENTARY 1 Examining Interest Convergence and Identity: A Commentary on Foote's Case 373
Robert Q. Berry III

COMMENTARY 2 Supporting a Teacher's Shift from Deficits to Funds of Knowledge: A Commentary on Foote's Case................... 379
Maura Varley Gutiérrez

COMMENTARY 3 A Commentary on Foote's Case 383
Nora G. Ramírez

17 Challenging Deficit Language 389
Imani Masters Goffney

COMMENTARY 1 Adjusting Perspectives: A Commentary on Goffney's Case.. 395
Joel Amidon

COMMENTARY 2 Supporting Strength-Based Perspectives and Understandings: A Commentary on Goffney's Case 401
Amy Roth McDuffie

COMMENTARY 3 Challenging Mathematics Teachers' Deficit-Language Use: A Commentary on Goffney's Case.......................... 407
Eugenia Vomvoridi-Ivanovlc

18 Moving from Addressing One's Target Identity to Addressing One's Nontarget Identities... 413
Beth A. Herbel-Eisenmann

COMMENTARY 1 Anticipating the Unexpected: Managing a Dilemma During Facilitation of a Social Justice Mathematics Task: A Commentary on Herbel-Eisenmann's Case 421
Lawrence Clark

COMMENTARY 2 Challenging PSTs' Views and the Inherent Subjectivity While Doing So: A Commentary on Herbel-Eisenmann's Case 427
Laura McLeman

COMMENTARY 3 On Denial and the Search for Explanation: A Commentary on Herbel-Eisenmann's Case 431
José María Menéndez

19 Learning About Students and Communities Using Data and Maps.. 435
Laurie H. Rubel

COMMENTARY 1 The Frog in the Pan: Developing Critical Awareness in Mathematics Teachers: A Commentary on Rubel's Case... 445
Rodrigo Jorge Gutiérrez and Alice Cook

COMMENTARY 2 Can Mathematics Pave the Road to Social Justice? A Commentary on Rubel's Case.. 451
Robert Klein

COMMENTARY 3 Being Students and Teachers of Math and Social Justice: A Commentary on Rubel's Case............................... 457
Cynthia Nicol

20 "Let Me Be Your Cultural Resource": Facilitating Safe Spaces in Professional Development... 463
Anita A. Wager

COMMENTARY 1 Opening Spaces in Mathematics Teacher Education: A Commentary on Wager's Case................................... 469
Corey Drake

COMMENTARY 2 *Nosotras* Spaces: Cobuilding Transformational Bridges: A Commentary on Wager's Case 475
Carlos A. LópezLeiva

COMMENTARY 3   Seeing the Problem Before Attempting to Solve It: The Role of Noticing Sociopolitical Narratives in Equity-Focused Work: A Commentary on Wager's Case 479
Jennifer M. Langer-Osuna

About the Editors 485

FOREWORD

Fran Arbaugh
The Pennsylvania State University
AMTE President (2013–2015)

Christine D. Thomas
Georgia State University
AMTE President (2015–2017)

> *As we ask teachers to display cultural sensitivity and competence (Ford, 2005),*
> *we must ask mathematics teacher educators to enact research and teaching practices*
> *that address the needs of all learners.*
> —Strutchens et al., 2012, p. 1

With the statement above, the editorial board for the Special Equity Issue of the *Journal of Mathematics Teacher Education* (JMTE) challenged our field not only to conduct needed research about equitable practices in mathematics teacher education, but also to *enact teaching practices* in our work with preservice and inservice teachers—practices that support *all learners*. As the most recent Presidents of the Association of Mathematics Teacher Educators (AMTE), we are pleased to contribute the foreword for this important and timely resource for mathematics teacher educators.

AMTE has a strong, albeit relatively short, history of supporting a focus on equity in mathematics teacher education, beginning in 2007 when AMTE President Jennifer Bay-Williams established the first Equity Task Force. Co-chaired by Rochelle Gutiérrez and Edd Taylor, this task force was charged

Cases for Mathematics Teacher Educators, pages xvii–xix
Copyright © 2016 by Information Age Publishing
All rights of reproduction in any form reserved.

with designing "a strategic plan for intentional ways that the organization can advocate for equitable practices, including to support and increase the diversity of mathematics teachers and mathematics teacher educators" (AMTE Board of Directors, 2012). As a result of the recommendations of this task force, the Board of Directors amended the AMTE goals to include the promotion of "equitable practices in mathematics teacher education, including increasing the diversity of mathematics teachers and teacher educators" (http://amte.net/about/mission). This task force was also instrumental in setting the stage for the JMTE special issue on equity. The members of the editorial board for this special issue, entitled *Foregrounding Equity in Mathematics Teacher Education,* consisted of AMTE members with AMTE President Marilyn Strutchens (2011–2013) serving as editor.

Seeing that more work needed to be undertaken by AMTE with regard to equity and mathematics teacher education, a second Equity Task Force was appointed by President Marilyn Strutchens and was active for multiple years under the leadership of Beth Herbel-Eisenmann, Rochelle Gutiér-rez, Julia Aguirre, and Anita Wager. The work of this task force included planning and implementing a "learn and reflect" strand focused on equity at two AMTE annual conferences as well as advising the Board during the writing of the AMTE Position Statement on Equity in Mathematics Teacher Education (published on the AMTE website in November, 2015).

This brief overview of the history of AMTE's focus on equity and mathematics teacher education brings us to the publication of this book. When AMTE entered into an agreement with Information Age Publishing to establish the AMTE Professional Book Series, the AMTE Board of Directors brainstormed ideas for the focus of the first three books of the series. Very quickly, *equity and mathematics teacher education* rose to the top of our list, and we went to work to identify book editors who would guide the content of the book and serve as stewards of the AMTE mission. As a Board, we felt that the important work that AMTE members had undertaken in the area of equity needed to be supported and shared with the field, and we were very pleased when Dorothy Y. White, Sandra Crespo, and Marta Civil undertook the Board's charge to produce a book focused on equity and mathematics teacher education that would serve our membership as well as serve mathematics teacher educators globally.

As representatives of the AMTE Board of Directors, we are very proud to present *Cases for Mathematics Teacher Educators: Facilitating Conversations About Inequities in Mathematics Classrooms* as the inaugural book in the AMTE Professional Book Series. The publication of this book is the latest link in an ever-growing chain of AMTE projects that focus on *equity and mathematics teacher education.* Many of the chapters in this book were contributed by former members of the AMTE Equity Task Forces and by AMTE members who participated in the equity learn-and-reflect- strand sessions at our annual conferences. We are excited to present this book to the field, and strongly believe that it makes an important contribution by supporting AMTE members and mathematics teacher educators around the world to *enact equitable teaching practices.*

REFERENCES

Association of Mathematics Teacher Educators Board of Directors. (2012). *A history of the Association of Mathematics Teacher Educators: The beginnings through January 2012.* Retrieved from http://amte.net/sites/all/themes/amte/resources/AMTEHistory_Jan2012.pdf

Ford, D. Y. (2005). Welcoming all students to room 202: Creating culturally responsive classrooms. *Gifted Child Today, 28,* 28–30, 65.

Strutchens, M., Bay-Williams, J., Civil, M., Chval, K., Malloy, C. E., White, D. Y., D'Ambrosio, B., & Berry, R. Q. (2012). Foregrounding equity in mathematics teacher education. *Journal of Mathematics Teacher Education, 15,* 1–7.

FOREWORD

In the fall of 1992, the Association of Mathematics Teacher Educators' (AMTE) first newsletter was published, marking the beginning of the AMTE Publications "history." In that first issue, AMTE President Mark Spikell (1992–1993) indicated a primary goal of the association was to,

> Provide a national forum... to discuss issues of mutual professional concern [and to] share ideas on effective ways of promoting the NCTM Standards, NCSM and MAA recommendations on teaching school mathematics and developing programs to improve the mathematics education of practicing and future teachers. (AMTE, 1992)

A purpose of AMTE publications is to help facilitate that goal and since 1992, the organization has thoughtfully considered various types and venues in which to publish and share the collective and individual work of its membership.

Twenty-four years later, AMTE's collection of publications include, in addition to the *Connections* newsletter, the *Contemporary Issues in Technology and Teacher Education (CITE)*-Math online journal (in conjunction with the Society for Technology and Teacher Education), the *Mathematics Teacher Educator* online journal (in conjunction with the National Council of Teachers of Mathematics), and seven monographs. With this first book, *Cases for Mathematics Teacher Educators: Facilitating Conversations About Inequities in Mathematics Classrooms*, AMTE is launching its latest publication venture, a Professional Book series (in conjunction with Information Age Publishing), providing yet one more arena for its members to share findings and recommendations from their scholarly endeavors as they relate to "improving the mathematics education of practicing and future teachers."

Unlike AMTE's monographs with a call for papers on a chosen topic, book topics were initially chosen by the AMTE Board, targeting projects that were currently underway through a task force or working group. Thus, a book focusing on equity issues was a natural choice given the work of the Equity Task Force and other related committees focusing on equitable practices in mathematics teacher education. The editors, Dorothy Y. White, Sandra Crespo, and Marta Civil, working with many, many different authors crafted the substance of this current book focusing on various cases that share real experiences of mathematics teacher educators as they have attended to issues of equity arising "in the moment" in their practice. I see this book as being helpful to mathematics teacher educators as they learn from the cases and commentaries themselves but also a useful tool for them to use in their classrooms with preservice teachers or in their professional development of inservice mathematics teachers. The cases provide insight into the author's instructional goals and what can happen in classrooms while the commentaries provide thoughts on what might have or could have happened given different reactions or responses. I encourage you to read the cases that follow and grow in your understanding of what constitutes equitable teaching practices.

—**Christine A Browning**
AMTE Publications Director

REFERENCES

Association of Mathematics Teacher Educators. (1992). *AMTE Newsletter, 1*(1).

PREFACE

This book is intended for all educators who are committed to improving the mathematics learning of every student. This book is specifically designed for Mathematics Teacher Educators (MTEs) to facilitate conversations about inequities in mathematics classrooms across various instructional settings (i.e., content and methods courses for prospective teachers, as well as graduate courses and professional learning experiences). The idea for this book emerged at the PrOMPTE (Privilege and Oppression in the Mathematical Preparation of Teacher Educators) conference in October 2012. This conference brought together about 40 MTEs whose practice and research center on addressing inequities in mathematics education. In particular the conference focused "on racism and classism as simultaneous systems that oppress some people while granting privileges to others" (Herbel-Eisenmann et al., 2013, p. 10).

In the conversations that took place during the conference we noted how difficult it is to engage ourselves and engage others in work aimed at addressing issues of oppression and privilege, as they interact with the teaching and learning of mathematics. We realized that few MTEs are equipped to do this kind of work on their own. From our collective experience, we surmised that this work requires ongoing reflection and conversations with colleagues. We also noted that very few resources (if any) were available to support MTEs in initiating and maintaining these needed conversations. We came up with the idea of writing cases for mathematics teacher educators in order to extend these conversations beyond the PrOMPTE group. We presented a few sample cases at the Association of Mathematics

Cases for Mathematics Teacher Educators, pages xxiii–xxiv
Copyright © 2016 by Information Age Publishing
All rights of reproduction in any form reserved.

Teacher Educators (AMTE) annual meeting in 2014 and invited members to join the conversation by contributing to the book as either a case or commentary author. It was at that AMTE annual meeting that we began conversations with the AMTE Board to sponsor this book project.

This book includes 19 cases and 57 corresponding commentaries that provide a needed resource for MTEs to engage prospective teachers, practicing teachers, and future teacher educators in discussions about inequities, privilege, and oppression in society, in schools, and in the mathematics classroom. It is the product of the thinking and experiences of 87 individuals who are committed to the improvement of mathematics teacher education. We are very much indebted to all the authors who have contributed to this book, and we appreciate their patience and willingness to share their stories. We especially want to thank the series editor, Christine Browning, for her support and encouragement. Finally, this book would not have been possible without the insights and hard work of the section editors, Imani Masters Goffney, Mathew D. Felton-Koestler, and Joi A. Spencer.

—**Dorothy Y. White**
Sandra Crespo
Marta Civil

NOTE

1. The conference was funded by CREATE for STEM Institute through the Lappan-Phillips-Fitzgerald CMP 2 Innovation Grant program at Michigan State University.

REFERENCE

Herbel-Eisenmann, B., Bartell, T. G., Breyfogle, L., Bieda, K., Crespo, S., Domínguez, H., & Drake, C. (2013). Strong is the silence: Challenging interlocking systems of privilege and oppression in mathematics teacher education. *Journal of Urban Mathematics Education, 6*(1), 6–18.

CHAPTER 1

FACILITATING CONVERSATIONS ABOUT INEQUITIES IN MATHEMATICS CLASSROOMS

Dorothy Y. White
University of Georgia

Sandra Crespo
Michigan State University

Marta Civil
University of Arizona

Not everything that is faced can be changed,
but nothing can be changed until it is faced.

—James Baldwin

Teaching mathematics is a complex endeavor. Preparing and supporting mathematics teachers requires educators who recognize that mathematics classrooms are spaces where complex interactions happen between the teacher and students and among peers. Nasir, Hand, and Taylor (2008) have argued that "mathematics classrooms are inherently cultural spaces

Cases for Mathematics Teacher Educators, pages 1–5
Copyright © 2016 by Information Age Publishing
All rights of reproduction in any form reserved.

where different forms of knowing and being are being validated" (p 206). Whose knowledge and who is being validated are equity concerns. Teachers across all grade levels are constantly challenged to make their classrooms equitable learning environments for their students. As mathematics teacher educators (MTEs) who prepare the next generation of mathematics teachers and support current classroom teachers, we have a responsibility to acknowledge and challenge these inequities. This means that we must engage with this work in our teacher education contexts. We have to articulate how we prepare teachers to take on the challenge of believing in the competence and potential of every student. We also need to be able to support them in making their classrooms into equitable places for everyone to learn mathematics.

A growing number of MTEs are explicitly foregrounding equity in their research and teaching and are deliberately designing course experiences to prepare teachers to tackle issues of inequities within their future mathematics classrooms (e.g., Foote, 2010; Jacobsen, Mistele, & Sriraman, 2012; Rodriguez & Kitchen, 2005; Strutchens et al., 2012; Wager & Stinson, 2012). Mathematics education professional organizations have also made strides towards making equity a priority. For example, the National Council of Teachers of Mathematics (NCTM, 2014) has suggested that creating an equitable environment for all students in every school requires teachers who "understand and use the social contexts, cultural backgrounds, and identities of students as resources to foster access, motivate students to learn more mathematics, and engage student interest" (p. 115). More recently, the Association of Mathematics Teacher Educators (AMTE, 2015) has called for MTEs to "advocate for equitable practices for all mathematics teacher educators, P–12 mathematics teachers and students" (p. 2). This book provides responses to these calls through this collection of cases centered on equity as a vehicle to prepare MTEs to discuss inequities in their courses.

Cases are widely used in our field as pedagogical tools to engage educators in analyzing dilemmas of practice. Uniquely, our work focuses on the preparation of mathematics teacher educators. Although cases have served to engage prospective and practicing teachers in conversations about mathematics instruction and pedagogy, very few are explicitly about equity issues (e.g., Crockett, 2008) and particularly focused on the *teacher educator's* role in engaging with these issues. This book is specifically designed for MTEs to facilitate conversations about inequities in mathematics classrooms across various instructional settings (i.e., content and methods courses for prospective teachers, as well as graduate courses and professional learning experiences).

This collection of cases reflects broad theories related to educational reform. For example, we can see connections with Oakes's (1992) three dimensions of change in the educational context: "technical, normative, and political" (p. 12). *Political change* refers to changes in power structures and the distribution of resources, whereas *technical change* focuses attention on the details of curriculum and instructional practice (e.g., how to implement the Common Core standards). *Normative change* attends to the conceptions, dispositions, and ideologies (e.g., beliefs about student intelligence and capability) that undergird an educational context. Here, we foreground the work of MTEs as a normative undertaking. How might teacher educators work to invoke normative change amongst the next cadre of mathematics teachers? How could they challenge prospective and practicing teachers' ideas about who is intelligent and who is mathematically capable? Because normative inequities are so embedded in our practices, they are often invisible. As such, it is essential that MTEs help their students consider ways in which they participate in, recreate, and potentially interrupt these inequities. We realize that equity cannot simply be a hoped-for goal. Rather, its promotion must be embedded in the *norms* as well as *practices* of those who wish to bring it about. This book of cases provides an opportunity for MTEs to problem solve with fellow educators, and to learn strategies for engaging future and curent mathematics teachers in the courageous work of bringing about equity in their mathematics classrooms.

ORGANIZATION OF THE BOOK

The book includes twenty chapters. This first chapter introduces the history and rationale for the book. The remaining chapters are organized across three sections: (1) *Conversations About Inequities in Mathematics Methods Courses,* (2) *Conversations About Inequities in Mathematics Content Courses,* and (3) *Conversations about Inequities in Graduate and Professional Development Contexts.* Each section includes five to eight chapters with each chapter including a case and three commentaries. The section begins with an overview of the chapters and guiding questions to support conversations about the cases and commentaries.

For each case, the authors identify an equity-related dilemma from their practice. Examples of these dilemmas include addressing issues such as low expectations, white privilege, and deficit discourse. The authors also provide the context for their case and share their responses and reflections as MTEs. Each case is accompanied by three reflective commentaries in order to broaden the interpretive lens on the dilemma. We asked the commentary authors to share their positionality and how they interpreted and responded to the case author's dilemma to allow the reader to understand their responses and to highlight the different ways the dilemma can be perceived and handled. The commentaries provide the reader with alternative perspectives and interpretations of the case dilemma and offer solutions based on the commentator's experiences, context, and positionality.

The cases and commentaries are intended to provide MTEs with a resource to engage in conversations around inequities in mathematics classrooms. The authors are diverse and represent a variety of personal and educational backgrounds and work in a variety of contexts and educational settings. These authors shared their experiences, challenges, and wonderings. Their cases and commentaries reflect multiple perspectives, each recognizing and underscoring the complexities of engaging people in conversations around inequities in mathematics classrooms. We envision the book being used as course readings in university courses, professional development sessions, and book clubs among MTEs. We encourage MTEs to discuss the cases with colleagues, try the strategies in their own contexts, and reflect on their experiences.

AN INVITATION

The cases and commentaries provided in this book represent the start of conversations among MTEs and mathematics education researchers to engage prospective and inservice teachers about inequities in mathematics classrooms. We see this work as an invitation for other MTEs to join the conversation, because it is through the sharing of our practice that we all learn and grow. This kind of conversation is what Singleton and Hays (2008) have referred to as a *courageous conversation*. These authors define courageous conversation as a "strategy for breaking down racial tensions and raising racism as a topic of discussion that allows those who possess knowledge on particular topics to have the opportunity to share it, and those who do not have the knowledge to learn and grow from the experience" (p. 18). In this book we extend the notion of courageous conversation to all contexts of inequity that MTEs may have to confront. Following AMTE's (2015) position paper on equity in mathematics teacher education, "this means actively working toward a more just and equitable mathematics education free of systemic forms of inequality based on race, class, language, culture, gender, age, sexual orientation, religion, and dis/ability" (p. 1).

However, in order for MTEs to find the courage to engage their students in these types of conversations, they must be aware of potential issues that could arise and "have thought through these issues from multiple angles in order to steer the conversation in a positive direction" (Singleton & Hays, 2008, p. 19). In this book, the authors have courageously shared their dilemmas and responses when attempting to address inequities in their work. The case commentary authors have joined the case authors by sharing their professional and personal views, contexts, and suggestions on how they would have handled the dilemma in their specific contexts.

As teacher educators, we prepare the next generation of teachers and support the current teaching force. We must be active participants alongside our prospective and inservice teachers in the work of dismantling systems of inequities in our schools and mathematics classrooms. We believe the *Four Agreements of Courageous Conversation* offered by Singleton and Hays (2008) can support MTEs in these efforts: "stay engaged, expect to experience discomfort, speak your truth, and expect and accept a lack of closure" (p. 22). As the opening quote by James Baldwin reminds us, "nothing can be changed until it is faced." But we must make changes in informed and strategic ways, not on our own but with trusted colleagues. We hope that this book serves as a resource to open conversations within and across mathematics teacher education communities to work to identify and face the inequities that we want to change.

REFERENCES

Association of Mathematics Teacher Educators. (2015). *Position: Equity in mathematics teacher education.* Retrieved from http://amte.net/position/equityin mathematics

Crockett, M. D. (2008). *Mathematics and teaching: Reflective teaching and the social conditions of schooling.* New York, NY: Routledge.

Foote, M. Q. (Ed.). (2010). *Mathematics teaching & learning in K–12: Equity and professional development.* New York, NY: Palgrave Macmillan.

Jacobsen, L. J., Mistele, J., & Sriraman, B. (Eds.). (2012). *Mathematics teacher education in the public interest.* Charlotte, NC: Information Age.

National Council of Teachers of Mathematics. (2014). *Principles to actions: Ensuring mathematical success for all.* Reston, VA: Author.

Nasir, N. S., Hand, V., & Taylor, E. V. (2008). Culture and mathematics in school: Boundaries between "cultural" and "domain" knowledge in the mathematics classroom and beyond. *Review of Research in Education, 32,* 187–240.

Oakes, J. (1992). Can tracking research inform practice? Technical, normative, and political considerations. *Educational Researcher, 21*(4), 12–21.

Rodriguez, A. J., & Kitchen, R. S. (2005). *Preparing mathematics and science teachers for diverse classrooms: Promising strategies for transformative pedagogy.* Mahwah, NJ: Erlbaum.

Singleton, G., & Hays, C. (2008). Beginning courageous conversations about race. In M. Pollock (Ed.), *Everyday antiracism: Getting real about race in school* (pp. 18–23). New York, NY: The New Press.

Strutchens, M., Bay-Williams, J., Civil, M., Chval, K., Malloy, C. E., White, D. Y., D'Ambrosio, B., & Berry, R. Q. (2012). Foregrounding equity in mathematics teacher education. *Journal of Mathematics Teacher Education, 15,* 1–7.

Wager, A. A., & Stinson, D. W. (Eds.). (2012). *Teaching mathematics for social justice: Conversations with educators.* Reston, VA: National Council of Teachers of Mathematics.

PART I

CONVERSATIONS ABOUT INEQUITIES IN MATHEMATICS METHODS COURSES

Imani Goffney
University of Maryland–College Park

Sandra Crespo
Michigan State University

Conversations about educational equity and diversity begin early in many university-based teacher preparation programs but taper off as courses and experiences turn the focus onto subject-matter content knowledge for teaching. Methods courses, typically offered in the last year or so in most teacher preparation programs, tend to foreground learning about ambitious teaching and about students' thinking with few conversations about educational equity. Yet as Feiman-Nemser (2001) argued, high-quality programs of teacher preparation are characterized by the coherence and continuity of the experiences that are offered to teacher candidates. This suggests that the study of educational equity needs to be integrated and sustained in order to make a lasting impact on future teachers.

Cases for Mathematics Teacher Educators, pages 7–11
Copyright © 2016 by Information Age Publishing

7

Mathematics methods courses are a key component of teacher preparation programs. These courses are often designed to review seminal mathematics concepts and ideas and to teach the corresponding pedagogical practices that are central to the elementary, middle, and secondary school grade bands. However, they are also contentious courses as these are places where educational theory and practice collide and where the complexity of teaching mathematics becomes visible to prospective teachers. As such, mathematics methods courses can be a fruitful site for integrating attention to equity, social justice, and diversity within the preparation of mathematics teachers.

The work of a mathematics methods course is often conceptualized as providing experiences for prospective teachers to "unlearn" the mathematics they experienced in K–12 schools (Ball, 1988). Infusing a focus on educational equity adds yet another layer of complexity to an already challenging course to teach for mathematics teacher educators. Cochran-Smith (2001) wrote about the responsibilities of the teacher educator to "unlearn" alongside with their students and to interrogate the assumptions deeply embedded within the teacher education curriculum and pedagogies that may also serve to reproduce educational inequities. She asked us "to own our complicity in maintaining existing systems of privilege and oppression, and to grapple with our own failures to produce the kinds of changes we advocate" (p. 158).

This section includes case scenarios and analytical commentaries that will support mathematics teacher educators to facilitate discussions of privilege and oppression within their teacher preparation classrooms. We anticipate that readers of this section will learn from the scenarios and dilemmas faced by mathematics teacher educators as they work to address issues of social justice and equity within their mathematics methods courses. We hope that readers of this section identify strategies they could use in their local contexts and institutional settings as well as identify resources that will support how they engage in discussions around privilege, oppression, social justice, and equity.

The eight cases in this section invite us to grapple with serious challenges that arise when prospective teachers are asked to consider what it might mean to teach ambitious mathematics to all students, including those who they consider or have been labeled as struggling students of mathematics. The collection of cases in this section are situated in the mathematics methods course but their grade level and contexts are not identical. Each case includes information about features of the context and of the course goals that are relevant to the situation. As a collection, these cases foreground two challenges for mathematics teacher educators who make equity a priority in their mathematics methods courses: (1) the challenge of designing learning tasks for prospective teachers to learn to notice and analyze social inequities within and outside the mathematics classroom, and (2) the challenge of addressing or not addressing the deficit language and discourses about students of mathematics, especially those traditionally marginalized and underserved.

In her case, "Addressing Deficit Language in Math Methods: Providing Critical Feedback to Preservice Teachers," Julia Aguirre shares the dilemma of offering course projects that elicit prospective teachers' deficit discourse about students of color. The case foregrounds the issues that arise when prospective teachers are asked to consider the cultural wealth of students who are different from them and the importance of the critical feedback that is provided by the teacher educator to help them revise their ideas. In "Understanding White Privilege: When a Good Task Is Not Enough," Kristen Bieda shares a mathematics for social justice task she used with her prospective teachers hoping to model how the mathematics class could be a site for exploring relevant social issues related to White privilege. When the task did not elicit the kind of analysis about White privilege she had expected, she found herself telling prospective teachers what they should have been able to get out of this lesson. In reflecting back on this situation, Bieda offers new insights and strategies to address this dilemma.

In the next case, Sandra Crespo writes about an e-mail she received from a prospective teacher who dared to question: "Why are you asking for these impossible math lessons?" But this was more than a prototypical question about the relevance or applicability of a mathematics methods course assignment. This e-mail argued that those impossible lessons were not appropriate for the "low students" she was teaching in the partner school where she was completing her internship year. This case highlights multiple ways of reading and responding to prospective teachers' concerns with the rhetoric and the reality of ambitious mathematics teaching. In their case, "Problematizing Gender: Learning to Embrace Uncertainty," Carlos (Nick) Gomez and Eric Siy share a lesson they designed where mathematics could be used to interrogate normativity about gender. Reflecting on their task and classroom interactions, they wonder if they played it too safe and by doing so undermined the potential impact and outcomes for their prospective teachers.

In "Challenging and Disrupting Deficit Notions in Our Work with Early Childhood and Elementary Teachers," Courtney Koestler helps us to realize that deficit labels about children's mathematical competence start when they are really young, even before they have had formal mathematics instruction. She shares the ways in which she seeks to disrupt deficit language about children and support her early childhood mathematics educators to notice and talk about their students' mathematical potential. In "Case X: Opportunities for America's Youth," Kimberly Melgar and Dan Battey share a mathematics task focused on exploring the meaning of fairness and of equity in society. The task includes data and statistics about life outcomes related to social justice for different ethnic groups. They expected that this kind of task would provide a transformational experience for the prospective teachers in their course and were dismayed that the task did not achieve their goals. They frame this as a dilemma for the teacher educator to learn how to challenge their prospective teachers' privilege in a mathematics methods course.

In "Hearing Mathematical Competence Expressed in Emergent Language," Judit Moschkovich shares how she seeks to address problematic beliefs prospective teachers hold about learners who are bilingual or are learning English as a new language. She addresses this dilemma by designing learning activities that both challenge limited conceptions about teaching mathematics to English Language Learners and to support prospective teachers to gain new perspectives and practices that will help them rethink their stereotypical preconceptions. She adopts what she calls a "cognitive compassion" approach to listening for the emerging shifts in prospective teachers' conceptions and understandings of teaching mathematics to English Language Learners. In the final case of this section, Dorothy White focuses on "Tracking in a Local Middle School: Do You See What I See?" She shares a field-related experience she has incorporated into her mathematics methods course in order to invite prospective teachers to make connections between what they are studying in her courses and the realities of mathematics teaching in a local school. As the title suggests, she invites us to wonder with her about what it will take for prospective teachers to see how structural and systemic inequities are actualized in very real ways in schools and inside the mathematics classroom.

Each of these cases and their commentaries offer images about the work of teaching prospective teachers to question and disrupt systems of oppressive education and to learn the practices of "teaching to transgress" (hooks, 1994). This collection of cases offers us an invitation to open the conversation and create "communities of struggle" (hooks, 1994) who will learn together the challenges and the possibilities of enacting a vision of mathematics education as a practice of liberation and freedom. Designing and enacting mathematics methods courses that deliberately integrate mathematics content, pedagogy, with social justice issues is inherently complex. As such, the cases in this section address issues related to creating and using tasks for teacher education students, responding to tasks and work that is produced by prospective teachers, and structuring discussions around these issues. As you read through the cases and commentaries in this section, we invite you to consider the following questions:

- How do you connect with the dilemma highlighted by the case author and commentators?
- What issues of educational equity are foregrounded and discussed in each of these cases?

- What insights and questions were raised for you about how and why infuse conversations about educational inequities in and out of the mathematics classroom in a mathematics methods course?
- What new insights and questions do you now have about how oppression and privilege operate inside and outside the mathematics classroom?
- What new goals, tasks, and tools are you ready to try in your own mathematics methods courses in your own local context and communities of practice?

REFERENCES

Ball, D. (1988). Unlearning to teach mathematics. *For the Learning of Mathematics, 8*(1), 40–48.

Cochran-Smith, M. (2001). Blind vision: Unlearning racism in teacher education. *Harvard Education Review, 70*, 156–190.

Feiman-Nemser, S. (2001). From preparation to practice: Designing a continuum to strengthen and sustain teaching. *Teachers College Record, 103*, 1013–1055.

hooks, b. (1994). *Teaching to transgress: Education as the practice of freedom.* New York, NY: Routledge.

CHAPTER 2

ADDRESSING DEFICIT LANGUAGE IN MATH METHODS

Providing Critical Feedback to Preservice Teachers

Julia Maria Aguirre
University of Washington Tacoma

INTRODUCTION OF THE CASE

This case focuses on supporting preservice teachers (PSTs) to move away from using deficit discourse about children, families, language, and culture in relation to mathematics learning through critical feedback (Black, Harrison, Lee, Marshall, & Wiliam, 2004). I teach in a small public urban-serving state university. Our K–8 teacher preparation program is a 1-year post-baccalaureate licensure (K–8 general education credential) with a second-year option to obtain a master's degree. Our demographics trend toward older students with a mean age of 28. The racial/ethnic and gender breakdown generally mirrors our state's teacher workforce demographics:

Cases for Mathematics Teacher Educators, pages 13–21
Copyright © 2016 by Information Age Publishing
All rights of reproduction in any form reserved.

85% White, female, and mono-lingual English. Our partner schools that host placements for our PSTs are highly diverse with most serving poverty-impacted and working class families and ethnic communities that speak multiple languages including Spanish, Vietnamese, Russian, Ukranian, Korean, Amheric, and Marshallese.

At my institution PSTs take a two-quarter sequence of mathematics methods. A major assignment is called the Mathematics Learning Case Study (MLC). Developed by the National Science Foundation-funded TEACH MATH (Teachers Empowered to Advance Change in Mathematics) project1, the MLC is designed to help preservice teachers gain knowledge about children's mathematical learning and funds of knowledge through interacting with one student (Bartell, Foote, Drake, Roth McDuffie, Turner, & Aguirre, 2013; Carpenter, Fennema, Franke, & Levi, 1999; Civil, 2007; Foote, Roth McDuffie, Aguirre, Turner, Drake, & Bartell, 2015; Gonzalez, Moll, & Amanti, 2005). The assignment consists of three parts. In Part 1, PSTs conduct an interview with a focal student about their interests, outside school activities, and views of mathematics. The student must be different from the PST in one or more sociocultural ways (e.g., gender, language, culture, race). The PST also shadows the student during a school day to see possible participation differences across academic and nonacademic settings. Part 2 is a classic cognitively guided instruction (CGI) problem-solving mathematics interview involving operations with whole numbers (Carpenter et al., 1999). Part 3 is a final reflection on what the PST learned across these different settings and how the PST will adjust mathematics instruction to support the student's learning. The reflection subprompts guide PSTs to discuss specific student strengths and competencies. PSTs need to support specific claims with evidence from the MLC and course readings to justify how they would help this student learn mathematics. Yet, to be a realistic writing exercise that strengthens professional communication, the reflection format contains three options (a) student progress report (to be sent to parents); (b) teacher–parent conference discussion report (to be used in a conference setting); (c) grade-level team or IEP (Individualized Education Program) discussion report.

This case study focuses on the dilemma of addressing PST deficit discourse about students in a way that allows the PST to recognize the deficit discourse and alter their characterizations of student math learning to a more strength-based resource orientation. The case centers on the written feedback given to a PST who elected to revise her MLC-Part 3 submission formatted as a student progress report to parents. The PST's original submission included instances of deficit language about her focal student's attitude and capacity to learn mathematics. The PST also acknowledged her own challenges to help the student communicate mathematical thinking. On the other hand, the PSTs noted efforts were made to affirm differences in cultural approaches to learning mathematics and the use of L1 (home language, Spanish). The case highlights different ways critical feedback addresses deficit discourses as well as acknowledges or pushes the PST's constructive attempts to communicate their evaluation of student competencies and promote mathematics learning of students.

THE DILEMMA (WHAT HAPPENED?)

The introductory paragraph of the final reflection focused on what the PST learned about her focal student, a 4th-grade student of Mexican descent named Angela. The PST conveyed that Angela felt math was "her least favorite subject." According to the PST, Angela indicated that math was a "series of drills and worksheets" and that she "never used mathematics outside of school." The PST concluded, "Angela may believe math is important, but she does not seem to believe that it has a purpose outside of the classroom." Next, the PST discussed Angela's struggles during the problem-solving interview:

> There were a variety of story problems I had her solve, and she had been confident in solving most of them. The one story problem she seemed to lack self-assurance when solving had been the "measurement division" problem that dealt with remainders. I found it difficult to encourage Angela to describe what she was thinking not only when she was experiencing difficulty but also when I tried to get her to explain why she thought her answer was right. She wrote the correct operation to use in this problem, but her final answer and her appearance conveyed uncertainty. When she wrote down her answers, she would write the equation she thought correctly depicted the situation, and her answer right afterwards. Any other operations she may have performed were not explained, despite my attempt with guiding questions meant to help her explain.

> Something else I noticed about Angela was that she used one method throughout the interview, and avoided using the unifix cube manipulative provided or any other technique, like drawing out the problem, to solve. She would write down the math operation she thought represented the situation, and her answer immediately afterwards. This leads me to believe that previous math classes she may have been in were highly teacher-directed and procedural focus, and this can be very limiting to her understanding of the methods she uses to solve math problems. (Van de Walle, Karp, & Bay-Williams, n.p.)

The PST noted her own struggles in helping Angela communicate her mathematical thinking. She also conjectured that Angela's past math learning experiences may have affected her capacity to solve problems in different ways. However, this excerpt also highlights a strong focus on Angela's limitations such as solving a problem in only one way, avoiding the use of manipulatives, or making a drawing to solve the problem.

The reflection also included commentary related to the role home language plays in learning mathematics. Knowing that Angela spoke Spanish at home, the PST affirmed the use of L1 when she stated, "Although my own fluency in Spanish is limited, and bilingual education is not featured in our school, I would like to find a way to incorporate Spanish into her math assignments." The PST went on to provide brief examples of altering instruction such as "simplifying teacher talk" to "reduce the linguistic complexity" of the math lessons for ELLs as well as "draw upon home and community based knowledge and experiences as a resource to support their math understanding."

The PST also offered ideas aimed at how Angela's parents could work with Angela to support her math learning. The PST conveyed that Angela did her work at home by herself and "does not ask for help when she doesn't understand math problems." The PST recommended that the parents work with Angela in Spanish to do homework. This would allow Angela to see "the value in the math she works on even though it may different from your own (parent) school experiences learning math."

In addition, the PST indicated she wanted mathematics lessons to be more "culturally relevant" and wondered about the connection between Angela's hesitancy in explaining her thinking and cultural ways of doing school mathematics in her home country of Mexico. She wanted to create lessons that were more "authentic" to her focal student. Although solid connections to course concepts and readings were made, specific examples for lesson ideas that might have come from her interviews and shadowing experiences were not included to illustrate her points.

Evaluating the original submission was a dilemma for me because it contained both deficit- and resource-based views. As a mathematics teacher educator, I want to explicitly challenge PST deficit discourses about children and families in a way that leads to a transformative learning opportunity for the PST rather than defensive disengagement. At the same time, I want to provide feedback that highlights the resource-based views I want PSTs to develop.

AUTHOR'S RESPONSE TO THE DILEMMA

With this PST's reflection, there was a strong focus on what the student, Angela, could not do mathematically. Although she briefly acknowledged that Angela solved most of the problems with confidence, the PST chose to highlight Angela's struggles with math problems. My feedback to the PST made this point explicit and asked her to consider how Angela's parents might react reading this description of their daughter as a math learner. In a comment bar attached to the paragraph describing the problem-solving interview I wrote:

> Again, what point do you want to convey here to Angela's parents? This paragraph depicts what she did wrong in both her solving the problem and her inability to communicate to you about her thinking. How do you think her parents will react to this description of their daughter? What do you want them to do or think?

In addition, I wanted to push PSTs to illustrate and justify points with specific examples. Here, the PST's positive view of the use of Spanish to support Angela's math learning is an important step forward in building a resource-oriented view of mathematics. Yet, the brief descriptions of how this would be done in the classroom demanded more specific details. My feedback on this point was:

> You are condensing several findings here. Simplifying teacher talk and reducing linguistic complexity are strategies for discourse. Connecting problems that draw on community based funds of knowledge is a separate and equally important strategy used by the teachers in the study by Turner & Celedón-Pattichis (2011).

In terms of the PST's suggestion of how parents could help Angela at home with homework, I wanted to push the PST's thinking beyond language, while important, to include mathematical knowledge and experiences that parents may have that might also be a resource to support Angela's math learning. The PST made the suggestion that helping Angela with homework in Spanish would be valuable even if the mathematics were different from their own schooling experience. It was unclear to me what the PST meant by "different." However, the PST knew that Angela's parents immigrated from Mexico. I wanted to check to see if her recommendations to the parents were or could be connected to a reading we had that focuses on the symbolic notation practices of immigrant students (Perkins & Flores, 2007). In my comment for this paragraph, I stated:

> Push your thinking here. Is it just about language or is it about math experiences, procedures, math knowledge that parents may have that can support her math development? Are you connecting to the Perkins & Flores (2007) piece here?

It is important to also point out that this PST connected to course concepts and readings linking to cultural and linguistic funds of knowledge. Her acknowledgement of wanting to make mathematics lessons more "intriguing" for this student through culturally relevant and authentic contexts is a positive step toward a resource view of instructional practice. The PST felt that making connections to Angela's interests would give the student more "ownership" of the mathematics. My comments acknowledged this point and pushed for more details that connect her instructional approach back to her interviews with Angela.

> Many big ideas here. Focus on the Turner and Font Strawhun (2007) piece on authentic problems. What is more important to convey to parents, is what kind of authentic issues might be important to Angela. Do you have an idea based on your interviews?

The PST received a B– for her grade on this assignment. The grade reflected a combination of content and writing issues. But in my course a B– does not require revision. However, this PST decided to revise her reflection in light of my comments. In an email accompanying her revision the PST wrote:

> After looking at the comments you wrote about my final reflection, I realized how poorly written and negative my paper had been. Although it was not my intention to be so negative about my MLC student, I'm sort of embarrassed at how I organized and worded this reflection. I decided to re-write the paper, and included your edited version in this email just in case. I tried to incorporate your edits on my previous paper into my new version. I attempted to limit the points I discussed in this version, and hopefully was able to go more in-depth about the ones I decided to re-write about.

This PST took responsibility to revise her work. And, she recognized the negativity in her original submission. This to me is a positive step toward helping PSTs take notice about the power of their language and how they can change their orientation toward a more positive strength-based view of children and their families.

RETHINKING MY APPROACH

In my experience designing and implementing a K–8 mathematics methods course that privileges mathematics, mathematical thinking, and cultural/community-based funds of knowledge, one of my professional growth areas is trying to provide substantive critical feedback that promotes teacher learning (Black et al., 2004). I am sensitive to deficit discourses commonly used in various educational settings that PSTs pick up and use liberally to describe children and families, especially those from working class/poverty-impacted communities and communities of color. I take these comments personally because they cut to my core values related to equity and they perpetuate negative stereotypes about youth of color that, in my own K–12 teaching experience as well as my research, I seek to eradicate. In addition, the deficit comments negatively impact me as a member of a sociocultural group often referenced in that deficit discourse (i.e., parent of elementary-age children and a Latina with working-class family history).

Looking back, the multiple goals of this assignment may have added confusion and complexity for beginning teachers to engage. I was asking them to write for at least two audiences, myself as their instructor who wanted to see evidence of critical thinking and connection to readings, as well as a professional-related audience (e.g., teacher colleague or parent). Particularly in light of professional communication with parents, citing literature in a student progress letter may have compromised the authenticity of the assignment and complicated the writing for the PSTs who selected this format.

In consultation with my TEACH MATH colleagues, I made several adaptions to this assignment. Now, PSTs must

- explicitly discuss the next instructional moves they would make with this student including a specific problem-solving task.
- describe what they would communicate to the parent/guardian about the child's mathematical learning providing "examples of strengths/areas for growth."
- complete a Math Action Plan to help structure their parent communication and use in a mock teacher–parent conference activity (Aguirre, Mayfield-Ingram, & Martin, 2013).

I feel these changes clarify the importance of PSTs taking the responsibility and justifying their next instructional moves with students on the basis of evidence from the MLC and course readings. Using a separate prompt focusing on communication with parents and the PST's plan to support the child's learning also emphasizes an important routine practice that PSTs must develop as part of their professional repertoire. Lastly, I also use my exchange with this PST as an exemplar for future PSTs taking math methods. The original submission with my evaluative comments and the revised copy are available to PSTs for review.

FINAL COMMENTS

I will continue to work on how to give constructive critical feedback to PSTs that promotes teacher learning and challenges deficit discourses. I believe the more explicit I am in pointing out how language used in communicating about mathematical learning can reinforce or challenge deficit views of children and families is an important part of PSTs' professional growth. At the same time, I want to explicitly acknowledge PSTs' attempts to incorporate a more strength-based perspective about students and their families as resources for mathematics instruction. I want to continue to work on how to balance my comments so that I keep PSTs engaged to think critically about their developing practice and how to support children's mathematical learning.

NOTE

National Science Foundation Award No. DRL #1228034

REFERENCES

Aguirre, J., Mayfield-Ingram, K., & Martin, D. (2013). *The impact of identity in K–8 mathematics learning and teaching: Rethinking equity-based practices.* Reston, VA: National Council of Teachers of Mathematics.

Bartell, T. G., Foote, M. Q., Drake, C., Roth McDuffie, A., Turner, E. E., & Aguirre, J. M. (2013). Developing teachers of Black children: (Re)orienting thinking in an elementary mathematics methods course. In J. Leonard & D. B. Martin (Eds.), *The brilliance of Black children in mathematics: Beyond the numbers and toward new discourse* (pp. 341–365). Charlotte, NC: Information Age.

Black, P., Harrison, C., Lee, C., Marshall, B., & Wiliam, D. (2004). Working inside the black box: Assessment for learning in the classroom. *Phi Delta Kappan, 86*(1), 9–21.

Carpenter, T., Fennema, E., Franke, M., & Levi, L. (1999). *Children's mathematics: Cognitively guided instruction.* Portsmouth, NH: Heinemann.

Civil, M. (2007). Building on community knowledge: An avenue to equity in mathematics education. In N. Nasir & P. Cobb (Eds.), *Improving access to mathematics: Diversity and equity in the classroom* (pp. 105–117). New York, NY: Teachers College Press.

Foote, M. Q., Roth McDuffie, A., Aguirre, J., Turner, E. E., Drake, C., & Bartell, T. G. (2015). Mathematics learning case study module. In C. Drake et al. (Eds.), *TeachMath learning modules for K–8 mathematics methods courses.* Teachers Empowered to Advance Change in Mathematics Project. Retrieved from: www.teachmath.info

Gonzalez, N., Moll, L. C., & Amanti, C. (Eds.). (2005). *Funds of knowledge: Theorizing practices in households, communities, and classrooms.* Mahwah, NJ: Erlbaum.

Perkins, I., & Flores, A. (2007). Mathematical notations and procedures of recent immigrant students. *Mathematics Teaching in the Middle School, 7,* 346–352.

Turner, E., & Celedón-Pattichis, S. (2011). Problem solving and mathematical discourse among Latino/a kindergarten students: An analysis of opportunities to learn. *Journal of Latinos and Education, 10*(2), 1–24.

Turner, E., & Font Strawhun, B. (2007). Posing problems that matter: Investigating school overcrowding. *Teaching Children Mathematics, 13*, 457–463.

Van de Walle, J. A., Karp, K. S., & Bay-Williams, J. M. (2010). *Elementary and middle school mathematics: Teaching developmentally* (7th ed.). Boston, MA: Pearson Education.

COMMENTARY 1

PROVIDING RESPECTFUL AND ABILITY-ORIENTED FEEDBACK TO PARENTS

A Commentary on Aguirre's Case

Monica Gonzalez
University of Houston

MY POSITIONALITY

I draw from multiple sets of experiences when I write this commentary. I come from an ethnically diverse family where my mother is Puerto Rican and my father is a mixture of European descent. I grew up with a wonderful blend of language, tradition, and viewpoints that helped me appreciate and accept the diversity around me. However in school this was not always the case. I once overheard my first-grade teacher say, "We're in America! They need to learn English!" when referring to a group of students in an ESL class. She made it sound like speaking Spanish would be a hindrance to these students. Knowing that teachers might have deficit opinions of their students helped to shape my views as a graduate student and emerging researcher with a deep interest in equitable mathematics teaching.

Cases for Mathematics Teacher Educators, pages 23–26
Copyright © 2016 by Information Age Publishing

23

I also draw from my experiences as an elementary school teacher and then later an assistant principal in a linguistically, culturally, and ethnically diverse elementary school in the greater Houston area. As an assistant principal, I realized that many new teachers in my school lacked the practice discussing students' progress, grades, and behavior in a respectful and productive way with parents. For me, Dr. Aguirre's case highlights the fact that activities connected to new teacher practice, like providing feedback to parents about students' mathematics progress, are valuable in teacher preparation (Feiman-Nemser, 2001) and take time to learn. The layers of the task used in this case provide opportunities for the scaffolding needed by new teachers to develop these skills. I will further unpack the idea of taking into consideration the parents' perspective of the feedback that will be given. For the preservice teachers, I will specifically focus on explicitness of expectations in the grading rubric as well as providing opportunities for revisions of the assignment.

INTERPRETING THE DILEMMA

The new teachers with whom I have worked would oftentimes forget the difference between the way they looked at students and the way in which the parents of the students would see their children. Parents see their child as an individual and not as a collective group. Dr. Aguirre brings up an important point about considering the parents' point of view when thinking about how to describe students' mathematical strengths and areas of improvement to parents. Listing topics that the student struggles with and providing no areas of strength will leave parents feeling discouraged and oftentimes defensive. It is important for teachers to build a trusting partnership with parents in order to help children learn mathematics. That trust can be violated if the parents do not believe that the teacher wants their child to be successful. The Math Action Plan (Aguirre, Mayfield-Ingram, & Martin, 2013) that was suggested in Dr. Aguirre's revisions to the assignment is a nice guideline for preservice teachers to use when they practice writing feedback to parents. It would require the preservice teacher to consider strengths, areas for improvement, and ways in which the student, teacher, and parents can all work towards the student's mathematical success. All of this information is more helpful than just telling a parent that their child is struggling in mathematics class. A mathematics methods course provides a safe context for a preservice teacher to practice this skill before they encounter the possibility of upsetting a child or parent.

The preservice teachers in this case were given reflection prompts to guide their discussion of specific strengths and competencies of the students. This suggests that the preservice teachers had access to explicit expectations for the assignment. As someone who would potentially implement a similar activity with my preservice teachers, I would find it helpful to know more about how the expectations were communicated and later assessed. On the basis of the feedback Dr. Aguirre gave the preservice teacher, I assume that there was an expectation that the language used to describe students, parents, and groups of people should be respectful and ability oriented (Featherstone, Crespo, Jilk, Oslund, Parks, & Wood, 2011). If a preservice teacher uses deficit language on the assignment, will it affect that person's grade or will it only be attended to in the written feedback? Said another way, I wonder if a preservice teacher could pass this assignment if only deficit language were used. Since the preservice teacher wrote in a way that combined ability-oriented language and deficit language, my next point has to do with the revisions to the assignment.

The preservice teacher in this case took it upon herself to revise the assignment even though the grade she received did not require revisions. She felt like she did not convey what she intended, and she wanted to fix that. From my own experiences working with new teachers, many of them had difficulty communicating their feedback to parents in a way that was productive, just like the preservice teacher from this case. I would suggest that all preservice teachers be required to revise their assignments based on the teacher educator feedback. It takes time and practice to develop the skill of providing ability-oriented feedback, and completing revisions gives the preservice teacher the needed practice. I had a preservice teacher who completed a similar assignment to the one in this case during the fall semester, and she made an assertion that the student she interviewed could not multiply 3 digits by 3 digits and therefore needed to talk to the parent about attending summer school. This particular preservice teacher needed to be reminded that a 10-minute student interview is not enough time to come to that sort of conclusion about a student. She was challenged to think about how the parents might react to this news so early in the school year and what she might be able to do before summer to help the student develop the needed mathematical skills. From an assistant-principal point of view, I want these types of mistakes to be made and addressed before the first teaching job. Methods courses do not have a lot of time to spend on appropriately providing feedback to parents, however the assignment that Dr. Aguirre describes in her case is worth spending time on.

RESPONDING TO THE DILEMMA

What I hope readers take away from this commentary about Dr. Aguirre's case is that providing ability-oriented feedback is a skill that will take time and practice for preservice teachers to develop. It is not a skill that is intuitive and therefore should not be assumed that preservice teachers or even new teachers will do well automatically. A mathematics methods course is a safe place in which preservice teachers can practice providing feedback about children's mathematical progress. Teacher educators themselves need to be well versed in recognizing and providing ability-oriented feedback in order to develop this skill in their preservice teachers. The value of practicing this skill is the potential of saving the preservice teacher from difficult situations with parents, building a supportive network that can improve student success, and giving administrators peace of mind that their teachers will communicate with parents in a helpful and respectful way.

REFERENCES

Aguirre, J., Mayfield-Ingram, K., & Martin, D. (2013). *The impact of identity in K–8 mathematics learning and teaching: Rethinking equity-based practices.* Reston, VA: National Council of Teachers of Mathematics.

Featherstone, H., Crespo, S., Jilk, L., Oslund, J., Parks, A., & Wood, M. (2011). *Smarter together! Collaboration and equity in the elementary math classroom.* Reston, VA: National Council of Teachers of Mathematics.

Feiman-Nemser, S. (2001). From preparation to practice: Designing a continuum to strengthen and sustain teaching. *The Teachers College Record, 103,* 1013–1055.

MISSING IN THE NUMBERS: EXAMINATION OF TEACHER AND RACIAL IDENTITIES

A Commentary on Aguirre's Case

Ebony O. McGee
Vanderbilt University

I enjoyed reading this case study, which focused on a White female preservice teacher as she attempted to infuse her mathematics teaching pedagogy within an anti-deficit, culturally affirming frame for a 4th-grade female student of Mexican descent named Angela. With that said, I would like to offer some strategies and suggestions to potentially improve the experience of the preservice teacher and her racially and culturally diverse students.

MY POSITIONALITY

For more than 10 years, I have been researching the lives and experiences of Black STEM (science, technology, engineering, and mathematics)

Cases for Mathematics Teacher Educators, pages 27–31

students, from high school all the way up through the pipeline to Black STEM faculty status. Having endured the challenges associated with being Black, female, and a "doer of mathematics" (Martin, 2006), I have come to understand that my experiences are similar to those of the students I research and teach. My research has been influenced by theories of race and racism, which have helped me recognize that issues around power, privilege, race, class, and sexual oppression are at the root of many of the academic barriers these students face and of the devaluation of their essential human attributes based on race and its intersection with class and gender (McGee, 2014).

INTERPRETING THE DILEMMA—INTERROGATING TEACHER IDENTITY AND ASKING TOUGH QUESTIONS

Researchers have traditionally assessed teacher efficacy with reductionist measures, such as how many courses a teachers has in a subject area, where they graduated from, and the number of years accumulated in the field (Spencer et al., 2012). Significant research has suggested that teachers are not generally well prepared to teach students whose cultural values and beliefs differ from the mainstream (e.g., Adair, Tobin, & Arzubiaga, 2012; Delpit, 2012; Milner, 2010). Schools with high populations of students of color are commonly stigmatized, regarded as hard-to-staff, and are often understaffed or staffed with inexperienced teachers (Gay, 2014). Delpit, referring to teaching in multicultural contexts, claimed it is necessary "to really see, to really know the students we must teach" (p. 182). Recently, critical educators have expanded notions of what constitutes effective pedagogy and learning in the classroom (Milner & Howard, 2013; Sleeter, 2012). Therefore, in order to really know and teach students of different cultural and racial backgrounds, teachers need to understand the nature of their students' racial identities, that is, their cultural practices, values, and beliefs and how these shape them as learners and members of ethnic communities (Lee, Sleeter, & Kumashiro, 2015).

Thus, I wondered what preparation was in place for preservice teachers (PSTs) to develop these skills of understanding their students' racial identities and cultural practices as a Part of their teacher preparation program. Did these courses or learning opportunities also include highlighting the linkages between teacher identity, self-reflection, and critical cultural consciousness – all of which are imperative for improving the educational opportunities and outcomes for students of different cultural and racial backgrounds? Most teacher education programs provide far too few opportunities for self-reflection and introspection, resulting in many education students not being prepared to confront their own racial identities, a fundamental first step in teaching students of color (Smith, Smith-Bona-hue, & Soutullo, 2014).

In order for PSTs to develop the knowledge and skills for instilling a cul-turally affirming anti-deficit mathematics pedagogy, they should be encour-aged to ask critical questions, not only about teaching techniques and skills, but also to pursue inquiries that encourage a self-conscious consideration of what can lead them to an enhanced understanding of themselves and their students, questions such as

- What biases may I have against students of cultures, skin tones, reli-gions, sexual orientations, SES, and racial identities different from mine?
- Do I unintentionally privilege certain types of students over others who I deem as different?
- What assumptions do I hold with students who share my same skin tone?
- How well does my pedagogy reflect my knowledge regarding racism, colorism, classism, sexism, and students whose first language is not English?
- What kinds of coping strategies do I employ to better understand and, if necessary, improve the racial and cultural dynamics in the classroom?
- Are there aspects of my students' identities that I am ignoring out of fear of change or lack of knowledge?
- As a teacher, what can I learn from teachers of diverse backgrounds who teach students of different cultural and racial backgrounds from their own?
- How have my beliefs about learning and pedagogy changed as a result of interacting with students of different cultural and racial backgrounds in the classroom?

These questions serve as a starting point for more robust examinations of race, culture, and other key facets of one's identity and can encourage a heightened consciousness with regard to teachers' own held ideologies and assumptions that may be influencing their perceptions of their students.

RESPONDING TO THE DILEMMA

I contend that the deficit language in the preservice teacher's analysis was an opportunity for growth as opposed to how I think the author might have viewed the PST, that is, missing the linkages between the scholarly readings and how these theoretically dense frameworks should be taken up in real classroom interactions. I also think there was some potential to unpack some of Angela's ideologies. When Angela proclaimed that her least favorite subject was mathematics, I am reminded of how girls of color are socialized to feel this way (McGee & Spencer, 2014). In other words, her disdain towards mathematics may not be just an internally driven ideology. Furthermore, Angela most likely never learned school-based mathematics in ways that validated her culture or affirmed her identity as a female. Possibly a series of home visits would have allowed the PST to see more connections to at-home mathematics learning and practices that often go underappreciated in education and learning contexts. I also wondered if Angela's uncertainty about the story problem or her avoidance of using what the PST thought would be an ideal strategy in the cube manipulation had more to do with the PST's novice teaching techniques than with Angela's lack of understanding.

Also, Angela's hesitance in offering authentic cultural connections to mathematics in her narrative could be due to the institution's lack of bilingual education affirmation which, as the piece contends, is "not featured" in Angela's school. When the institution itself is culturally repressive, this sends the message that no one in the school is to be trusted, including the teachers. However, teachers who align their pedagogy with what is best for their students are often able to establish trust and a justice-orientated classroom (Spencer et al., 2012). Also, some forms of culturally robust mathematics that provide high mathematical thinking and reasoning competencies (e.g., dominoes, sophisticated card games) are looked upon negatively in that they are stereotypically associated with activities such as drinking and gambling. I wished the author would have brought up these important aspects of mathematics learning and participation when responding to the PST.

Lastly, I feel that the grade on the assignment could be dependent on what the PST learned in the dialogue with the author, as opposed to the PST's initial responses to the assignment. Otherwise PSTs like this will be receiving lots of B– or lower grades with insufficient opportunity to learn and grow from their experiences and to more critically reflect on their own identities in addition to those of their students. Inasmuch as the PST appeared to be gradually learning, it is unclear if there should, subsequently, be deliberate attempts to revise and resubmit her assignments and, more importantly, her thinking in order to achieve the standards set by the author. What happens when we let a B– teacher into a diverse classroom, where students are faced with multiple forms of marginalization? Who really suffers—the PST, Angela or both?

REFERENCES

Adair, J. K., Tobin, J., & Arzubiaga, A. (2012). The dilemma of cultural responsiveness and professionalization: Listening closer to immigrant teachers who teach children of recent immigrants. *Teachers College Record, 114*(12), 1–37.

Delpit, L. D. (2012). *"Multiplication is for White people": Raising expectations for other people's children.* New York, NY: The New Press.

Gay, G. (2014). Teachers' beliefs about cultural diversity. In H. Fives & M. G. Gill (Eds.), *International handbook of research on teacher beliefs* (pp. 436–452). New York, NY: Routledge.

Lee, J., Sleeter, C., & Kumashiro, K. (2015). Interrogating identity and social contexts through "critical family history." *Multicultural Perspectives, 17*(1), 28–32.

Martin, D. B. (2006). Mathematics learning and participation as racialized forms of experience: African American parents speak on the struggle for mathematics literacy. *Mathematical Thinking and Learning, 8*(3), 197–229.

McGee, E. O. (2014). When it comes to the mathematics experiences of Black preservice teachers...race matters. *Teachers College Record. 116*(6), 1–50. Retrieved from http://uex.sagepub.com/content/early/recent

McGee, E. O., & Spencer, M. B. (2014). The development of coping skills for science, technology, engineering, and mathematics students: Transitioning from minority to majority environments. In C. C. Yeakey, V. L. Sanders, & A. Wells (Eds.), *Urban marginality: Youth, cities and neighborhoods in transition* (pp. 351–378). Lanham, MD: Lexington Books.

Milner, H. R. (2010). What does teacher education have to do with teaching? Implications for diversity studies. *Journal of Teacher Education, 61*(1–2), 118–131.

Milner, H. R., IV, & Howard, T. C. (2013). Counter-narrative as method: Race, policy and research for teacher education. *Race Ethnicity and Education, 16*, 536–561.

Sleeter, C. E. (2012). Confronting the marginalization of culturally responsive pedagogy. *Urban Education, 47*, 562–584.

Smith, S. C., Smith-Bonahue, T. M., & Soutullo, O. R. (2014). "My assumptions were wrong": Exploring teachers' constructions of self and biases towards diverse families. *Journal of Family Diversity in Education, 1*(2), 24–46.

Spencer, M. B., Dupree, D., Tinsley, B., McGee, E. O., Hall, J., Fegley, S. G., & Elmore, T. G. (2012). Resistance and resiliency in a color-conscious society: Implications for learning and teaching. In K. R. Harris, S. Graham, T. Urdan (Editors-in-Chief), C. B. McCormick, G. M. Sinatra, & J. Sweller (Associate Editors), *APA educational psychology handbook: Vol. 1. Theories, constructs, and critical issues* (pp. 461–494). Washington, DC: American Psychological Association.

CRITICAL DIALOGUES TO PROMOTE TRANSFORMATIVE LEARNING

A Commentary on Aguirre's Case

Beatriz Quintos
University of Maryland

The case study "Addressing Deficit Language in Math Methods: Providing Critical Feedback to Preservice Teachers" is an opportunity for math educators to examine the possibilities to promote transformative learning within their courses. The dilemma addresses how to support teachers in recognizing deficit discourses and alter their characterizations to a more strength-based resource orientation through critical feedback.

MY POSITIONALITY

I can relate to this case study because I have taught elementary mathematics methods for 6 years in a large public state university. I have been the

Cases for Mathematics Teacher Educators, pages 33–37

instructor of this course in the undergraduate elementary program and in a 2-year graduate program in which interns serve as instructional assistants in an elementary school. In the last revision of this course I included as a major assignment a modification of the Mathematics Learning Case Study (MLC) developed by the National Science Foundation-funded TEACH MATH (Teachers Empowered to Advance Change in Mathematics) project1, as the author in the case study did. As an uncomfortable admission, I must say I have held and continue to identify deficit discourses and practices in me as an individual and mathematics educator. I sometimes experience fear of the other, cultural ignorance, ambivalence, or a preference for those who I initially perceive are "like me." It is this awareness that guides my equity framework as well as my responses to the students in my course and to this case study.

INTERPRETING THE DILEMMA

Aguirre identifies the dilemma as addressing deficit discourse through critical feedback. For this commentary, I use Valencia and Black's (2002) definition of *deficit discourse* as a way educators explain learning gaps or underachievement by locating a deficit within students, families, and communities (p. 83). Other scholars have also identified common examples of deficit thinking in schools; for example, García and Guerra (2004) pointed to overgeneralizations about family backgrounds, caring at the expense of academics, absence of a cultural lens, monocultural view of child-rearing practices and success, and assumptions that children and families need to change to reflect the dominant culture. In this case, Aguirre identifies negative descriptions of students' learning as examples of deficit discourse. Her use of the term sometimes seems to diverge from the definition by Valencia and Black (2002) and García and Guerra (2004), but most importantly I suggest identifying deficit discourses is a complex activity that merits problematizing.

Deficit thinking permeates society, and schools and teachers mirror these beliefs (García & Guerra, 2004, p. 154). These authors have alerted educators that a common misguided effort to deconstruct deficit thinking is to focus on teachers as the problem. They have advocated for a broader perspective towards a critical examination of the systemic factors that perpetuate deficit thinking and reproduce educational inequities within classrooms. It is evident that the readings and discussions that the author includes in her course are critical to support preservice teachers. Without these supports, the critical feedback provided by the instructor would be isolated and hard for preservice teachers to consider. The second note of caution from García and Guerra points to the risk of assuming that discourse and cultural sensitivity automatically results in equity practices. This warning does not diminish the importance of language, but it does point to the fact that language is part of a bigger problem. This case highlights the role of language and ponders its power to create change.

RESPONDING TO THE DILEMMA

After I read the case, I reviewed the responses from my own students to their Mathematics Learning Case Study and tried to identify "deficit discourses." Then I wondered if there is such a thing as phrases or vocabulary that suggest "deficit thinking"? According to Bakhtin (1968), words in language are half someone else's. On one hand, the speaker or writer shares with an expectation of a response, and on the other, the discourse and words have a history of usage to which it responds. It is without doubt that language is powerful and is important to address. However, instead of focusing on annihilating certain discourses, we have the opportunity to focus on language-in-action as a source of insight and regeneration. There is a need to deconstruct assumptions, explore intentions, and listen to personal and group histories. Dialogue as a deepening activity that allows each of us to see ourselves as in-process, rather than a discourse that is taken as an end in itself.

As educators we can see teachers' discourses as learning opportunities—spaces to listen to others' perspectives and experiences that have shaped them, to deconstruct and problematize the meanings embedded in utterances. This perspective seems to parallel the mathematics education reform emphasis to move away from evaluating right and wrong answers towards exploring the reasoning behind students' arguments. Mathematics educators know right answers can hide misconceptions or superficial understanding, while wrong answers can hide strong sense making through superficial errors. Of course, there is a danger of not addressing deficit thinking. There is also a risk at identifying deficit language. Bakhtin (1968) has called for dialogue that allows for creative opportunities for renewal and the possibility for individuals to speak back to dominant discourses. Is it possible for us to model a process of dialogue to supplant efforts to eradicate specific discourses that we determine to hold deficit views? As an example, let's consider parent communication inasmuch as it is the realistic context given in the assignment of Aguirre's case. Sharing reports with parents is a practice that can have many marginalizing or empowering shades. The students in my course included several comments in their cases that involved their views and assumptions of parents' roles in relation to children's mathematical learning. The following are excerpts from four graduate students in response to the prompt, "What was your learning about yourself from the case study?" In sharing these responses, I want to challenge the reader, as I would future students, to go beyond an immediate label or judgment and consider other ways of engaging with the resource-based and deficit thinking present in each of them.

Prompt: "What was your learning about yourself from the case study?" Responses:

1. Some of my assumptions were confirmed. For example, I believed that Natalie (African American) and Sebastian (Latino) did not have enough support at home to do their homework. I thought that their parents probably did not have time to help them with their homework. I also thought that their parents did not understand many topics that teachers are teaching at school. I also think that not being fluent in English is a barrier for some parents to help their children with homework.
2. Another erroneous assumption I made was that they did not receive a lot of academic support at home since their parents both worked and may not have graduated college or even high school in their native countries. This definitely was not true, especially in Alejandra's case. In fact, her positive attitude and perseverance are a direct result of her mother's creativity and encouragement. Despite the fact that Alejandra's father works long hours as a cook, he makes time to check her homework and her mother encourages her to practice math over the summer.
3. In order to solidify this strong foundation, practicing math in school isn't enough. A child's parents must push them, practice with them, and learn to incorporate math into every day. . . . As a teacher of first or second graders, I will be sure to provide math materials and games for my students to keep at home, so that they will have a way of practicing with their families outside of school.
4. It is important to note that a students' race/ethnicity has directly no correlation in how he/she performs academically, therefore it can't predetermine a student's success or failure.

These statements are filled with historical vestiges, struggles with dealing with difference, hopes, biases, as well as deep understandings. They address ideas that are critical to discuss from different perspectives when working with families of culturally and linguistically diverse backgrounds. As an educator I am interested in promoting critical dialogues, modeling a process of open and vigorous discussion in which we aim to make sense of what makes sense to the other (Bakhtin, 1968). A dialogical methodology (Padilla, 1993) brings text for others to respond, deconstruct, and problematize. The goal is to move away from a focus on truth and individual creation. Instead it is possible to acknowledge the deep social and political nature of language and therefore the need to understand each other's histories, assumptions, and beliefs. As critical educators we need to create open spaces to explore persuasive assumptions or discourses. To supplement critical feedback to push students' discourse, I suggest bringing selected ideas from students for everyone to ponder as another way to create possibilities of transformative learning.

NOTE

National Science Foundation Award No. (DRL #1228034)

REFERENCES

Bakhtin, M. M. (1968). *Rabelais and his world* (H. Iswolsky, Trans.). Cambridge, MA: Massachusetts Institute of Technology.

García, S. B., & Guerra, P. L. (2004). Deconstructing deficit thinking: Working with educators to create more equitable learning environments. *Education and Urban Society, 36,* 150–168.

Padilla, R. (1993). Using dialogical research methods in group interviews. In D. Morgan (Ed.), *Successful focus groups* (pp. 153–166). Newbury Park, CA: SAGE.

Valencia, R., & Black, M. (2002). "Mexican-Americans don't value education!" On the basis of the myth, mythmaking, and debunking. *Journal of Latinos and Education, 1*(2), 81–103.

CHAPTER 3

UNDERSTANDING WHITE PRIVILEGE

When a Good Task Is Not Enough

Kristen Bieda
Michigan State University

INTRODUCTION OF THE CASE

The racial demographics of the U.S. teaching force have changed very little over time and continue to fall short of reflecting the diversity of the U.S. student population. In 2011, 84% of U.S. teachers identified as non-Hispanic Whites, a decrease of only 7 percentage points from the data released in 1986 (Feistritzer, Griffin, & Linnajarvi, 2011), whereas only 55% of K–12 students identified as non-Hispanic Whites (Davis & Bauman, 2013). While it is critical that we better understand the lack of diversity in the pool of U.S. teachers, it remains a challenge for teacher educators to prepare a predominately White contingent of prospective teachers with the knowledge, beliefs, and perspectives they need to teach in equitable ways for all students. Despite issues of equity and social justice not yet being a central focus of my scholarship in mathematics education, it is a core commitment in my teaching of mathematics and my teaching of future mathematics teachers.

Cases for Mathematics Teacher Educators, pages 39–46
Copyright © 2016 by Information Age Publishing
39

One way to prepare White teachers for diverse classrooms is to directly address their understanding of their racial privilege, which can often be a topic that they do not willingly wish to acknowledge or will be quick to downplay in its significance. The dilemma presented in this case is one of how to help prospective teachers see or be aware of inequity when their own privilege or lack of understanding about a situation hampers their ability to do so. This case illustrates this dilemma from my own attempts as a teacher educator to use a mathematics task to spark a conversation among prospective secondary teachers in my mathematics methods course about the manifestation and impact of White privilege in U.S. society. This particular course is the second in a four-semester sequence of methods courses that students at Michigan State University must take for certification. Students enroll in the course in their senior year of completing a Bachelor's degree in Mathematics, and they will engage in a full-year student teaching experience in the following year to fulfill requirements for licensure. Prior to this course, students have learned about social foundations of schooling and issues of urban education in coursework they completed as freshmen and sophomores.

THE DILEMMA: WHEN THE TASK FAILS
TO UNCOVER WHITE PRIVILEGE

One goal of my secondary mathematics pedagogy course is to help my students experience how mathematics can be used as a tool to understand societal issues of race, class, and oppression. Prior to taking this course, the prospective secondary teachers have had opportunities in their freshman and sophomore years of their degree program to explore issues of White privilege in the context of public schooling through foundations of education courses required for certification. However, now as seniors, my students are immersed in real classrooms through field experiences and can reconsider issues of privilege within the context of the subject matter they are preparing to teach—mathematics. One approach we take in the course is to discuss how motivating it can be for students to see how mathematics can be used to understand the world they live in, especially for those students who otherwise are not motivated by grades, academic status, or parental approval. To support this goal, I use several activities from www.radicalmath.org, a great resource of tasks for teaching mathematics for social justice.

In this class, I decided to use the South Central task (Gutstein, 2005), inspired by the implementation of another teacher educator, Andrew Brantlinger (2005), to help students confront not only their understanding of ratios and proportional relationships but also their own White privilege—something they might not have done before. As is typical of the secondary mathematics pedagogy courses I teach, all of the students in the class where this dilemma arose identified as White.

The task (Gutstein, 2005, pp. 101–102) consists of several parts that my students completed in groups of 4.

1. Three types of places that exist in communities in the U.S. are movie theaters, community centers, and liquor stores. If you were an urban planner and had the responsibility to design and plan a small city or large town, what seems to be a reasonable ratio of each of these to people, that is, how many people to each movie theater (i.e., individual screens), etc., would seem to make sense to you? What are you basing your estimate on?

2. Think about an "average" city in the U.S.—what does that mean to you? Now consider a city in the U.S. that is "average" in some sense, in that it is relatively dense but has a lot of one- or two-family homes with some three-flats as well as a mix of apartment buildings. In this type of city, *using your ratios from #1 above*, what would be a reasonable number of community centers, movie theaters, liquor stores? (You will need to think about how many people are in this "average" city and work from there.)

3. The year 2002 marked the tenth year anniversary of the Rodney King verdict in the community of South Central, Los Angeles. When the verdict was announced, and South Central broke out in rebellion/rioting (depending on your perspective), National Public Radio reported on the actual number of movie theaters, community centers, and liquor stores in a "3-mile radius" from the corners of Florence and Normandy (the epicenter of the rebellion) in South Central.

    Using your work from #1 and #2 and the map of Chicago (if you want), but *without* using other background knowledge—what would you estimate the number of movies theaters, community centers, and liquor stores to be in South Central? South Central is slightly less dense than Chicago, and you can use your ratios from problem #1 as well (Chicago's population is about 3,000,000).

4. As reported on the radio, there were 0 movie theaters, 0 community centers, and 640 liquor stores in the 3-mile radius area. What is the *density* of liquor stores in the area? How far, on average, assuming the liquor stores are evenly distributed, would you have to walk from any house in South Central to find a liquor store?

I first had groups consider Parts 1 and 2, and then brought the class together for a discussion about the decisions they had made regarding the number of movie theaters, community centers, and liquor stores in an "average" city. The prospective teachers in this class had generated the following ratios in response to #1:

Group Number	1 Movie Theaters per ...	1 Liquor Stores per ...	1 Community Centers per ...
1	27,500 people	2,500 people	50,000 people
2	5,000 people	750 people	5,000 people
3	5,000 people	5,000 people	100,000 people
4	3,500 people	1,500 people	2,500 people
5	25,000 people	500 people	50,000 people
6	5,000 people	2,000 people	20,000 people

The discussion of their responses to #1 and #2 elicited a lot of debate, but I was especially surprised at how the majority of groups had approximated ratios of movie theaters to community centers to liquor stores that were similar to what was found in South Central at the time of the Rodney King riots, given that they were using their hometowns (which tended to be predominately White, midsize suburban cities) as references for an "average" city. All of the groups expected far fewer community centers per capita than movie theaters, and in general expected many, many more liquor stores per capita than either movie theaters or community centers. Moreover, many of them questioned the utility of community centers. Few of the prospective teachers had ever been to a community center, and many had questions about the purpose of community centers.

My goal for this lesson was to have students use proportional reasoning to shed light upon inequitable distribution of tangible resources using the case of South Central LA at the time of the Rodney King riots. However, I did not expect that most of the students in my class thought it was completely appropriate and acceptable—that it was *normal* or *average*–to have such a skewed proportion of liquor stores in relation to movie theaters and community centers. At the beginning of Part 3, I had planned to show a montage of video footage depicting the Rodney King beating, as well as the aftermath of the verdict, so as to situate the context of the upcoming tasks. Nearly all of my students were toddlers during the riots, so I knew I would need to provide some context to help them understand the situation in South Central. I was now faced with the dilemma of how to problematize the lack of movie theaters and community centers for my students, so that they would understand the role of the mathematical work in making sense of the real-world situation.

AUTHOR'S REFLECTION ON THE DILEMMA

There are different approaches a teacher educator could take to help prospective teachers examine an issue of teaching practice. The approach I took for this class was to *model a lesson* where prospective teachers would, as students, use quantitative reasoning to examine the inequities present in a neighborhood quite different from their own. Another approach could have been to engage in a dialogue using the Socratic method, or an instructor-centered lecture about the inequities like a "sage on the stage." One advantage to the modeling approach is that prospective teachers experience a lesson as students, while reflecting upon the experience (both in-the-moment and afterward) from a teacher's perspective. Being able to exercise this dual perspective is a key habit of mind of experienced practitioners. However, there were particular missteps I made in modeling the task that resulted in the implementation morphing from student-centered to instructor-centered.

The challenge I faced heading into the reveal of the actual numbers of liquor stores, movie theaters, and community centers was that the experience they were having as students as they developed the ratios in Parts 1 and 2 was not promoting the kind of reflection from a teacher's perspective I had hoped for. More specifically, as students believed a high density of liquor stores was typical of an average city, my reveal of the number of liquor stores, movie theaters, and community centers in South Central at the time of the Rodney King riots would not elicit the appropriate surprise to generate discussion about wealth disparity, cultural capital, and other related issues of White privilege. Without this element of surprise, the prospective teachers would not get to experience as students how their quantitative reasoning helped them to consider deeper societal issues at play in this particular news story.

So, what did I do as the lesson unfolded? After showing a video montage to give them some context of the riots, I decided to go directly to #4 and reveal the actual numbers of liquor stores, community centers, and movie theaters in South Central at the time of the riots. As I had expected, this did not elicit any reaction from the class. Unfortunately, this reveal came with only minutes left in class, so I posed two questions for them to reflect upon for homework—*Given the actual ratios of liquor stores, community centers, and movie theaters per capita in South Central L.A. at the time of the riots, what concerns do you have? How do these ratios help you to better understand why riots broke out after the verdict was announced in the Rodney King trial, if at all?* At the next class period (4 days later), we briefly discussed their reflections about these questions. I was frustrated by the shallowness of their reflections. For instance, some said that the lack of movie theaters contributed to angst about the verdict because residents maybe did not have an outlet like going to the movies to take their minds off the news media. Others said that they did not see the relevance of considering the number of community centers, as they still did not understand the role of community centers and what they might indicate about the health of a community. To address these reflections, I resorted to telling—telling the prospective teachers about the role of movie theaters and community centers for providing recreation activities that are family-oriented for a community, as well as telling them how community centers often provide after-school activities for children. The fact that many of my students did not know what community centers provided to residents illustrates White privilege in action; my students did not need a place to go if their parent or guardian could not pick them up from school, so they had not utilized the offerings provided by a community center.

Reflecting upon this class, I still have many unresolved issues regarding how I modeled the South Central task, how to better engage my students in thinking about issues of White privilege, and how mathematics can be used to help unpack issues of white privilege. Brantlinger (2005) also experienced how a good task might not be enough. Although his ninth-grade students initially thought that South Central should have a roughly equal number of liquor stores and community centers, he still found that the task ended up reinforcing stereotypes about residents of low-income neighborhoods rather than exposing issues of income inequality that promote lack of resources in these areas. Both Brantlinger and I struggled to balance telling our students what we wanted them to know with allowing the discussion to naturally follow their own reasoning, in part because we felt pressed for time.

Brantlinger's (2005) reflection upon his teaching of the South Central task has helped me consider ways I could have improved my facilitation of this task. First, I wish I would have given an assignment prior to modeling the task in class where the prospective teachers had to obtain actual numbers of liquor stores, community centers, and movie theaters in their hometown to build my students' *community knowledge* (Gutstein, 2005). They could also have investigated what kinds of services community centers in their hometowns offer to residents so that they better understood the community center as a cultural resource. Then, the prospective teachers would likely not only have a more accurate accounting for the numbers of these resources but could also identify what community centers are, and what they offer, prior to class.

My experience suggests that work to support the prospective teachers' community knowledge may be more critical for this task than work to develop other types of knowledge for doing mathematics for social justice. For example, I asked the prospective teachers to read Brantlinger (2005) as homework for the second class session after our work with the South Central task. The prospective teachers reflected upon how Brantlinger helped develop the three types of knowledge Gutstein (2005) discussed for mathematical understanding that promotes social justice: *classical, community,* and *critical.* Surprisingly, no prospective teacher mentioned how different their own estimations of the ratios of movie theaters, community centers, and liquor stores in a typical community were from the estimations made by Brantlinger's ninth graders, raising a question for me as to whether my students recognized why the ratios of movie theaters, community centers, and liquor stores have a bearing upon the cultural capital of a community.

Second, I wish I would have done more to build what Gutstein (2005) called *critical knowledge* by having them unpack various reasons for citizens' riots. This could have been an excellent opportunity to engage in interdisciplinary work and would have helped students better understand the situation in South Central at the time of the Rodney King trial without having had any memory or personal experience of the riots. Finally, I would plan for more time to discuss why ratios are a useful tool for using mathematics to understand the sociopolitical dimensions of the riots. Because I was so focused on how I was going to deal with my students' perceptions that the resource allocation in South Central around the time of the riots was unproblematic, I neglected to have them consider how the task illustrates ratios as a useful mathematical tool for unpacking sociopolitical issues. Specifically, in this particular task, ratios are more useful than just comparing total number of liquor stores, community centers, and movie theaters between communities because equitable resource use depends upon both number of resources and potential population to use those resources.

As a concluding comment, I wish to note that modeling a lesson that promotes prospective teachers' understanding of teaching mathematics for social justice is more demanding on my knowledge and skills as a teacher educator than if I were modeling a lesson to teach a particular piece of content or mathematical practice. In planning for the class, I not only had to do the task and anticipate various strategies the prospective teachers would use to solve the task, but I also had to anticipate their reactions to the societal issues they might confront in discussing the context of the task. As a result, I had to question my own understanding of the issues surrounding the Rodney King riots and how my reactions might reveal my own White privilege. One misunderstanding I had as a high schooler during the time of the riots was that the rioting was gang-related. I was aware that the LAPD were being vilified for the treatment of Rodney King, and my high-school self attributed the violence of the riots to general mistrust of the police. When I initially read the task as a graduate student, I realized how little I understood about the events surrounding the riots (my community knowledge) as well as how little I understood about how the number of community centers says something about the cultural "health" of a community (my critical knowledge). As a White female growing up in a predominately White, middle-class community, I had a relative wealth of community resources, and community centers were places where my friends and I volunteered as an outreach experience. So, as a teacher educator, you must interrogate how your life experiences have shaped your beliefs and the privileges you enjoy if you want your prospective teachers to do so as well. Teacher education that works against privilege and oppression is a practice of laying bare your own assumptions and being willing to recognize ways that you benefit from particular privileges provided to you regardless of race or creed.

REFERENCES

Brantlinger, A. (2005). The geometry of inequality. In E. Gutstein & B. Peterson (Eds.), *Rethinking mathematics: Teaching social justice by the numbers* (pp. 97–102). Milwaukee, WI: Rethinking Schools.

Davis, J., & Bauman, K. (2013, September). *School enrollment in the United States: 2011.* Retrieved from http://www.census.gov/prod/2013pubs/p20-571.pdf

Feistritzer, C. E., Griffin, S., & Linnajarvi, A. (2011). *Profile of teachers in the US, 2011.* Washington, DC: National Center for Education Information.

Gutstein, E. (2005). South Central Los Angeles: Ratios and density in urban areas. In E. Gutstein & B. Peterson (Eds.), *Rethinking mathematics: Teaching social justice by the numbers* (pp. 101–102). Milwaukee, WI: Rethinking Schools.

COMMENTARY 1

ACKNOWLEDGING PERSONAL PERSPECTIVES TO BUILD MATHEMATICAL UNDERSTANDINGS

A Commentary on Bieda's Case

Cynthia Oropesa Anhalt
The University of Arizona

MY POSITIONALITY

This case elicits thoughts of issues beyond "White privilege." I want to acknowledge that educating teachers who are predominantly White (and non-Hispanic) for educating non-White students is a challenge, especially when exploring serious social issues. As a guest commentator, I first give my background so the reader gets an understanding of my positionality as a Hispanic female whose family roots are of a mixture of Mexican and Spanish-European cultures. As a child, I learned from listening to adult family conversations about the high status given to my father's side of the family, whose ancestry is from Spain, and my mother's Mexican ancestry

Cases for Mathematics Teacher Educators, pages 47–51
Copyright © 2016 by Information Age Publishing
All rights of reproduction in any form reserved.

being addressed as lower. I recall heated discussions among aunts and uncles from both sides of the family that caused grief. Growing up in a small community in a border city, interaction between families occurred regularly through social events.

One topic of conversation that produced the most discomfort for me was skin color because there is a wide range of lightness and darkness of skin among family members. Firsthand experience allowed me to witness statements of discrimination and status within my extended family that caused conflict in my beliefs regarding skin color for much of my K–12 education. As an undergraduate, I began to question and challenge this status that was "given" and in most cases, accepted, among my extended family. It is important for me to give this background information, rather than simply say that I am a Hispanic female, because my questioning of family perspectives has shaped my perspectives over the years.

The following are my viewpoints as a mathematics educator in working with K–12 prospective and practicing teachers in content, curriculum, and assessment courses as well as professional development and research projects.

INTERPRETING THE DILEMMA

In summary, *White privilege* is an implicit social advantage that lives in peoples' subconscious biases that systematically benefits White people. From Williams's (2004) perspective, White privilege is a way of "conceptualizing racial inequalities that focus on advantages that White people accrue from their position in society as well as the disadvantages that non-White people experience" (p. 429). This bias can manifest itself on paper, such as giving privilege to "White-sounding" names on résumés or applications, or in personal interactions, such as teacher-student exchanges. Blum (2008) talked about "unjust enrichment" privileges, in which White people benefit from the injustices done to persons of color. From her perspective of being White, McIntosh (1988) conceptualized racism "as something which puts others at a disadvantage, but had been taught not to see one of its corollary aspects, white privilege which puts me at an advantage" (p. 1). In general, it is difficult to address White privilege with prospective teachers without initially acknowledging and discussing personal perspectives because sometimes the beneficiaries of White privilege may not think that this systematic unfairness applies to them. This unawareness is difficult to uncover personally and may be difficult in a group setting.

A concern I have about the case presented is that discussions about racism did not seem to take place before analysis of the case by the students. The author writes that the dilemma she sees is "one of how to help prospective teachers see or be aware of inequity when their own privilege or lack of understanding about a situation hampers their ability to do so." Presenting a case with serious complex issues, such as race, privilege, and economic disparity, needs ample time in a classroom for honest discussion, especially if the students know little of the historical tensions that led to violent crimes. If I were to use this task, I would ponder how to build empathy and understanding of the context, that is, the social inequities and other factors that surrounded the riots that followed the announcement of the verdicts in the 1992 Rodney King case. Critical social issues are implicitly embedded in the history and facts of the case, and a clear understanding of these issues should be part of the discussion.

As presented, the mathematical task of ratio and proportions of community centers, movie theaters, and liquor stores to the number of people in communities appears to be secondary to the primary issues of race and social inequities across communities with varying resources. Although a case for the connections between the number of community centers, movie theaters, and liquor stores to the number of people in a community *could* be created by building background knowledge about the "health of a community," the task itself leaves much for interpretation, extrapolation, and assumptions about the factors leading to the breakage of the riots. The reflection questions posed after the reveal of the actual numbers of liquor stores, community centers, and movie theaters in South Central LA at the time of the 1992 riots imply a causal relationship, for which no evidence was presented. The situation is much more complex. As Manuel Pastor, professor of American Studies and Ethnicity, stated,

> Economic distress caused by the departure of manufacturing industries and high unemployment and widespread distrust of the police department set the stage for the outrage following the King case verdict... There was lots of economic frustration; there was racial tension in the air. Then the word of the acquittals set it off. (Associated Press, 2012, para. 8)

RESPONDING TO THE DILEMMA

Beginning the task with a discussion of background information on the roles of community centers and movie theaters in a community and possible connections between these and acts of unrest might have provided a better frame for the activity. The discussion could also allow the teacher to know the students' points of view and develop a more accurate expectation of their possible reactions to the topic and data. The prospective teachers may bring up factors that could be related to riots, such as wages, that could help set the stage for the tougher topics about racism, social inequities, White privilege, and the living conditions in poverty areas. These conversations could then lead to discussion of the purpose for movie theaters, community centers, and liquor stores. The prospective teachers were expected to take a "big leap" in understanding social inequities and issues of race and discrimination in order to make connections between concrete examples of community cultural capital and White privilege.

After discussions take place, the activity could be modified to present information for prospective teachers to establish clear connections between the data and the social outcomes (riots). Establishing connections allows for discussing deeper mathematical ideas, such as changes in the data that would produce different outcomes. For example, working under the assumption that the ratio of liquor stores to community centers is an indicator for civil unrest, one could ask the question, "How low would that ratio need to be in order to avoid riots?" This kind of approach, which makes the task mathematically rich, is often taken in mathematical modeling tasks in which students are asked to pose questions about a social situation and translate to mathematical language to make deep connections between the mathematics and the situation that is being described (see Anhalt & Cortez, 2015a, 2015b).

RECOMMENDATIONS FOR MATHEMATICS TEACHER EDUCATORS

Learning from these experiences is fundamental to teacher education, especially if we approach teaching by modeling rich mathematical lessons with embedded social issues. I close with six points to keep in mind as we do the work in mathematics teacher education that addresses social inequities.

- Establish mathematical goals for activities and tasks related to social issues so that the mathematics informs and questions beliefs.
- Get to know your students and learn what they care about, so you have realistic expectations of their reactions to critical social topics.
- Build background information about the social issues in the contexts of problems; additional mathematics data can add to background knowledge.
- Allow prospective teachers' points of view to add to the discussion, and don't assume they will be similar to those of peers or the instructor.
- Acknowledge prospective teachers' background experiences.
- Raise critical social topics that permeate society because students come from diverse backgrounds, and we need teachers that will bring empathy to diverse student backgrounds different than their own.

REFERENCES

Anhalt, C., & Cortez, R. (2015a). Mathematical modeling: A structured process. *Mathematics Teacher, 108*, 446–452.

Anhalt, C., & Cortez, R. (2015b). Developing understanding of mathematical modeling in secondary teacher preparation. *Journal of Mathematics Teacher Education*, 1–23. doi: 10.1007/s10857-015-9309-8

Associated Press. (2012, April 29). Rodney King, key figure in L.A. riots, dead at 47. *Herald Net*. Retrieved from http://www.dailyherald.com/article/20120617/news/706179885/

Blum, L. (2008). "White privilege": A mild critique. *Theory and Research in Education, 6*(309), 309–321. doi: 10.1177/1477878508095586

McIntosh, P. (1988). *White privilege: Packing the invisible backpack*. Retrieved from http://www.uakron.edu/centers/conflict/docs/whitepriv.pdf

Williams, L. (2004). *The constraint of race: Legacies of white skin privilege in America*. University Park, PA: Pennsylvania State University Press.

THE IMPORTANCE OF CONTEXT AND NUANCE IN DESIGNING LEARNING EXPERIENCES FOR TEACHERS

A Commentary on Bieda's Case

Erica N. Walker
Teachers College, Columbia University

MY POSITIONALITY

I have been a mathematics educator (teacher, researcher, and professor) for roughly 20 years and was born and raised in the South. As an African American student in racially mixed public schools, and then as an African American teacher in similarly diverse schools, I have always noticed issues of equity and access in mathematics classrooms. As a professor and researcher, my work is influenced by my own experiences as a teacher and student in these settings. One of my most vivid—and favorite—memories of high school mathematics teaching is when a White male student pointed out that it was as important for him to have had me as a math teacher and

Cases for Mathematics Teacher Educators, pages 53–56
Copyright © 2016 by Information Age Publishing
All rights of reproduction in any form reserved.

role model as it was for the Black students in the class who had said that I was their first Black math teacher. My point here is that teachers often influence students in ways we cannot always predict—and we are role models to all of our students, whether or not we share their backgrounds.

INTERPRETING THE DILEMMA

This case highlights the importance of nuance and context when educators construct meaningful tasks for instruction for use with students in classrooms as well as teachers in teacher preparation programs. Questions we might consider when designing instruction include, What are the purposes of the activity? What do we expect students (in this case, teachers) to learn from it? What "residue" (Hiebert et al., 1996) from the activity might we hope teachers draw from the experience and later build on when they design instructional tasks and teach mathematics? We usually think of residue as having a mathematical meaning, that is, what kinds of mathematical ideas and foundational thinking do we want students to take from our lessons? But I think the idea of residue is valuable also when we think of teaching, as an inherent sensemaking activity from which we, as teachers, are continually learning.

I commend the author as a teacher educator interested in helping preservice teachers understand how to effectively implement mathematics lessons for social justice, and for her thoughtful reflections on the task itself. My first impression upon reading the case was that this activity may have been inaccessible to White teachers from affluent socioeconomic backgrounds. One possible adaptation might have been to ask students to first reflect on their own cities (which the author notes as being "predominantly White, midsize suburban cities") and the perceived proportions of movie theatres, liquor stores, and community centers relative to each other; then to examine similar proportions in other locations. Their own experiences and perceptions of cities may have led them to think that these ratios were "appropriate" without any comparison to other cities of similar size but with different demographics. It is also important to consider that South Central is a *neighborhood*, not a city. Thus, looking at the proportion of these amenities in other neighborhoods, within the same cities, is important to provide some context. As a person who grew up in a predominantly Black, middle-class neighborhood in a diverse Southern city, the experience of growing up in South Central or Chicago does not resonate with me. However, despite my neighborhood being comparable in terms of housing quality and proportion of professionals to other "desirable" city neighborhoods, for many years we had fewer community amenities than our neighbors in the "Whiter" sections of town.

As Brantlinger (2005) discovered, this case could reify stereotypes rather than ameliorate them. Further, it did not uncover White privilege for the students he taught either. What do the riots (despite a caveat about the use of the terms *riots* versus *rebellions* earlier in the essay, the author defaults to "riots" later) in South Central have to do with White privilege? If the point is that certain communities are underserved in terms of amenities that develop human capital, there may be a better way to make that point. The issue that may most underscore the existence of White privilege, and the issue most relevant to its existence, is that Rodney King was beaten by White police officers, this was captured on camera, and the police were acquitted. People were angry about that, which is understandable. If the officers—even one—had been convicted, it is doubtful the riots would have happened. That seems to me to be salient, but it is masked here.

That cultural capital, or "health" of a neighborhood, is defined here as the ratios of movie theaters, community centers, and liquor stores to the number of neighborhood residents or even to assume that these ratios have significant bearing upon the cultural capital of a community seems reductive. This does not take into account the other important cultural institutions of any given neighborhood, such as religious institutions or libraries, nor does it account for the valuable knowledge and experiences of the residents. In Brantlinger's (2005) original description of the lesson, he clarified that this was the focus of the NPR piece, but this point is not made as clear in the classroom activity presented here. When I think about my own predominantly Black neighborhood growing up, and the fact that by similar narrow metrics it would be considered to have little cultural capital when there were neighbors of varied professions who were actively involved in the education of the community's children, the assumption that it could in any way be considered culturally bereft is maddening.

RESPONDING TO THE DILEMMA

Another approach might be, as Bieda states, to have teachers consider "how mathematics can be used to understand the world they live in," or in this case, teach in. The current debates and discussions around "stop and frisk" and overzealous policing are much more timely, and studying the ratios of who gets stopped, for what reasons, and who is convicted for similar offenses is, I would argue, much more relevant to the recent memories and lives of new teachers. Another option is to present teachers with a task directly relevant to school mathematics teaching: Teachers could reflect on their own schooling or student teaching experiences to consider how mathematics might illuminate certain situations. For example, students might say that they've noticed fewer Black and Latino/a students versus Asian and White students in a Calculus course, or that there is differential availability of advanced mathematics courses across high schools within the same district.

Regardless of the approach taken, as the author notes, it is important to consider how deep and nuanced students' and teachers' knowledge and understanding of the issues at play are, so that appropriate and timely scaffolding activities can take place. For example, a very important point to consider is how discussions of White privilege might be different if all of the teachers in the course were not White. If the purpose of an activity is to document mathematically the existence of White privilege so that (White) teachers can understand it, resources and examples (such as home buying and school allocation of resources) drawn from Brantlinger's (2005) and Gutstein's (2005) excellent work, among others, could be used carefully as cultural reference points that may have more meaning and resonance for teachers. I'd also suggest some critical background reading for teachers, such as *Being Black: Living in the Red* by Dalton Conley (2009), and *Black Picket Fences: Privilege and Peril Among the Black Middle Class* by Mary Pattillo-McCoy (1999). It is important to understand that societal messages about value and worth of students, their communities, and their education are so pervasive that despite evidence to the contrary—and our careful crafting of educational experiences designed to ameliorate them—preservice teachers may still hold fast to stereotypical notions about particular students (Walker, 2007). Finally, as a reminder for us all, it is important that teacher education activities devoted to preparing teachers to teach for social justice do not contribute to "teachers' essentializing students' experiences to one shared experience based on their group membership" (Walker, 2012, p. 1450).

REFERENCES

Brantlinger, A. (2005). The geometry of inequality. In E. Gutstein & B. Peterson (Eds.), *Rethinking mathematics: Teaching social justice by the numbers* (pp. 97–102). Milwaukee, WI: Rethinking Schools.

Conley, D. (2009). *Being Black, living in the red: Race, wealth, and social policy in America.* Los Angeles, CA: University of California Press.

Gutstein, E. (2005). Home buying while brown or black. In E. Gutstein & B. Peterson (Eds.), *Rethinking mathematics: Teaching social justice by the numbers* (pp. 47–52). Milwaukee, WI: Rethinking Schools.

Hiebert, J., Carpenter, T. P., Fennema, E., Fuson, K., Human, P., Murray, H., Olivier, A., & Wearne, D. (1996). Problem solving as a basis for reform in curriculum and instruction: The case of mathematics. *Educational Researcher, 25*(4), 12–21.

Pattillo-McCoy, M. (1999). *Black picket fences: Privilege and peril among the Black middle class.* Chicago, IL: University of Chicago Press.

Walker, E. N. (2007). Preservice teachers' perceptions of mathematics education in urban schools. *The Urban Review, 39,* 519–540.

Walker, E. N. (2012). Mathematics, teacher preparation for diversity. In J. Banks (Ed), *Encyclopedia of diversity in education* (pp. 1449–1452). Thousand Oaks, CA: Sage.

SUPPORTING NOVICE MATHEMATICS TEACHERS' RACIAL CONSCIOUSNESS

A Commentary on Bieda's Case

Craig Willey
Indiana University Purdue University–Indianapolis

MY POSITIONALITY

As a White man, I struggle to understand how White folks can deny that racism is rampant in the United States. Bonilla-Silva (2014) helped us understand how contemporary racism takes on different forms than the overt racism of the past, but I still work hard to understand the prevalence/prominence of color-blind and post-racial perspectives in our communities. I operate from a place where I acknowledge that I contribute to racist discourses, acts, and systems of oppression everyday through my White privilege, which includes the luxury of not having to grapple with racist microaggressions (and overt aggressions) on a daily—or even hourly—basis. I also acknowledge that I will not reach a finite point of racial consciousness where I am not contributing to racist structures and discourses; but I,

Cases for Mathematics Teacher Educators, pages 57–61
Copyright © 2016 by Information Age Publishing

like all White folks, have the agency to try to "see" racism, and "hear" the narratives of people of color, so as to lessen the effects of racism incrementally over time. Admittedly, there is a lot of structural undoing and systemic rethinking to do, as Martin (2015) has recently reminded us, specifically referencing the institutional enterprise of mathematics education:

> The concerns for equity expressed in *Principles to Actions*, like earlier documents, make note of the need to ensure mathematics success for all students with particular expressions of concern for African American, Latin@, Native American, and poor students; that is, those who have been the least well-served by school-based mathematics education. This is a 26-year-old message, couched in a 400-year-old quest for equity in the U.S. (pp. 4–5)

My comments below represent one White perspective, a perspective based largely on my developing ability to "hear," an intentional process through which we can understand misrepresentative, dominant narratives about people of color and honor and elevate counterstories (e.g., Fernández, 2002; Howard, 2008).

INTERPRETING THE DILEMMA

The dilemma here lies in what, if asked, we think is appropriate for Others—in this case, Black communities. This task presents an indirect way of getting at loaded questions. As White power brokers, if given the opportunity, how would we reshape communities of color? Would we make them more like our own? How well do we understand the cultural values, desires, and ambitions of communities of color? It is interesting, yet not entirely surprising, that White preservice teachers (PSTs), through their proposed ratios, would elect to re-create the same social conditions that existed in South Central at the time of the Rodney King riots. This suggests that there is work to be done in helping White PSTs (a) understand the issues facing urban communities, (b) understand their implication in these issues, and (c) leverage these understandings to be a part of meaningful solutions, both in mathematics classrooms and more broadly (Gutstein, 2012).

This task forces us to grapple with the questions above. It seems challenging, to say the least, to engage a class of White PSTs in questions of this magnitude. Nonetheless, we have to understand that we *did* create these communities of color, and thus, the conditions within which they live. And, here lies the largest myth of urban communities: that Black folks (or other residents of color) are responsible for their own social arrangements and status within the larger society.

RESPONDING TO THE DILEMMA

In this case, Bieda indicates that the PSTs struggled to understand the depth of racial oppression. Given the dominant White reaction of disdain and confusion to recent protests in Ferguson and Baltimore, this is not surprising. It seems imperative that *real* issues need to be spotlighted, from the voices of urban residents of color. It needs to be clear that, in the cases of South Central, Ferguson, and Baltimore, this is not about poor Black people acting out because of some arbitrary inequity. It's about seemingly immovable oppression at the hands of White power and interests. It's about being misnamed, misunderstood, misrepresented, and unheard *intentionally* for centuries. It's about feeling hopeless and skeptical that tomorrow will be a better day for their communities.

There seems to be some ambiguity around the context of the lives of Black folks in South Central, Ferguson, Baltimore, and countless other places. I continue to think that there is promise in sharing counterstories—in their entirety, so as not to settle for simple sound bites that fail to reveal the whole picture—such as the NPR story sharing the perspectives of 16 residents of the West Baltimore neighborhood (Inskeep, 2015); these opportunities might shed light on the socially contrived situations from which eruptions happen. It is once we start to see Other People's reality that we can ask ourselves worthy questions pertaining to this South Central task: What might be included in a community center? Who might go there? How might it be used (beyond recreational uses)? How might small investments like this change the culture of a community? My assumption is that the answers of these questions will be qualitatively different following an in-depth look into, and meaningful interactions within, a community of color. Indeed, there has been movement in urban teacher education programs to create sequences of clinical experiences that allow a "deep dive" into the social construction of communities of color and where critical themes, such as psychosocial distress and the role of community spaces, can be explored. I look forward to more concentrated attention and analyses around the impact of these experiences on PSTs' outlook and approaches to mathematics teaching.

TAKE-AWAYS FOR MATHEMATICS TEACHER EDUCATORS

It is important to underscore Bieda's point that working to develop culturally relevant mathematics teachers, or social justice mathematics teachers, demands much from the mathematics teacher educator. And, as this case illustrates, (White) justice-oriented mathematics teacher educators often know what they want to hear from PSTs in terms of acknowledging systems of privilege and oppression, but misunderstand the personal sense-making journey that needs to take place in order to fully understand Whiteness, White privilege, and the corresponding racial order. As it stands, we are collectively ill-equipped to strive and account for incremental progress, especially the kind of meaningful progress that translates into a wholly different form of mathematics teaching and learning with students of color, or the collective Black (Martin, 2015).

To work towards more effective teacher education that will lead to culturally relevant mathematics teachers, two things are centrally important. First, as Bieda alludes to, White teacher educators—even those who see themselves as critically and racially conscious—likely can benefit from engaging in critical identity work (e.g., Applebaum, 2010; DiAngelo, 2012). What exactly this looks like and how we "know" increased consciousness around our Whiteness and the implications of our White privilege on people of color cannot be easily outlined. Nonetheless, there is a need to continually call into question our assumptions about individuals and communities of color, as well as how our words and actions manifest to either perpetuate or dismantle stubborn systems of racial oppression (Leonardo, 2004).

Second, we can continually help PSTs contextualize current and historical events, and understand this broader context as the one within which their students live and learn mathematics. This means that we are constantly processing current (racialized) events through our own evolving lenses and refining our focus on White privilege, race-based oppression, and micro- and macro implications. Related, we can be resourceful consumers of "new" knowledge or evidence of privilege and oppression. For example, a recent publication articulating the relationship between poverty and shortened life expectancy among African Americans (Chae et al., 2015) might shed some light on the factors leading to psychosocial distress in Black communities and, thus, help PSTs situate acts of resistance and protest, such as riots, in the larger context of income inequality and community disinvestment/neglect.

Indeed, as critical mathematics teacher educators, we aim to help develop a particular disposition, skill set, and level of consciousness that, in the past, has not been explicitly required of teacher candidates and practicing teachers, nor are these likely to be reflected in formal documents outlining professional standards for teachers (Martin, 2015). Therefore, as Bieda suggests, we can work thoughtfully and collaboratively through these dilemmas, as it is these tenuous spaces, like the implementation of the South Central task, where we might realize our proclaimed goals of helping teachers see the political nature and utility of mathematics teaching and learning.

REFERENCES

Applebaum, B. (2010). *Being White, being good: White complicity, White moral responsibility, and social justice pedagogy.* Lanham, MD: Lexington Books.

Bonilla-Silva, E. (2014). *Racism without racists: Color-blind racism and the persistence of racial inequality in the United States* (4th ed.). Lanham, MD: Rowman & Littlefield.

Chae, D. H., Clouston, S., Hatzenbuehler, M. L., Kramer, M. R., Cooper, H. L., Wilson, S. M., Stephens-Davidowitz, S. I., Gold, R. S., & Link, B. G. (2015). Association between an Internet-based measure of area racism and Black mortality. *PLoS ONE, 10*(4), e0122963. doi:10.1371/journal.pone.0122963

DiAngelo, R. (2012). *What does it mean to be White? Developing White racial literacy.* New York, NY: Peter Lang.

Fernández, L. (2002). Telling stories about school: Using critical race and Latino critical theories to document Latina/Latino education and resistance. *Qualitative Inquiry, 8*(1), 45–65.

Gutstein, E. (2012). Connecting community, critical, and classical knowledge in teaching mathematics for social justice. In S. Mukhopadhyay & W.-M. Roth (Eds.), *Alternative forms of knowing (in) mathematics* (pp. 300–311). Dordrecht, The Netherlands: Sense Publishers.

Howard, T. (2008). Who really cares? The disenfranchisement of African American males in preK–12 schools: A critical race theory perspective. *The Teachers College Record, 110*, 954–985.

Inskeep, S. (2015). *Baltimore is not Ferguson. Here is what it really is.* Retrieved from: https://news.wbhm.org/npr_story_post/baltimore-is-not-ferguson-heres-what-it-really-is/

Leonardo, Z. (2004). The color of supremacy: Beyond the discourse of 'white privilege'. *Educational Philosophy and Theory, 36*(2), 137–152.

Martin, D. B. (2015). The collective Black and principles to actions. *Journal of Urban Mathematics Education, 8*(1), 17–23.

CHAPTER 4

WHY ARE YOU ASKING FOR THESE IMPOSSIBLE MATH LESSONS?

Sandra Crespo
Michigan State University

In this case I share an all-too-common dilemma for mathematics teacher educators teaching university-based courses—that of prospective teachers questioning course assignments that run counter to dominant teaching practices
they are observing in their school field placements. In this case I invite the
reader to think about how they might respond to prospective teachers' push
back on course assignments that require them to design and teach lessons
that embody principles of ambitious and equitable mathematics teaching.

At the time of the particular incident I share here I had a collection of
well-rehearsed teacher educator responses, such as: "*You do not need a degree
from MSU to teach students how to complete worksheets,*" "*We are purposefully focusing on practices that are hard to learn alone and on the job,*" and other variations
of these kinds of statements. However, these responses seemed insufficient
to address the questions from a prospective teacher who seemed really dis-
tressed about my course's expectation that they plan and teach high-quality
mathematics lessons regardless of who and where they were teaching.

Cases for Mathematics Teacher Educators, pages 63–69
Copyright © 2016 by Information Age Publishing

Unlike the push backs I had come to expect from prospective teachers who questioned the effectiveness of mathematics instruction that shares intellectual authority with students, this prospective teacher's concern did not seem to be about the desirability of this kind of mathematics teaching but rather about the feasibility of doing so in the context of an inner-city school with students who are considered to be struggling with mathematics. This incident challenged me to articulate more explicitly my commitment to equity in mathematics education and how to communicate it in my teacher education courses. I share this case in order to invite other mathematics teacher educators to do the same.

CONTEXT

During their yearlong internship, teacher interns in my institution's teacher preparation program take an advanced mathematics methods course that supports their teaching in their field classroom and also extends their ideas about designing mathematics instruction that is equitable to all students. The internship year is the 5th year of this program, and during that year the interns spend 4 days of the week teaching in a local school alongside a mentor teacher and once a week meet on campus with course instructors of Master's-level courses focused on developing teachers' professional vision and best practices in each content area.

In the internship year mathematics course, the interns are asked to design high-quality mathematics lessons that give students opportunities to show their mathematics smarts as they work together in group-worthy tasks (Cohen, 1994; Lotan, 2003) and during productive whole-class discussions (Chapin, O'Connor, & Anderson, 2009). More often than not, the interns who are placed in Detroit public schools push back at the course's expectations that they design and teach lessons that engage students in challenging mathematical work and that address issues of unequal participation during either whole-group or small-group mathematics lessons.

It is important to note that discussions and principles of ambitious and equitable mathematics teaching are not just sprung out of the blue into this internship course. These ideas are studied throughout our teacher preparation program starting in the first year of the program with a foundational course about diversity in schools and teachers' roles in addressing inequities in their classrooms. This thread continues throughout the program, culminating in the senior-year mathematics methods course where prospective teachers learn about students' mathematical funds of knowledge and how to design lessons that bridge students' home and community knowledge with the mathematics they are studying in school. They also then learn about *Complex Instruction* as an approach to designing productive and equitable group work (Cohen, 1994; Featherstone, Crespo, Jilk, Oslund, Parks, & Wood, 2011) and implement lesson studies with their peers to practice and explore some aspect of complex instruction methods in their field-placement classrooms. Finally, in their internship year course they are required to design and teach these kinds of lessons (that attend to children's mathematical funds of knowledge and that use complex instruction methods) as part of their Unit Planning and Reflection project, which they are required to implement in their field-placement classroom.

It is also important to consider the historical and political context of our teacher preparation program, especially in relation to school placements in the city of Detroit some 70 miles from the MSU campus. Detroit experienced economic boom in the early 1900s when it became known as the car industry headquarters. With this industry also came the rise of a unionized workforce and their fight for fair wages and working conditions for plant workers. Racial tensions in mid-1960s compounded with the economic tensions between Whites and Blacks led to what is known as the "White flight," or the move of the White middle class to the suburbs of Detroit. The city's population has steadily declined, creating huge economic problems for the city in terms of its ability to collect enough taxes from the largely unemployed remaining population to the point where in 2013 the City declared bankruptcy.

The dominant narrative of Detroit as a city in economic decline and of racial segregation and tension is constantly present in our teacher preparation program in spite of our program's efforts to disrupt it and offer counternarratives to the story of the city and its public schools. Our students, many of whom are from anywhere but the city of Detroit, have grown up with this narrative of decline and come to the internship with varied understandings and approaches to doing equity work in education, and in this particular context.

The teacher candidates in the MSU program tend to be White middle- and working-class females who grew up in small towns and in suburbs in Michigan. Teacher candidates who request internship placements in Detroit public schools or near Detroit often do so because they intend to live with their parents and ameliorate the expense of a full-year internship by saving on housing, and not necessarily or explicitly because they are committed to working with underresourced city schools and with a diverse student population. Nonetheless our teacher preparation program has become increasingly more explicit over the years about working with teacher candidates and interns about unpacking their assumptions about teaching mathematics to students who are culturally, racially, and linguistically different from themselves.

THE DILEMMA

I received the following e-mail from one of my interns questioning how realistic my expectations were for them to be able to teach such "impossible" kinds of lessons to "these low kids." As a mathematics teacher educator with more than 15 years teaching prospective teachers of mathematics, I am very familiar with the interns' push back on course assignments that seem irrelevant or disconnected from their field placement. As a mathematics educator committed to doing equity work, I was challenged by this e-mail as I could see multiple issues to take up and address in my response. However, I also considered that I could not respond to all of those issues at once and therefore was pressed to consider how to productively respond, how to be strategic about which issue I was going to take on and which issues I was not going to address in my immediate response. I understood this problem as requiring me to figure out how to productively support my interns to fully embrace teaching high-quality mathematics not only in the wealthy schools but more so in underresourced schools and classrooms.

Hi Sandra,

I am writing up my math lessons today, and trying to get a lot done and I cannot help but be frustrated. I am teaching first grade and I happen to have the lowest first grade class out of the three. My mentor teacher and I have been struggling because we are meant to stick to a Unit plan that my school paid a lot of money to have done. The principals also expect us to stay with the unit lesson schedule. However our students are too low and they need a lot of extra work. We have recently been doing addition, and only 4 out of 17 were able to complete the work with understanding. The following day we were supposed to move on to the next lesson and there is no time to catch them up. So each day, more and more of our class gets confused and I hate that there isn't much I can do. It is a horrible time for me to be taking over the plans for three weeks because starting Monday, all the teachers are going to sit down together and discuss strategies to separate the class or veer from the unit (I'm not sure what they are thinking yet). I am happy we are going to change things in order to help our lower students and the majority of the class but I am worried about my planning and teaching. I cringe when I read the lessons I am supposed to teach because they are so high above my children's learning levels. I need advice, should I continue writing these impossible lessons and turn them in to you? I feel like that is my only option but then when I teach it, I won't experience success and neither will my students. Sorry for putting this all on you, since I was absent the last two days (your class and school the following day) my mentor teacher just e-mailed me mentioning the meeting we are going to have on Monday to talk about how to make math better for everyone but that it is extremely difficult to explain via e-mail.

I hope I explained that well and that it makes sense. Please give me any advice that you have. And thank you for your help in advance.

Melissa (Pseudonym)

AUTHOR'S REFLECTIONS

Some version of this particular scenario has since replayed itself every year that I teach in my institution's teacher preparation program. Predictably the interns that push back on the course's requirements to teach ambitious mathematics lessons are those who are placed in economically disadvantaged and working-class neighborhood schools in the cities of Detroit, Flint, and Lansing. In my many years teaching in this teacher preparation program, I almost never hear the interns placed in the affluent suburban schools consider their students "too low" to teach them in the ways they are learning about in our university classes. This does not mean that all is well with these other interns or that they do not experience the theory-practice tensions widely documented in the teacher education literature (Cochran-Smith & Zeichner, 2009). However, these interns' reasoning and justifications tend to be framed quite differently, such as citing parents' pressures on what they will or will not support teachers to do with their students.

In considering this e-mail now many years later, I can see that how I read and interpret this case scenario will suggest different ways to respond. I can associate this particular episode with three interrelated issues of equity that operate at different grain sizes—individual, societal, and structural. The first focuses on the teacher (in this case the teacher intern's) expectations about their students and how their expectations influence not only their students' academic achievement but also the teachers' moment-to-moment interactions with their students. At the level of student-teacher interactions, framing some students as "low" and others as "advanced" has consequences in how a teacher designs learning opportunities and evaluates their learning.

The literature and empirical research is very clear on this problem of classroom practice and was first documented by Rosenthal and Jacobson (1968), who conducted an experimental study whereby teachers were led to believe that some of their students were at the verge of a dramatic growth in their IQ. This piece of information was enough to influence these teachers' expectations, and over the following 2 years these students' academic achievement dramatically improved in comparison to the other students in their class. Studies of gender differences in the classroom (e.g., Sadker & Sadker, 1994) have also documented differences in teacher expectations for boys and girls in the mathematics classroom and how these affect their teaching practices and ultimately in how different students come to see themselves as competent or not math students. Ladson-Billings's (1995) research has also documented the interplay of teachers' practices and beliefs and especially so for teachers of students of color. Reading this episode as a case of teacher education students expressing their low expectations for their low-income students and students of color suggests that the work for the mathematics teacher educator here is to respond in ways that challenge and disrupt the deficit language of low expectations about students of color to help them understand the consequences of teachers' low expectations for their students.

At a larger grain size this episode reminds me of Paulo Freire's (1970) discussions about oppressive forms of education that serve to reproduce rather than challenge social inequalities. The language in this intern's e-mail is suggestive of the dominant language and practices that characterize Freire's *banking* concept of education, where students are considered emptied of any worthy knowledge and are filled up with the wisdom and knowledge from the official curriculum dispensed by the teacher. I can read in this intern's note some of Freire's words: "projecting an absolute ignorance onto others, a characteristic of the ideology of oppression, negates education and knowledge as processes of inquiry" (p. 58). Reading this episode as an instantiation of what Freire called a *banking* approach to education helps me imagine a response that questions and brings into focus the absoluteness of the statements the intern is making about her students and about the philosophy of education that is associated with those kinds of statements about students, the curriculum, and the teacher.

A third issue that can be read in this case is related to *White Privilege* (McIntosh, 1989) and the ways in which this White student seems to recognize and take action when she considers herself to be placed at a disadvantage: "I won't be successful on this assignment." Blaming her students for her inability to enact any type of ambitious mathematics teaching rings close to entitlements those in power often express in their attempts to extricate themselves of any responsibility for causing others' sufferings. Another question I could raise about this episode is whether this teacher intern feels more entitled to push back on this course requirement because I am a faculty of color who is explicit about her commitment to equitable mathematics teaching. Would this intern use such forceful and demanding tone (as opposed to an asking "would it be okay" tone) if the instructor were a White teacher educator? Reading potential racial tensions into this episode brings yet another dimension and layer of complexity to be unpacked and addressed.

I wish I had a happy ending for this story or that I could say that I have since figured out how to swiftly respond and address this and similar "complaints" our teacher candidates raise about being required to have high expectations for students who have been constructed as struggling students of mathematics. Truth is I am still haunted by this e-mail and have no good answer for how to best address it. I leave this case open and inconclusive because the point of sharing it is to invite others to provide more lenses and strategies to read and respond to these and similar kinds of situations whereby our teacher preparation students challenge us to make explicit our values and commitments to equity in mathematics education.

REFERENCES

Chapin, S. H., O'Connor, C., & Anderson, N. C. (2009). *Classroom discussions: Using math talk to help students learn.* Sausalito, CA: Math Solutions.

Cochran-Smith, M., & Zeichner, K. M. (2009). *Studying teacher education: The report of the AERA panel on research and teacher education.* Mahwah, NJ: Erlbaum.

Cohen, E. G. (1994). *Designing groupwork: Strategies for the heterogeneous classroom.* New York, NY: Teachers College Press.

Featherstone, H., Crespo, S., Jilk, L., Oslund, J., Parks, A., & Wood, M. (2011). *Smarter together! Collaboration and equity in the elementary math classroom.* Reston, VA: National Council of Teachers of Mathematics.

Freire, P. (1970). *Pedagogy of the oppressed.* New York, NY: Continuum.

Ladson-Billings, G. (1995). Toward a theory of culturally relevant teaching. *American Educational Research Journal, 32,* 465–491.

Lotan, R. (2003). Group-worthy tasks. *Educational Leadership, 8,* 72–75.

McIntosh, P. (1989). White privilege: Unpacking the invisible knapsack. *Peace and Freedom* (July/August), 10–12.

Rosenthal, R., & Jacobson, L. (1968). *Pygmalion in the classroom.* New York, NY: Holt, Rinehart & Winston.

Sadker M., & Sadker D. (1994). *Failing at fairness: How America's schools cheat girls.* New York, NY: Scribner.

COMMENTARY 1

IDENTIFYING AND SUPPORTING THE NEXT SMALL STEP TOGETHER

A Commentary on Crespo's Case

Megan Franke
UCLA

MY POSITIONALITY

As a teacher educator in a program specifically focused on both social justice and meeting the needs of Los Angeles communities, I resonate with
this case. I too have puzzled about how to help my mathematics methods
students get to know their students as people and mathematically, to understand the ways the structures of schooling and society shape both their students' participation and other teachers' perceptions and practice. In considering my response to the case, I found myself drawn to the work of those
whose work has pushed me to consider race, identity, cultural practice, and
the political, economic, and social forces that shape schooling (see *Studying Diversity in Teacher Education*, Ball & Tyson, 2011). I want to argue that
in responding to the case we need to consider the value in having Melissa's

Cases for Mathematics Teacher Educators, pages 71–74
Copyright © 2016 by Information Age Publishing
All rights of reproduction in any form reserved.

perspectives voiced, understand Melissa, the spaces where her ideas are voiced and interrogated with her colleagues, and the ways to support her to take on a piece of the work being put forward in her methods course with the goal of generativity.

INTERPRETING THE DILEMMA

Melissa's reported concerns are not uncommon and I would argue expected, even in programs that take on issues of privilege and power. Melissa's email, while worrisome, can also be productive and leveraged to create a space for learning. Spaces where Melissa and her fellow classmates can learn together can be created (Gutiérrez, 1996; Horn & Little, 2010; Strutchens, Quander, & Gutiérrez, 2011) by asking the kinds of questions raised in Melissa's email, have them taken up, interrogated, and discussed in ways that allow students to consider themselves, their experiences, and their beliefs in relation to the students and schools in which they are working, and pressing students to consider in that moment the reasons for their perspectives and assumptions (Hollins, 2011). It is often in these moments, when doing the work of teaching, where our students are forced to make sense of themselves as teachers, in relation to students, the mathematics, other teachers, and so on. I recognize that this is easier to say than to do—and that who is in the discussion, how the discussion is framed, how participation is supported, the range of ideas voiced, the norms developed, and the structures of our universities and teacher education programs all shape what occurs. I also recognize that the conversations cannot be separate from practice or from those with whom our teacher education students work. As teacher educators we must then not only create the spaces for this to occur but also figure out who participates in the collective work and how to support continued learning in it.

RESPONDING TO THE DILEMMA

Understanding Melissa is paramount. I want to see how she sees herself in relation to the task she has been given and the structures in which she is trying to do this work (Olsen, 2011). It is hard to makes sense of the email without knowing Melissa. But I do recognize the complexity in what Melissa writes. I hear the worry in her comments—fear that she won't be able to handle all of what is being asked of her, concern that she is going to have to teach math in a way not accepted by her mentor teacher and the other first-grade teachers (and maybe herself), frustration or worry about not meeting the requirements or goals of the teacher education program. I can see how her concerns about her students reflect the pervasive view of a number of teachers and are different than those discussed in the teacher education program (Ladson-Billings, 2011). I want to understand the ways structures are shaping her response, her view of her students, and her need to write the email. I do not want to make excuses for her; quite the contrary, I want her to see what her students do know about mathematics. I want to find the space in which to productively engage her to see herself in relation to her mentor teacher and students, the program, and the mathematics and help her be explicit about how she is participating and managing the values and practices of the program with those of the teachers at her school.

I want to help Melissa see the situation she is in—to identify for herself—the concerns she has and where they may be coming from—from her perspective, the other teachers' perspectives, and the students' perspectives. To do so, I need to know her, I need to ask her questions, I need to know where the openings are to press on how she views the students and her situation and how she might take up an idea from methods in relation to her current situation. I want to position her as seeing herself as someone who is knowingly teaching against the grain, pushing back on established cultural practice, and challenging the status quo in an attempt to meet the needs of students. I also want to position her as someone who is capable of getting started on this work—that it does not need to wait until she is in her own class, or is a better teacher. I want her to see she can continue to make explicit her assumptions, find ways to challenge them, and continue to learn from her practice.

I recognize the difficulty in what I want to accomplish. What I have learned is that small steps matter. That waiting does not work. The work that occurs in the moment-to-moment teaching practice is exactly where we must attend to the issues of equity being raised here. We need Melissa to see herself as being able to make progress in teaching as a social justice mathematics educator (my term), to see this is a developing practice, something you can continually get better at, where the goal is generativity (Ball, 2009; Franke, Carpenter, Levi, & Fennema, 2001). I think one question we want to ask is what small step can we support Melissa to make, how do we support that step and tie it to the work that has occurred in the methods course, and engage her in thinking about it with her colleagues and the school community. Another question is how the step she takes is tied to a way for her to make sense of what she tries and the result of that in ways that are seen from the perspective of the students she is working with and the cultural, social, political, and economic context in which she is working.

REFERENCES

Ball, A., & Tyson, C. (Eds.). (2011). *Studying diversity in teacher education.* Lanham, MD: Rowman & Littlefield.

Ball, A. F. (2009). Toward a theory of generative change in culturally and linguistically complex classrooms. *American Educational Research Journal, 46,* 45–72.

Franke, M. L., Carpenter, T. P., Levi, L., & Fennema, E. (2001). Capturing teachers' generative growth: A follow-up study of professional development in mathematics. *American Educational Research Journal, 38,* 653–689.

Gutiérrez, R. (1996). Practices, beliefs, and cultures of high school mathematics departments: Understanding their influences on student advancement. *Journal of Curriculum Studies, 28,* 495–530.

Hollins, E. R. (2011). Teacher preparation for quality teaching. *Journal of Teacher Education, 62,* 395–407.

Horn, I. S., & Little, J. W. (2010). Attending to problems of practice: Routines and resources for professional learning in teachers' workplace interactions. *American Educational Research Journal, 47,* 181–217.

Ladson-Billings, G. (2011). Asking the right questions: A research agenda for studying diversity on teacher education. In A. Ball & C. Tyson (Eds.), *Studying diversity in teacher education* (pp. 385–398). Lanham, MD: Rowman & Littlefield.

Olsen, B. (2011). "I am large, I contain multitudes": Teacher identity as a useful frame for research, practice and diversity in teacher education. In A. Ball & C. Tyson (Eds.), *Studying diversity in teacher education* (pp. 257–274). Lanham, MD: Rowman & Littlefield.

Strutchens, M. E., Quander, J. R., & Gutiérrez, R. (2011). Mathematics learning communities that foster reasoning and sense making for all high school students. In M. E. Strutchens (Ed.), *Focus in high school mathematics: Fostering reasoning and sense making for all students* (pp. 101–114). Reston, VA: National Council of Teachers of Mathematics.

EQUITABLE MATHEMATICS TEACHING FOR *ALL* STUDENTS

A Commentary on Crespo's Case

Christa Jackson
Iowa State University

MY POSITIONALITY

As an African American female mathematics teacher educator, I have experienced similar push back on course assignments in which prospective teachers express that students are not capable of engaging with mathematics. Typically, the students whom prospective teachers most often identify as incapable are students of color. My prospective teachers have never explicitly expressed they were referring to students of color, because I, their instructor, am/was one of "those" students.

Cases for Mathematics Teacher Educators, pages 75–78

INTERPRETING THE DILEMMA

After reading the case, I agree with the author that there are multiple issues that need to be addressed in the prospective teacher's email. First, although Melissa (the prospective teacher) has been immersed in a teacher education program centered on equitable mathematics teaching, it appears she is only applying this equity-focused paradigm with students she believes are capable of doing the mathematics. Barlow and Cates (2006) have suggested, "Beliefs affect how teachers see their students...thereby impacting [their] instructional practices" (p. 64). It appears from Melissa's email that she has unproductive beliefs and extremely low expectations of her first-grade students because they are the "lowest" class. Additionally, she describes feeling constrained and restricted by the district/school's pacing schedule for the course. Her low expectations of her students are exemplified in two forms—the labeling of her class and the "ability" of her students based on an assessment/homework that was administered.

Melissa appears to feel trapped; no matter what she does, she will not "experience success and neither will her students." This is indicative of a *fixed mindset*. Although Carol Dweck (2006) and others (e.g., Boaler, 2013) have discussed fixed and growth mindsets in terms of students learning mathematics, I argue that these "mindsets" have also been applied to prospective and inservice teachers (Gutshall, 2013; National Council of Teachers of Mathematics [NCTM], 2014; Rattan, Good, & Dweck, 2012). The mindset embodies who can and cannot learn and do mathematics. A fixed mindset results in low expectations of students, particularly students of color and students from low socioeconomic backgrounds, whereas a teacher with a growth mindset has high expectations of his/her students and implements a variety of instructional strategies that draw on students' funds of knowledge (Dweck, 2010; NCTM, 2014).

RESPONDING TO THE DILEMMA

The literature documents that when teachers have high expectations of their students, they will excel in the learning of mathematics. In one study, Jamar and Pitts (2005) examined the high expectations of Mr. Lee, a White mathematics teacher in an urban school district. Mr. Lee believed his students were able to learn meaningful mathematics and responded to them in positive ways. He used his students' prior knowledge as stepping-stones to new knowledge and communicated to his students that they already had the foundation needed to learn. As a result, his students engaged in rigorous, challenging mathematics. Instead of Melissa focusing on what her students cannot do and what they do not understand, her instructor might encourage her to take a step back and focus on what her students *do* under-stand and build off of that understanding as she is preparing her lesson(s).

It is important that we, as mathematics teacher educators, realize that providing opportunities for our prospective teachers to engage in equitable mathematics teaching practices as well as the principles of ambitious teach-ing is critical, but it may not change deep-seated beliefs or expectations on who can and cannot learn and do mathematics. I anticipate that if Melissa were asked if she engages in the practice of equitable mathematics teach-ing, she would respond with a resounding, "Yes." However, her email does not appear to correlate with this belief. Although prospective teachers may have been immersed in a teacher education program that deliberately fo-cuses on equitable mathematics teaching, some prospective teachers may need to witness its implementation in practice and see ways to enact such practices. Providing an opportunity for the prospective teacher to see this in action could open up a discussion on the nuances that exist when learn-ing to teach mathematics equitably to all students, specifically to students from underresourced communities.

Many prospective teachers express they have high expectations for all of their students, and all of their students can learn mathematics. Unfor-tunately, when prospective teachers are teaching in an environment where the schools are underresourced and it appears the students are not learning or understanding the mathematics, the notion of equitable mathematics teaching is halted. Their perceived high expectations have been countered with the reality of their environment. Because many of the prospective teachers seem to revert to a deficit perspective of the students and families when placed in these communities, it may be helpful to provide a copy of Melissa's email to prospective teachers prior to beginning their internship, and have them discuss the equitable practices or thoughts, if any, and the inequitable practices or thoughts that are prevalent in Melissa's email. As the prospective teachers notice and discuss some of the deficit language and thoughts in the email, they can share alternative views or solutions for Melissa that can help disrupt the deficit view of the students. For example, the interns could discuss how Melissa could better draw on her first grad-ers' funds of knowledge and use it within her mathematics instruction. It may also be helpful for the mathematics teacher educator to discuss why she only receives these types of emails from prospective teachers who are interns in the inner city and not from the interns who are placed in more affluent communities.

Sometimes prospective teachers have a deficit view of the community in which students of color and students from low socioeconomic status live, and this deficit viewpoint is evident in their instruction. In order to help prospective teachers better develop more of an asset view of the inner city and the students who live in this community, the mathematics teacher educator could engage the prospective teachers in a neighborhood asset mapping (Beaulieu, 2002). Neighborhood asset mapping involves going on a neighborhood "walk" to identify the assets within the community.

Although prospective teachers may have been immersed in a teacher education program focused explicitly on teaching mathematics to students who are racially, culturally, and linguistically different from themselves, there are nuances that exist in the practice of equitable mathematics teaching. It is not a process that can be learned in a semester, a year, or over the course of time they are enrolled in a teacher preparation program. But, it is a career-long process. We, as mathematics teacher educators, must prepare our prospective teachers to continuously reflect on their teaching practices to ensure their instruction is embodying the principles of ambitious and equitable mathematics teaching.

REFERENCES

Barlow, A. T., & Cates, J. M. (2006). The impact of problem posing on elementary teachers' beliefs about mathematics and mathematics teaching. *School Science and Mathematics, 106*(2), 64–73.

Beaulieu, L. J. (2002). *Mapping the assets of your community: A key component for building local capacity.* Mississippi State, MS: Southern Rural Development Center.

Boaler, J. (2013). Ability and mathematics: The mindset revolution that is reshaping education. *Forum, 55*(1), 143–152.

Dweck, C. S. (2006). *Mindset: The new psychology of success.* New York, NY: Ballantine Books.

Dweck, C. S. (2010). Mind-sets and equitable education. *Principal Leadership, 10*(5), 26–29.

Gutshall, C. A. (2013). Teachers' mindsets for students with and without disabilities. *Psychology in the Schools, 50,* 1073–1083.

Jamar, I., & Pitts, V. R. (2005). High expectations: A "how" of achieving equitable mathematics classrooms. *The Negro Educational Review, 56*(2 & 3), 127–134.

National Council of Teachers of Mathematics. (2014). *Principles to actions: Ensuring mathematical success for all.* Reston, VA: Author.

Rattan, A., Good, C., & Dweck, C. (2012). It's OK—not everyone can be good at math: Instructors with an entity theory comfort (and demotivate) students. *Journal of Experimental Social Psychology, 48,* 731–737.

TURNING DISAPPOINTING STUDENT EMAILS INTO TEACHABLE MOMENTS

A Commentary on Crespo's Case

David W. Stinson
Georgia State University

INTERPRETING THE DILEMMA

The interpretation of the dilemma outlined by Professor Crespo provides a solid foundation for theoretical discussions in graduate courses for future mathematics teacher educators. In her three-part analysis, Professor Crespo discusses the dispositions of teachers toward teaching "other people's children" (e.g., Delpit, 1995), the benefits of classroom practices that reflect a problem-posing pedagogy (e.g., Freire, 1970/2000), and the unique challenges of faculty of color (e.g., Stanley, 2006). There is a sound literature base on each of these topics to draw on for course readings and classroom discussions. Many of the readers of this book, I assume, address each of these topics in some form or another in their courses for future teacher educators. In my brief response to Professor Crespo's analysis, rather than

Cases for Mathematics Teacher Educators, pages 79–83
Copyright © 2016 by Information Age Publishing
79

respond to the case scenario from a theoretical perspective, I discuss it from more of a practical perspective, if you will. In that, my response discusses what you might consider if you were the teacher educator (or the university field supervisor) who taught the methods course(s) of the student who wrote the email.

RESPONDING TO THE DILEMMA

First, no matter what level of mathematics education courses you teach (i.e., elementary school, middle school, or high school), as a teacher educator, you will receive similar emails from one, two, or more of your students each year throughout your career. In fact, rather than being the exception, such emails will be more the rule. That is to say, no matter how many course readings you have students read, reflective journals or essays you have students write, or critical classroom discussions in whole group or small group you have students engage in around issues of diversity and equity, as a teacher educator, you will receive such emails. This acknowledgement is not meant to in some way declare defeat but rather to be prepared.

When preparing a response, understand that such emails will have unique nuances but somewhat the same sentiment: "the problem" of teaching those other people's children. The key aspect while preparing, I believe, is to approach such an email as a *teachable moment*: moments that "represent new emerging ways for [teacher educators] and students to learn *from each other* by temporarily ignoring institutional identities in order to become participating members in an interdependent teaching and learning community" (Hyun & Marshall, 2003, pp. 125–126; emphasis in original). In other words, just as we ask our students to engage their students as learners, we must, in turn, engage with our students as learners. When acknowledging prospective teachers as learners, we can then approach learning to teach as assisted performance (Mewborn & Stinson, 2007)—"learn[ing] with help what they are not ready to do on their own" (Feiman-Nemser, 2001, p. 1016). In so doing, we can take such emails that are counter to what we had hoped that our students had learned to be indicators of where we need to provide students with more assistance. Such emails can be somewhat positive in that rather than provide "lip service" (i.e., providing you with what you want to hear), in this case scenario, the student who sent the email is clearly stating what she believes and why she (and evidently, her mentor teacher) believes high-quality, rigorous lessons are "impossible" when teaching other people's children. Take the email as a teachable moment, a catalyst for a face-to-face conversation, and avoid responding through email; think about how you, as the teacher educator, might provide more assistance to this learner.

During the face-to-face conversation, ask the student to unpack certain phrases such as "lowest first grade class out of three," "our students are too low," "so high above my children's learning levels," and so on. During these conversations, it is important to refrain from judgments, learn to listen to the student's concerns, provide alternative explanations, and make references to the course readings and class assignments and discussions. Again, just as we ask our students to learn where their students are through listening so that they might begin instruction where the students are (e.g., Fennema, Carpenter, Franke, Levi, Jacobs, & Empson, 1996; Leonard, 2008), as teacher educators, we must do the same: Learn where our students are through listening and begin instruction where the students are.

Second, given the demands on time, especially if you have several students who send similar emails, think about how you might take one case scenario back to the whole class or to future courses (with the student's permission, of course). You and the student can present the email to the class, the student can explain her or his "frame of mind," so to speak, at the time of writing the email, and then the two of you can discuss how the two of you, approaching learning to teach as assisted performance, worked through the dilemma in the collective goal of achieving equity in the mathematics classroom. As a teachable moment, such emails provide an excellent opportunity to engage in Freirian dialogical discussions on the promises and challenges of achieving an equitable mathematics education for all students even within the current regimes of student testing and teacher accountability. Dialogical discussions reflect a "horizontal relationship between persons ... [a] relation of 'empathy' between two 'poles' who are engaged in a joint search" (Freire, 1969/2000, p. 45).

Third, throughout these dialogical discussions, remember that more times than not beginning teachers' espoused beliefs are too often not reflective of and, in many ways, might be counter to their actual classroom teaching practices (see, e.g., Raymond, 1997; Wilson & Goldenberg, 1998). Therefore, even if beginning teachers demonstrate growth in embracing diversity and creating equitable classrooms in class assignments and during class discussions, when they encounter difficulty in enacting these new emerging beliefs they often resort to "blaming" the students or the structures of schools, not their practices. This blaming can be (is) further exacerbated when the mentor teacher's classroom practices are counter to equitable learning environments (see, e.g., Achinstein & Barrett, 2004). Acknowledgement of the incongruence between beliefs and practices is, again, not meant to in some way declare defeat but rather to be prepared. Think and rethink the field experiences so as to provide students with more or, we might say, different assistance: Assign prospective teachers to multiple mentor teachers, reconfigure full-day student teaching with half-day student teaching and half-day campus meetings, or include coteaching or coplanning among the prospective and mentor teachers and university supervisor (cf. Mewborn & Stinson, 2007).

In the end, emails received from students during their field experiences that express ideas that are counter to the goals of diversity and equity reflected in your mathematics education courses are always disappointing. Often your first reaction is one of anger, as you ask: "Did they not learn anything in my class?" However, if you have been teaching teachers for any length of time, you learn to expect such emails, no matter how explicitly issues of diversity and equity were addressed during your mathematics education courses or throughout their entire teacher education program. Nonetheless, although disappointing, such emails also can provide moments for practical assistance in learning to teach in equitable ways if we approach learning to teach as assisted performance. When we open "ourselves to the idea that teacher *education* is about structuring learning opportunities for future teachers, using the same principles we use to design educational opportunities for children" (Mewborn & Stinson, 2007, p. 1483; emphasis in original), such emails are no longer merely disappointing but rather catalysts for uncharted trajectories for us—teacher educator and prospective teacher—to explore as we jointly learn to teach.

REFERENCES

Achinstein, B., & Barrett, A. (2004). (Re)framing classroom contexts: How new teachers and mentors view diverse learners and challenges of practice. *The Teachers College Record, 106,* 716–746.

Delpit, L. (1995). *Other people's children: Cultural conflict in the classroom.* New York, NY: The New Press.

Feiman-Nemser, S. (2001). From preparation to practice: Designing a continuum to strengthen and sustain teaching. *The Teachers College Record, 103,* 1013–1055.

Fennema, E., Carpenter, T. P., Franke, M. L., Levi, L., Jacobs, V. R., & Empson, S. B. (1996). A longitudinal study of learning to use children's thinking in mathematics instruction. *Journal for Research in Mathematics Education, 27,* 403–434.

Freire, P. (2000). *Education for critical consciousness.* New York, NY: Continuum. (Original work published 1969)

Freire, P. (2000). *Pedagogy of the oppressed* (M. B. Ramos, Trans., 30th anniv. ed.). New York, NY: Continuum. (Original work published 1970)

Hyun, E., & Marshall, J. D. (2003). Teachable-moment-oriented curriculum practice in early childhood education. *Journal of Curriculum Studies, 35,* 111–127.

Leonard, J. (2008). *Culturally specific pedagogy in the mathematics classroom: Strategies for teachers and students.* New York, NY: Routledge.

Mewborn, D., & Stinson, D. (2007). Learning to teach as assisted performance. *The Teachers College Record, 109,* 1457–1487.

Raymond, A. M. (1997). Inconsistency between a beginning elementary school teacher's mathematics beliefs and teaching practice. *Journal for Research in Mathematics Education, 28,* 550–576.

Stanley, C. A. (2006). Coloring the academic landscape: Faculty of color breaking the silence in predominantly White colleges and universities. *American Educational Research Journal, 43,* 701–736.

Wilson, M., & Goldenberg, M. (1998). Some conceptions are difficult to change: One middle school mathematics teacher's struggle. *Journal of Mathematics Teacher Education, 1,* 269–293.

CHAPTER 5

PROBLEMATIZING GENDER

Trepidation and Uncertainty

Carlos Nicolas Gomez
Clemson University

Eric Siy
University of Georgia

Teacher education programs can be described as a planned series of interventions with the goal of challenging a prospective teacher's vision of self as a teacher-of-mathematics. Teacher education programs are in the business of identity development. Through coursework, field experiences, and other experiences, the prospective teacher is pushed to question who they want to be as a teacher-of-mathematics. An individual's narrative of future self is referred to as one's *designated identity* (Sfard & Prusak, 2005). Prospective teachers go through a series of interventions in their teacher education program that influences their designated identity as teachers-of-mathematics. Consequently, a prospective teacher's identity as a teacher-of-mathematics fluctuates as he or she progresses through the program. In other words, one's identity is constantly being constructed, deconstructed,

Cases for Mathematics Teacher Educators, pages 85–94
Copyright © 2016 by Information Age Publishing
All rights of reproduction in any form reserved.

and reconstructed (Flores & Day, 2006). Thus, one's identity as a teacher-of-mathematics is a fluid part of being, not a static object.

One aspect of the prospective teachers' identity is their gender. Scholars define the concept of gender in different ways, but it is commonly differentiated from *sex*, one's categorization based on genitalia whereas *gender* lies within the social (Esmonde, 2011). Gender is more than one's sex, it is socially constructed and performative (Butler, 1990). "The contemporary use of the term *gender* incorporates certain distinctions that play out in the concepts of gender expectations, gender expression, gender attribution, gender assignment, and gender identity" (Rands, 2009a, p. 420). How one performs their gender is their *gender expression* or the collection of actions, activities, gestures, and discourses that express "how traditionally 'masculine' or 'feminine'" the individual is (Gay, Lesbian & Straight Education Network [GLSEN], 2014, p. 12). *Gender roles* are the socially accepted ways of acting that are appropriate for males and females (Rands, 2009a). The individual's culture will dictate the appropriate and inappropriate actions and activities that are attached to each gender category. We agree with Mendick (2006) that gender is a verb:

> Gender features traditionally as a noun, an aspect of the social world, and as an adjective, pinpointing a particular strand of identity. However, its most important use is as a verb. In other words, gender, as with all differences between people, is something that we do and are done by not something that we are. (p. 10)

We see having a deeper understanding of gender as a necessity to better support lesbian, gay, bisexual, transgender, questioning, and intersexual (LGBTQI) youth and families. The Gay, Lesbian, & Straight Education Network (GLSEN) found that 38.7% of LGBT students felt unsafe at school due to their gender expression and 55.5% due to their sexual orientation (GLSEN, 2014). LGBTQI students who felt unsafe at school had higher number of absences than other students (GLSEN, 2014). Understanding gender identity can help with the complexities of gender nonconforming children and prevent issues that may arise due to a limited meaning of gender. For example, a teacher with a limited understanding of gender identities can potentially create a hostile learning environment for LGBTQI students. Teachers can lower the status of LGBTQI students by perpetuating stereotypical gender roles in the classroom if unaware of how gender is enacted and learned. This can be done by the use of contextualized problems featuring traditional gender roles or by calling out students who transgress established gender roles. We do not want to be misunderstood; we do not believe that ignoring gender is the solution. Instead, teachers must be aware of the influence of gender in the mathematics classroom. For prospective elementary teachers, being able to critically analyze gender roles can provide a different way to discuss the influence of culture, power, and other larger social issues in the classroom (see Rands, 2009b).

We wanted to promote an antigenderist approach to mathematics education. "An anti-genderist approach strives to combat the use of the gender binary (or other simplified understandings of gender categories) to structure schools and society" (Esmonde, 2011, p. 30). Therefore, we created a lesson to problematize the meaning of gender and the idea of gender as a dichotomy to a cohort of prospective elementary teachers. In doing this, we hoped to better prepare them for teaching mathematics in a more equitable manner. We also developed this unit in response to previous calls to action by mathematics education researchers to further explore issues of heteronormativity, genderism, and sexism in the mathematics classroom (Damarin & Erchick, 2010; Esmonde, 2011; Fennema & Hart, 1994) and to address the low emphasis of teacher education programs that explore issues of gender and sexual orientation (Jennings, 2007). We hoped to model mathematical inqu[ee]ry by problematizing meanings of gender in the mathematics classroom that are often taken for granted in order to develop prospective teachers' critical eye (Rands, 2009b). In addition, it is important to make prospective teachers aware of master narratives (in this case, narratives of gender) in society that shape mathematics education (DiME, 2007), especially the role of the teacher who might perpetuate these narratives.

One dilemma in developing this lesson arose from our institutional positions. In our position as graduate students and future researchers, our own professional identities are being developed throughout this process. We became more aware of the politics and power structure involved in academia. Developing this lesson uncovered our *in-between* positions as students and educators. Though we felt supported by our superiors, we found reluctant and closed-door support for our ideas that instilled a sense of trepidation and uncertainty in going through with the lesson. We felt we had to play it extra safe and were concerned with pushing the students into an uncomfortable discussion. This led to our second dilemma, the unproblematized prospective teachers. We found that the prospective teachers had already espoused a more developed view of gender. Consequently, we believe the impact of the lesson was compromised. We begin this case study by describing the overall theory that informed our lesson design and the activities constructed. Thereafter, we describe some of the hidden narratives that arose while discussing the planned lesson with others. We end this case with a reflection on our experience teaching a constructed equity lesson for the first time.

DESIGNING THE LESSON

To expand prospective teachers' understanding of gender, we chose to use mathematical inqu[ee]ry (Rands, 2009b) as our framework to inform our development of the instruments and corresponding lesson. According to Rands (2009b), *mathematical inqu[ee]ry* is teachers and students pushing each other to "interrogate normativity in mathematics, classrooms, and society" (p. 189). It involves problematizing the taken-for-granted meanings that signify what is and is not appropriate in society. Additionally, it is a way to push students to think about the ways that heteronormative discourse is used in institutions like schools and in mathematics. Using mathematics to challenge meanings follows Gutstein's (2003) ideas that mathematics should be used as a way to develop the critical-thinking skills students (in our case, prospective teachers) need in developing a sense of social justice.

We believe that prospective teachers need to be aware of how gender will affect their classroom. In order to do this, Rands (2009a) argued that teachers need to practice gender-complex education. In comparison to the other forms (e.g., gender-stereotyped, gender-blind), gender-complex education "takes into consideration the complex sets of privilege and oppression that students and teachers experience based on their gender categories, gender expressions, and the gender attributions others make of them" (p. 426). Exposing prospective teachers to different ways that gender can influence instruction is one of the objectives of our instruments and lesson. We felt this was important because gender-complex perspectives embrace mathematical inqu[ee]ry and expose the danger of a dichotomous view of gender.

Table 5.1 shows the five major activities that the students participated in during two 70-minute classes. A lesson on culture and the mathematics classroom based on a unit described by White, DuCloux, Carreras-Jusino, Gonzalez, and Keels (2016) was conducted prior to the lesson on gender. After having discussed culture as a larger entity, we found it important to then focus on a particular social construction culture influences, that of gender, and seeing how it infiltrates our mathematics classrooms.

The final discussion focused on the influence of gender in the mathematics classroom and how a teacher may perpetuate a heteronormative discourse. Prospective teachers need to understand the power and influence they have on students' identities as doers-of-mathematics. Mendick (2006) argued that doing mathematics is aligned with masculinity and "we need to re-make maths starting by recognizing that it is a social activity, and is negotiated not absolute" (p. 68). We suggested the prospective teachers to take two actions. The first is to alter given gender roles in tasks and problems. We asked the students to construct contextual mathematics problems where the character transgresses traditional gender roles (e.g., Billy is using his EZ-Bake oven). This aids in normalizing these actions, which are therefore no longer seen as transgressions. Secondly, we push the prospective teachers to include invisible populations. For example, word problems that include family context and consider problems that involve other family structures such as families with a single parent, same-sex parents, or children who live with

TABLE 5.1	Lesson Phases	
Activity	**Description**	**Objectives**
Talking about culture	Activities inspired by White et al. (2016) were conducted. This includes a culture toolkit, discussion on mathematics stereotypes, and sharing narratives of being othered.	• Raise awareness of "the role of culture in the teaching and learning of mathematics" (White et al., 2016, p. 165) • Awareness of students' own multicultural mathematics dispositions (White et al., 2016)
Gender Interrogation Checklist (see appendix)	List of objects and events for which one is forced to decide whether it is more suitable for a female or male.	• Question the binary view of gender • Expand gender to a spectrum
Walk like a man (Berkowitz, Manohar, & Tinkler, 2010)	Volunteers are asked to portray their opposing genders in a series of tasks (walking and sitting).	• Explore gender as performance • Begin exploring the social and cultural ways gender is constructed
Suot-Baro Scenarios (suotbaro.weebly.com)	Participant is given a gender (Suot or Baro) that is not based on one's gender identity. The participant then goes through six scenarios exploring the consequences of the decisions they make.	• Exploring the ways that gender roles are constructed socially and culturally • How gender roles affect emotions and decision making • Better understanding children who are gender nonconforming
Questionnaire	Reflection on the instruments and discussion that followed.	• Reflect on the influence of gender roles and gender identity in the mathematics classroom • Encourage committing to practice gender sensitive or gender-complex education (Rand, 2009a)

grandparents. By including these invisible populations in contextual problems, these populations and cultures are normalized. The act of normalization is not limited to the LGBTQI community but can be used to normalize various other cultures, races, and those with disabilities.

RISING TREPIDATION

In the days leading up to the lesson, we began to share and discuss our goals with others. As graduate students and first-time instructors, this was going to be our first equity-based lesson that we designed and implemented. We found it important to get feedback from those who were more experienced. However, most whom we spoke to, though supportive of the lesson, felt necessary to warn us of the backlash from the prospective teachers. They critiqued our selected framework of mathematical inqu[ee]ry as "intentionally making students uncomfortable." One individual told us to send an e-mail to the administration about our plans. Others had wanted to make sure that we were prepared for emotional reactions that prospective teachers may have, pointing to the region's political and religious backgrounds of the students commonly held at our institutions. We recognized both the political and the personal space issues of gender may take up, and promised that we would be cautious. We found value in pushing the prospective teachers to question as perturbation, which could potentially open them up to new perspectives, and while it may be "uncomfortable," it is likely to be a valuable learning experience. Additionally, the discussion of social issues should make one uncomfortable because the veil being lifted forces one to realize their position within the issue.

On the day prior to the lesson, both of us began to feel more trepidation. The tensions grew, and we began to doubt whether this was a good idea considering our position as graduate students. We found our concerns directly tied to the fear of the political space that issues of gender exist in, but also the political nature of our position as graduate students. We felt we were putting more at risk, and we worried if our choices would be supported publicly. Bower and Klecka (2009) found similar tensions within a group of practicing teachers who were afraid that including LGBT families in the curriculum would offend parents and administration. The teachers were concerned that the push back from these populations would lead to serious consequences and little support from others. Our dilemma going into the lesson was like a high-roll bet. We thought the lesson could either empower the students to proactively consider gender in the mathematics classroom or encounter great resistance against the perceived dominant culture of heteronormativity, which could compromise our roles as instructors.

The narratives of fear and uncertainty from the feedback had made us nervous and overly cautious. We no longer felt safe conducting the lesson and feared that our closed-door support would fall silent. We comforted ourselves in knowing that push back would not be due to our own gender identities as a straight male (first author, instructor) because we did not foresee students seeing this lesson as propaganda. Only the ideas could be problematic.

ALREADY PROBLEMATIZED

As the lesson began, we asked prospective teachers to discuss the meaning of gender and what it means to be female and/or male. During the discussion, we realized that all of the prospective teachers had discussed gender in previous classes. They already considered gender as a spectrum and did not see a binary as a useful way of making sense of the world. They also seemed blasé about the lesson. The prospective teachers complained of having seen this before and asked why every class had to do this. We were nonplussed. We began to worry that the students had become oversaturated with discussions of gender and culture, to the point where they no longer found value in these discussions. This raised two pedagogical concerns. First, our entire theory informing our lesson had been problematizing and challenging the notion of a binary meaning of gender, but it had already been problematized. Secondly, how do we motivate and engage students who no longer see the value of discussing larger social issues? We saw no option but to continue with the lesson as planned and attempt to lead the discussion to a productive space and turn awareness into action. Once the lesson was over, we waited for the anonymous questionnaire responses. Below are two responses from the questionnaire demonstrating the differing responses we received:

> The survey question seemed to demand certain answers and associations and demanding one to agree that gender is societal invented with no innate factors. Furthermore, the PowerPoint presentation and the fact that our teacher lectured on the topic of gender left less room for debate and personal interpretation of varying understandings of what gender is. I think that a girl is a girl whether she fits the stereotyped gender of feminine traits or female gender roles- therefore the physical attributes of a girl is defined by sex, but the features and characters of what makes a girl is a broader term that includes all aspects of her personality and desires. Girl and boy is more broad than it seems in my mind. (Student 5)

> With the Gender Interrogation Checklist, I knew that people tend to identify certain objects with certain genders, and I knew that I thought that was wrong. However, once I did the Suot-Baro Scenarios, I found myself trying to figure out which actions were more "female" and which ones were more "male." I found myself trying to figure out if I was a male or a female in this alternate universe. This made me realize that I look for this classification in actions myself and it caused me to rethink my stereotypes of male and female actions. (Student 15)

Even as we reflect today on the questionnaire responses of the students, we are left uncertain about our lesson. We are still unsure of how to make sense of this experience. We hope to continue to better conceptualize our lesson, and create a stronger intervention for our students.

REFLECTION: PLAYING IT SAFE
AND ACCEPTING UNCERTAINTY

At the end of the lesson, it seems like we were not able to problematize gender too much. Looking back, there were many opportunities to push the students to further problematize and tease out the influence of gender on the everyday and in our institutions (religious, social, and school related). Subconsciously, the warning discourse and the fears of others were influential. To be honest, we were disappointed that the lesson seemed to have gone *too* well. There was no uprising or conflicted stances, and the discussion was open. We feel like we played it safe, and turned our back to the critical nature of the lesson in order to keep those outside of our classroom and ourselves safe. If we had realized that the prospective teachers already viewed the gender-binary as problematic, we would have given more time for them to find its application in the mathematics classroom. In the future, we will take the opportunity to assess the prospective teachers' understanding of gender prior to the lesson. This will provide us with ample time to design a stronger intervention. We envision starting the conversation by giving them contextualized problems where a character transgresses gender norms. Such as the following:

> Johnny has 4 hair bows and Daniel gives him 6 more. How many hair bows does Johnny have?

> Jessica has some toy trucks. Her fathers give her eight more trucks. Jessica now has 14 trucks. How many trucks did Jessica start with?

An unexpected result was how we realized the need to problematize the fear of the political space within our own institutions. Could we show to others that there is no need to fear a backlash? Like Bower and Klecka (2009) found, fear of the political space holds back teachers from being advocates for the LGBTQI population in schools. We imagine the teachers being afraid of not being supported if a backlash occurs. Inservice teachers would have much more to lose because they are already in the space that involves stakeholders like administrators and parents. How can we help them to not be afraid of being agents of change?

In our current position, we have to accept that we will probably not know the impact of our lesson. Will the veil remain lifted? Will they see how these issues of gender are perpetuated in the mathematics classroom? We have to accept the uncertainty of what we do as teacher educators when it comes to teaching issues of equity. The intervention planned for these students hopefully aided in the evolution of their identities as teachers-of-mathematics, where they shifted slightly to be more aware of issues of gender. In general, we hope that our prospective teachers make a commitment to resolving issues that are important to them. Most importantly, we hope they remember that silence is the most dangerous action that a teacher can make.

APPENDIX: GENDER INTERROGATION CHECKLIST

Using the following items, classify each as either male or female. You are not allowed to check both for any item. If it will help you, you can ask the following question:

If I had a son and daughter, who would it be more suitable for?

	M	F
Building blocks		
Rocking horses		
Dinosaurs		
Cars		
Magnetic letters		
Puzzles		
Picture books		
Stickers		
Sleepovers		
Dress up		
Chess		
Video games		
Basketball		
Volleyball		
Tennis		
Algebra		
Statistics		

	M	F
Swimming		
Twitter		
Instagram		
Sesame Street		
School musicals		
Disney		
CNN		
Grilling		
Baking		
Camping		
Hide-and-seek		
Green		
Yellow		
Emotion		
Active		
Geometry		
Problem-Solving		

REFERENCES

Berkowitz, D., Manohar, N. N., & Tinkler, J. E. (2010). Walk like a man, talk like a woman: Teaching the social construction of gender. *Teaching Sociology, 38*(2), 132–143.

Bower, L., & Klecka, C. (2009). (Re)considering normal: Queering social norms for parents and teachers. *Teaching Education, 20*, 357–373.

Butler, J. (1990). *Gender trouble: Feminism and the subversion of identity.* New York, NY: Routledge.

Damarin, S., & Erchick, D. B. (2010). Toward clarifying the meanings of "gender" in mathematics education research. *Journal for Research in Mathematics Education, 41*, 310–323.

DiME. (2007). Culture, race, power, and mathematics education. In F. Lester (Ed.), *Second handbook of research on mathematics teacher and learning* (pp. 405–433). Reston, VA: National Council of Teachers of Mathematics.

Esmonde, I. (2011). Snips and snails and puppy dogs' tails: Genderism and mathematics education. *For the Learning of Mathematics, 31*(2), 27–31.

Fennema, E., & Hart, L. E. (1994). Gender and the *JRME*. *Journal for Research in Mathematics Education, 25,* 648–659.

Flores, M. A., & Day, C. (2006). Contexts which shape and reshape new teachers' identities: A multi-perspective study. *Teaching and Teacher Education, 22,* 219–232.

Gay, Lesbian & Straight Education Network. (2014). *The 2013 national school climate survey: The experiences of lesbian, gay, bisexual and transgender youth in our nation's schools.* New York, NY: Author.

Gutstein, E. (2003). Teaching and learning mathematics for social justice in an urban, Latino school. *Journal for Research in Mathematics Education, 34,* 37–73.

Jennings, T. (2007). Addressing diversity in US teacher preparation programs: A survey of elementary and secondary programs' priorities and challenges from across the United States of America. *Teaching and Teacher Education, 23,* 1258–1271.

Mendick, H. (2006). *Masculinities in mathematics.* New York, NY: Open University Press.

Rands, K. (2009a). Considering transgender people in education: A gender-complex approach. *Journal of Teacher Education, 60,* 419–431.

Rands, K. (2009b). Mathematical inqu[ee]ry: Beyond 'add-queers-and-stir' elementary mathematics education. *Sex Education, 9*(2), 181–191.

Sfard, A., & Prusak, A. (2005). Telling identities: In search of an analytic tool for investigating learning as a culturally shaped activity. *Educational Researcher, 24*(4), 14–22.

White, D. Y., DuCloux, K. K., Carreras-Jusino, A. M., Gonzalez, D. A., & Keels, K. K. (2016). Preparing preservice teachers for diverse mathematics classrooms through a cultural awareness unit. *Mathematics Teacher Educator, 4*(2) 164–187.

COMMENTARY 1

USING MEDIA TO PROBLEMATIZE GENDER STEREOTYPES IN THE MATHEMATICS CLASSROOM

A Commentary on Gomez and Siy's Case

Katrina Piatek-Jimenez
Central Michigan University

In the case "Problematizing Gender: Trepidation and Uncertainty," Carlos Gomez and Eric Siy use the lens of identity to frame the design of their classroom lesson to problematize gender in the mathematics classroom. In their work with prospective teachers, Gomez and Siy explore how prospective teachers' gender identities and beliefs about gender influence their beliefs about their role as future teachers of mathematics.

Cases for Mathematics Teacher Educators, pages 95–99
Copyright © 2016 by Information Age Publishing

MY POSITIONALITY

In my work, I also use identity as a lens to study the interplay between gender and mathematics, though I explore the construct of identity from a different perspective. My work focuses on undergraduate women mathematics majors and their choices whether or not to pursue a career in the field of mathematics. I investigate women's mathematics identities, gender identities, and beliefs about mathematicians, and explore how women mathematics majors handle the dissonance that arises and how this influences their future career choices (Piatek-Jimenez, 2008, 2015). I also question how society, especially in the form of media, aids in shaping these identities and beliefs. It is with this perspective that I comment on this case.

INTERPRETING THE DILEMMA

In their classroom lesson described in this case, Gomez and Siy began by encouraging their students to problematize and challenge the notion that gender is binary, wanting the students to view gender as socially constructed and lying along a continuum, though they found that many of their students already held this perspective prior to the beginning of their lesson. They felt the lesson went "too well" and were disappointed that no "uprising or conflicted stances" arose. Although this may have been the case, I would argue that the results of their lesson did appear to make an impact on some of their students. For both Student 3 and Student 15, the Suot-Baro Scenarios activity and following discussion left them realizing that despite their knowledge of gender biases and discrimination, they too wanted to classify gender as "male" or "female" in this alternative universe. This insight into their own thinking problematized gender for them in that it made them discover their own prejudices and stereotypes that they did not even realize they had. Recognizing this dissonance between their intended beliefs and their actual perspective during the activity seems to have allowed them to move forward in their thinking about gender stereotypes. Dunn (2005) has discussed the importance of prospective teachers experiencing such a disequilibrium in their beliefs when assisting them to learn to teach for diverse populations.

RESPONDING TO THE DILEMMA

Historically, gender stereotypes in society were blatant and acceptable. Although in recent years they have become more subtle, such stereotypes still exist. Therefore, in order to assist prospective teachers to recognize the problem, we need them first to recognize their own biases and their unknowing acceptance of others' biases. A few years ago, a colleague and I developed a course for prospective K–8 teachers titled "Mathematics in Popular Culture." In this course, we explore how popular culture (such as television, comics, movies, music, and books) portrays mathematicians and the discipline of mathematics, and we discuss the impact that this has on society's beliefs of who can and should do mathematics and the influence this has on their future mathematics classrooms.

Though we explore many different media in this course, one of the first media we view is the 2004 movie *Mean Girls*. We have found this to be an excellent way to begin the semester because nearly all of our students have seen this movie previously, many of them having seen it multiple times. We ask them, however, to watch this movie again now with different eyes than before, taking note of the way that mathematics, and those who are good at mathematics, are portrayed in this movie. This tends to be an eye-opening experience for most of our students, especially given that this movie is intended to be about high-school cliques and not about mathematics at all. Our students are often taken aback by the fact that they had never before noticed all of the stereotypes and negative messages about mathematics infused within this movie. It is at this point that they begin to realize that stereotypes about gender, about race, and about mathematics are so engrained in society that even though they are already aware of the problems that exist, they had not previously recognized the prevalence of these messages in society.

In addition to helping students recognize the gender stereotypes specific to mathematics found in popular culture, it is also important for prospective teachers to become aware of the gender stereotypes specific to mathematics commonly found in the mathematics classroom. One such source is through mathematics textbooks. Research on gender bias in mathematics textbooks began in the 1970s. At that time, scholars found that in general, not only were males portrayed more often in mathematics textbooks, and that this distinction increased by grade level, but that males and females were depicted in very different gender-stereotypic roles (Kepner & Koehn, 1977; Northam, 1982). Although we may expect that the situation has changed dramatically since the 1970s, recent research suggests that we have not made as much progress as one would hope. For example, in a study I conducted recently with two of my colleagues, we found that substantial gender bias remains in the images found in current middle-school mathematics textbooks (Piatek-Jimenez, Madison, & Przybyla-Kuchek, 2014). Although the sheer number of males and females pictured were nearly equal, the activities and roles of the two genders were portrayed quite differently. Adult males were shown more often as active and in professional roles, whereas adult females were more often portrayed as passive and doing recreational activities. Furthermore, nearly twice as many males were shown as having careers than females, and males were depicted in a much larger variety of careers than females. Interestingly enough, there was a clear effort made in these textbooks to include select images of females working in male-dominated careers; however, very little attempt was made to portray males working in female-dominated careers. I find this last point particularly interesting because I believe that it demonstrates a willingness of our society to allow women to step outside of society's gender-defined roles, a willingness that we typically do not afford to men.

Relating to this, toward the end of their case, Gomez and Siy suggest providing their prospective teachers with contextualized problems where "the character transgresses gender norms" such as boys trading pink hair bows or girls playing with toy trucks. I am interested to learn what they find from this experience. Given society's current stance on gender, I would hypothesize that these scholars experience much more resistance from their prospective teachers on the boys-trading-hair-bows problem. In today's society, girls playing with toy trucks (or any other "boy toys" for that matter) is much more acceptable, is often encouraged, and does not elicit the same level of discomfort as boys being interested in traditional "girl things."

CONCLUDING REMARKS

It is important for prospective teachers, during their college education, to learn how to problematize gender and to understand society's influence on gender expectations and gender roles. It is also important for prospective teachers to reflect on how these stereotypes influence their behavior as teachers and the behaviors of their students, in order to provide them insight into developing more equitable classrooms. What we may find as mathematics teacher educators is that our students may be gaining these insights in other courses in their teacher education program. This reality may free us up to discuss gender at a deeper level and specifically in the context of the mathematics classroom with our students. Helping our prospective teachers more clearly recognize societal beliefs about gender specific to mathematics, found both in the media and within the mathematics classroom, can aid them in developing more welcoming classrooms.

REFERENCES

Dunn, T. K. (2005). Engaging prospective teachers in critical reflection: Facilitating a disposition to teach mathematics for diversity. In A. J. Rodriguez & R. S. Kitchen

(Eds.), *Preparing mathematics and science teachers for diverse classrooms: Promising strategies for transformative pedagogy* (pp. 143–158). Mahwah, NJ: Erlbuam.

Kepner, H. S., & Koehn, L. R. (1977). Sex roles in mathematics: A study of the status of sex stereotypes in elementary mathematics texts. *The Arithmetic Teacher, 24,* 379–385.

Northam, J. (1982). Girls and boys in primary math books. *Education 3–13, 10*(1), 11–14.

Piatek-Jimenez, K. (2008). Images of mathematicians: A new perspective on the shortage of women in mathematical careers. *ZDM Mathematics Education, 40,* 633–646.

Piatek-Jimenez, K. (2015). On the persistence and attrition of women in mathematics. *Journal of Humanistic Mathematics, 5*(1), 3–54.

Piatek-Jimenez, K., Madison, M., & Przybyla-Kuchek, J. (2014). Equity in mathematics textbooks: A new look at an old issue. *Journal of Women and Minorities in Science and Engineering, 20*(1), 55–74.

PROBLEMATIZING GENDER: LEARNING TO EMBRACE UNCERTAINTY

A Commentary on Gomez and Siy's Case

Kai Rands
Independent Scholar

In this commentary, I will identify the issues and dilemmas I read in the case. I will share my own experiences as a teacher educator teaching both "math methods" courses and "diversity" courses and consider some possible ways to respond to the issues.

MY POSITIONALITY

As I read about the authors' experience introducing the topic of gender and realizing that the prospective teachers had already had discussions in previous classes, the experience resonated with me. Although I remember this happening more than once, I will mention one example: My first

Cases for Mathematics Teacher Educators, pages 101–105

semester teaching at a particular university, a review of course descriptions and syllabi led me to the presumption that the current curriculum did not address language diversity. As it turned out, the information "on the books" did not reflect recent changes in course content, and the prospective teachers had more familiarity with linguistic diversity than I had anticipated.

In my interpretation of my own experiences and the case, I see a lack of communication and collaboration among faculty. Linda Darling-Hammond (2012) has identified this obstacle as "traditional isolationism and individualism" that "plague many university faculties" and interfere with creating a coherent approach to supporting developing teachers (p. 92). Darling-Hammond's research indicates that courses in "highly successful programs" are carefully sequenced, "designed to intersect with each other," and are "tightly interwoven with...students' work in schools" (p. 94). Perhaps the more obvious approach to addressing a lack of communication, collaboration, and coherence is a top-down, hierarchical approach to infusing diversity and social justice perspectives throughout teacher education programs. A second, perhaps less obvious, approach proceeds horizontally rather than vertically. In this approach, collaboration and communication stem from teacher educators who reach out horizontally to one another to find and create connections among courses.

COVERING GENDER AND "WHY DO WE ALWAYS..." AS A RESISTANCE MOVE

Mathew Felton and Courtney Koestler (2012) noted that "prospective teachers see themselves as largely passive receivers of a static and depoliticized body of knowledge" (p. 25). Perhaps it is not too surprising that prospective teachers view *mathematics* as a "static and depoliticized body of knowledge" inasmuch as many prospective teachers' experiences as students may have led them to this view. More surprising is that the prospective students in the case seem to have this same view of *gender*. The prospective teachers' reluctance to engage with gender because of "having seen this before" implies an image of curriculum as a set of topics to be covered. The presumption seems to have been that seeing something once or twice or having several discussions about gender means that one has "covered" that topic and is "prepared" to teach with gender diversity in mind. Such a view overlooks the complexity of gender.

INTERPRETING THE DILEMMA

The case authors' suggest that the lesson may have gone "*too* well." I will offer another possible interpretation of the prospective teachers' responses.

Resistance takes many forms, and while it is true that there was "no uprising or conflicted stances" and that the teachers were "open to discussion," it is also possible that the prospective teachers were using a different form of resistance when they "asked why every class had to do this." Reading about this response reminded me of times in which I have encountered similar responses from prospective teachers. In one example, I asked prospective teachers to participate in a class activity related to privilege and oppression. Groups of students sat in circles. Each student began with a sheet of paper that listed one form of privilege and a sentence stem related to the form of privilege. For example, one sheet listed "sexuality privilege," with the sentence stem "Straight people can count on..." Prospective teachers wrote down as many different manifestations of the form of privilege on their sheet as they could within several minutes, then passed the sheets around the circle until each prospective teacher had added to each sheet. In all, there were nine different forms of privilege addressed. However, after the activity, several prospective teachers wrote in their journals something to the effect of "Why do we always have to talk about race?" In actuality, race privilege was only one of nine forms of privilege addressed. In my interpretation, the journal responses indicated that this predominantly White group of prospective teachers were experiencing resistance, especially to the topic of White privilege, a form of privilege they experienced. In the case, "Why do we always...?" may be a move to avoid further discussion, resistance to going deeper with the topic of gender, or a way to induce silence (see, e.g., Gay & Kirkland, 2003; Mazzei, 2011).

RESPONDING TO THE DILEMMA

In addition to my own experience with the "Why do we always..." response, I am led to interpret this response as resistance based on some of the prospective teachers' statements in their reflections. Despite general openness to the idea of gender as a spectrum, prospective teachers' statements tend to include contradictions and tensions that may indicate that they are still in a state of productive resistance. For example, Student 5 said that the tasks are set up to "demand" acceptance that "gender is societal invented" but also seems to be struggling with conceptualizations of gender that may be new to the student: "Girl and boy is more broad than it seems in my mind." This student seems to be resisting the "demand" to view gender in a new way at the same time as considering it. Moreover, the students' new conceptualization ("Girl and boy is more broad...") still invokes a binary view of gender as consisting of boys and girls. Student 15 stated that the student entered the discussion with the perspective that it was wrong to "identify certain objects with certain genders," yet "noticed myself trying to figure out which actions were more 'female' and which ones were more 'male'" in the Suot-Baro Scenarios. Again, the student overlay an unacknowledged binary view of gender on the hypothetical scenarios. The task prompted the student to "realize that I look for this classification" and to "rethink my stereotypes." Ultimately, as the case authors point out, the process of interpreting prospective teachers' responses (both written and oral) is uncertain, and alternative interpretations are possible. Moreover, unintended consequences are always a risk in education, and I agree that part of being a (teacher) educator is teaching without knowing the full impact of our practice.

GOING FORWARD

The activities shared in the case provide resources from which teacher educators can draw in staging encounters with gender with prospective teachers. Considering how to interweave gender-complex teacher education throughout teacher education programs holistically rather than as separate, individual courses can provide numerous encounters with gender diversity in different contexts, prompting prospective teachers to consider gender diversity in deeper and more nuanced ways. For some teacher education programs, the activities suggested in the case may fit better into a diversity-focused class, allowing prospective teachers to consider gender-complexity within mathematics curricula and pedagogy in the math methods course, as the case authors suggested. Also important is moving beyond the idea of gender diversity as a spectrum to addressing issues of privilege and oppression, power dynamics related to gender category and gender transgression oppression, and the intersection of gender and other forms of social regulation (e.g., race, class, sexuality, language). Finally, teacher educators committed to social justice can benefit from a broad view of the context of teacher education. As Cornbleth (2014) pointed out, teacher education programs are subject to the dynamics of state regulations, accreditation standards and agencies, the cultures of their institutions and the surrounding communities, employer markets, and professional organizations, among others. Understanding these contextual factors makes it more likely that teacher educators can press past the boundaries of playing it safe while staying on the inside of teacher education.

REFERENCES

Cornbleth, C. (2014). *Understanding teacher education in contentious times: Political cross-currents and conflicting interests.* New York, NY: Routledge.

Darling-Hammond, L. (2012). Building a profession of teaching: Teacher educators as change agents. In M. Ben-Peretz, S. Kleeman, R. Reichenber, & S. Shimoni (Eds.), *Teacher educators as members of an evolving profession* (pp. 87–102). Lanham, MD: Rowman and Littlefield Education.

Felton, M. D., & Koestler, C. (2012). 'Questions and answers *can* mean something': Supporting critical reflection in mathematics education. In R. Flessner, G. R. Miller, K. M. Patrizio, & J. R. Horwitz (Eds.), *Agency through teacher education: Reflection, community, and learning* (pp. 25–35). Lanham, MD: Rowan and Littlefield Education.

Gay, G., & Kirkland, K. (2003). Developing cultural critical consciousness and self-reflection in preservice teacher education. *Theory Into Practice, 42*, 181–187.

Mazzei, L. (2011). Desiring silence: Gender, race, and pedagogy in education. *British Educational Research Journal, 37*, 657–669.

GENDER ≠ SEX ≠ SEXUAL ORIENTATION

A Commentary on Gomez and Siy's Case

Marcy B. Wood
University of Arizona

Schools can be uncomfortable places for students who fail to conform, especially students who enact alternative gender performances. If schools are to be safe places, the gender work of Carlos Gomez and Erik Siy is critical, and I thank them for taking on this challenge. In my commentary below, I describe what I learned from their case. I then offer suggestions for extending this work.

INTERPRETING THE DILEMMA

I was particularly intrigued by the feedback from preservice teachers (PSTs). While Gomez and Siy expressed disappointment with their lesson, I felt the PSTs gained important insights from the activities. The two reported PST responses described changes in their thinking about gender roles, including a determination to rethink how they reinforced gender stereotypes.

Cases for Mathematics Teacher Educators, pages 107–111
Copyright © 2016 by Information Age Publishing

Also, PSTs' feedback offered insights into features of successful activities. The PSTs seemed to benefit more from the Suot-Baro scenarios than from the gender interrogation checklist. As one PST noted, they had already problematized the checklist idea that certain activities were identified with certain genders. Interestingly, the PSTs seemed to miss the authors' primary goal for the checklist, problematizing the binary construction of gender. Perhaps the checklist could be modified to present a range of gender descriptions including intergender or genderfree. Asking PSTs to decide activities suitable for an intergender person might help them think beyond gender as a dichotomy.

In contrast, the Suot-Baro scenarios helped PSTs think more deeply about gender stereotypes. In the Suot-Baro scenarios (suotbaro.weebly.com), participants take on the gender of Suot or Baro and then select one of two gendered responses to a scenario. If the PST chooses the "correct" gender performance, they are rewarded with acceptance, praise, and admiration. If their choice is "wrong," they are harassed, ostracized, and shamed. It is not possible to determine your gender, and thus it is not possible to make the correct gender choice. Yet, it is painfully clear when your choice transgresses the gender norm.

As PSTs first engaged in this activity, they worried about not knowing which choice was appropriate for their Suot or Baro gender. However, through discussion, they realized that their gender identity was less important than experiencing the pain of the "wrong" gender performance. This consequence helped some PSTs consider the impact of their negative reactions to students' alternative gender performances. In contrast, other activities, such as the gender interrogation checklist, did not explore the shame or hostility connected to gender transgressions. Perhaps PSTs benefit more from activities that go beyond noticing gender norms and help them experience the negative consequences of nonconforming gender performances.

RESPONDING TO THE DILEMMA

This case addresses many important issues. However, I would like to push the authors (and all mathematics teacher educators) to be more precise in their use of gender terms. In particular, many people frequently entangle gender, sex, and sexual orientation, leading to additional confusion rather than to clarity.

Male and Female

Gomez and Siy defined gender, noting that it was different from ("more than") sex. However, they then described gender using the sex words *male* and *female*. Gomez and Siy are not alone in using these words as gender labels. However, when male and female refer to gender, it suggests that gender is connected to sex and has biological origins. Instead, gender is culturally derived and dependent upon interactions with others.

When male and female are used for gender, people may wonder whether the conversation is intended to refer to cultural expectations or biological features or both. For example, consider this statement: "In 2003, the average scale score of male 4th-grade students (236) was significantly higher... than that of female 4th-grade students (233)" (McGraw, Lubienski, & Strutchens, 2006, p. 135). This use of male and female allows an interpretation that biology might be involved in the difference in mathematics achievement. This is dangerous, as once sex becomes a reason for differences, the differences can be seen as expected and immutable, rather than as something that can (and should) be challenged by addressing cultural expectations.

Also, when we lack specific vocabulary to differentiate gender and sex, we suggest that gender and sex should align and that people whose gender expression differs from their sex are not normal. For example, female students who fail to act like females are seen as weird and may be pejoratively labeled as lesbians. To address these concerns, many authors reserve the words male, female, and intersex to refer to sex and use words like masculine, boy, man, feminine, girl, woman, genderqueer, androgynous, or intergender to refer to gender.

Gender and Sexual Orientation

In conversations about gender, it is also important to differentiate gender expression and sexual orientation. Sexual orientation describes whom a person is attracted to whether this person is someone of the same sex, opposite sex, both, or neither. Sexual orientation is defined in terms of sex and not gender. For example, a female person may enact a masculine gender, be primarily attracted to male people, and consider herself to be heterosexual.

Many people entangle gender, sex, and sexual orientation, assuming that people whose gender performances are nonnormative are also homosexual. For example, many people assume that a male student who wears pink and likes dolls is also gay. However, wearing pink and liking dolls are cultural gender performances and are not inherently expressions of sexual orientation. When teachers fail to distinguish gender expression and sexual orientation, their interventions in bullying situations may result in additional harm. For example, many boys are harassed for being gay when in fact they are not gay, but are instead gender nonconforming (Kimmel & Mahler, 2003). When teachers intervene, they may respond to the actual taunt (about sexual orientation) without addressing the discriminations attached to the underlying issue (gender nonconformity). This intervention fails to address stereotypes about gender expression while also reinforcing stereotypes about homosexual people as unable to adequately perform the normative gender. It also incorrectly assumes that gender performance includes sexual orientation.

The case presented by Gomez and Siy contains several instances in which the authors connect gender expression and sexual orientation. For example, the argument for caring about gender expression is framed in terms of support for LBBTQI youth and families. The label LBBTQI includes people with nonconforming gender performances, sexual orientations, biological sex characteristics, or a combination of these. However, LBBTQI does not mean that homosexuality, gender performance, and sex are inherently linked. It would be more accurate for Gomez and Siy to argue that work on gender expression is in support of transgender, queergender, or androgynous youth and families. Some of these youth may also have a nonconforming sexual orientation, but conflating gender and sexual orientation adds confusion rather than clarity to conversations about these topics.

GENDER, MATHEMATICS, AND PSTS

My commentary has not yet addressed the critical intersection of mathematics education and gender. While I appreciate Gomez and Siy's work on gender, I urge us to move beyond their suggestion to use gender nonconforming story problems as a way of addressing gender in our classrooms. These stories may reinforce gender binaries (as when boys are shown with stereotypically feminine objects). Also, these stories fail to address a larger issue: What counts as culturally desirable mathematical activity maps onto masculine activity (Mendick, 2006) and thus provides men or boys (as opposed to women or girls or genderqueer individuals) with more opportunities to see themselves in mathematical activity. This is not to say that mathematics must be a masculine activity. Instead, what we tend to value and reinforce in schools (abstract thinking and rational reasoning) tends to be the same activities we assign to masculine gender performances.

If we want students of all genders to see themselves as mathematical people, we must move beyond masculine constructions of mathematics. We must design tasks that embrace a full spectrum of engagement. For example, teachers at one high school were successful in helping all students, including students of all genders, see themselves as capable of doing mathematics (Boaler, 2006). These teachers used a pedagogical strategy called Complex Instruction (CI) (see Featherstone, Crespo, Jilk, Oslund, Parks, & Wood, 2011), which requires many different ways of engaging with mathematics. As we help students become comfortable with connections between their gender performance and their mathematical activity, we make our classrooms safer and more successful places for all students.

REFERENCES

Boaler, J. (2006). How a detracked mathematics approach promoted respect, responsibility, and high achievement. *Theory Into Practice, 45*, 40–46.

Featherstone, H., Crespo, S., Jilk, L., Oslund, J., Parks, A., & Wood, M. (2011). *Smarter together! Collaboration and equity in the elementary math classroom*. Reston, VA: National Council of Teachers of Mathematics.

Kimmel, M. S., & Mahler, M. (2003). Adolescent masculinity, homophobia, and violence random school shootings, 1982-2001. *American Behavioral Scientist, 46*, 1439–1458.

McGraw, R., Lubienski, S. T., & Strutchens, M. E. (2006). A closer look at gender in NAEP mathematics achievement and affect data: Intersections with achievement, race/ethnicity, and socioeconomic status. *Journal for Research in Mathematics Education, 37*, 129–150.

Mendick, H. (2006). *Masculinities in mathematics*. New York, NY: Open University Press.

CHAPTER 6

CHALLENGING AND DISRUPTING DEFICIT NOTIONS IN OUR WORK WITH EARLY CHILDHOOD AND ELEMENTARY TEACHERS

Courtney Koestler
OHIO Center for Equity in Mathematics and Science
Ohio University

This kind of mathematics teaching sounds great, but it wouldn't work with my students.

I would love to have my students really engage in problem solving, but they just don't know their facts.

My students don't come with any outside experiences to build on. All they ever do is play video games and watch TV outside of school.

Unfortunately, the sentiments above have been voiced by prospective and practicing teachers in various explicit and implicit ways at the beginning of almost every semester I have taught courses at the university level. Because the kind of pedagogy I advocate is often very different from what many of the teachers with whom I work have experienced in their own schooling, and what they may currently experience in the field, some teachers are

Cases for Mathematics Teacher Educators, pages 113–121
Copyright © 2016 by Information Age Publishing
All rights of reproduction in any form reserved.

quick to voice resistance to teaching approaches that are different from traditional, teacher-centered pedagogies. They challenge the idea that "progressive" kinds of pedagogy can be done in their classrooms with the students in front of them. In this case, I discuss the dilemma of how to confront, challenge, and disrupt certain deficit discourses, while at the same time trying to create an open, honest, and reflective classroom community where teachers can freely discuss their perspectives.

MY BACKGROUND AS A TEACHER EDUCATOR

The work I reflect on here is based on my 10 years as a university-based teacher educator where I have worked with prospective and practicing PreK–8 teachers at three distinct universities and their surrounding school districts. My 7 years as a classroom teacher and mathematics coach in public schools located in exceptionally diverse communities also has greatly influenced how and why I advocate critical and child-centered pedagogies for all students. As a K–8 teacher and coach, I was afforded many high-quality professional development experiences that supported me in learning about child-centered mathematics pedagogies (e.g., Cognitively Guided Instruction), community-based perspectives (e.g., Funds of Knowledge for Teaching), and critical approaches to teaching as well as putting these progressive pedagogies into practice in my work with culturally, linguistically, and economically diverse students.

MY APPROACH TO TEACHING TEACHERS

A long-standing challenge in our work as teacher educators is advocating pedagogical approaches that are often very different from what many prospective teachers experienced in their own schooling as children. In their K–12 classrooms, teachers have been socialized through an "apprenticeship of observation" (Lortie, 1975) regarding what counts as mathematics, mathematics teaching, and mathematics learning. Many of these experiences were traditional, teacher-centered pedagogies at odds with child-centered teaching practices and pedagogical approaches effective in meeting the needs of today's students (Ball & Cohen, 1999; Darling-Hammond, 2013). In my work as a university-based teacher educator, I have tried to foreground diversity, equity, and social justice throughout course readings, major assignments, class activities, and class discussions so that teachers are supported in thinking critically about their work as teachers in the social, political, and cultural spaces in which it takes place and they are prepared to more equitably and justly teach the diverse students in front of them.

However, I often start by introducing the idea of *child-centered approaches to teaching*.

To me, the idea of looking at children as competent mathematical thinkers with important background experiences on which to build is what I consider a child-centered approach to teaching. In my current work with teachers, I use this term as a way to focus teachers' attention on the importance of valuing children and connecting to student thinking. I also appreciate this term because it directly challenges teacher-centered pedagogies and may help teachers reconceptualize the way we think about the kinds of important mathematics that are discussed in classrooms. It also connects readily to ideas about community mathematics—what kinds of mathematical practices are children engaging in outside of school with their families and in their communities—and critical mathematics—what kinds of mathematics do these children need to use (and are interested in) to address relevant issues of social justice in their schools, community, and the world. I also believe it aligns well with other important concepts in early childhood and elementary teaching, such as emergent curriculum, constructivism, and sociocultural theories of learning.

An explicit goal of the courses I teach is to engage in critical reflection (see Felton & Koestler, 2012, 2015) and critical discussions to discuss and unpack the *practical theories* (Handal & Lauvas, 1987) that teachers hold based on their own personal experiences, knowledge and understandings they learn from others, and their own personal values and beliefs. Because these practical theories influence the way teachers frame mathematics teaching and learning and the students in their classrooms, it is important for teachers to have the space to discuss and unpack them. By highlighting equity, diversity, and social justice to critically reflect on their perspectives and practice, teachers may see the need for more child-centered and critical ways of teaching and be able to envision how to implement it in their own classrooms.

In order to support a broader understanding of different kinds of important mathematics, I have adapted Eric Gutstein's (2006, 2007) "3Cs" framework—classical mathematics, community mathematics, and critical mathematics—to think about the kinds of mathematics I want to include and the kinds of mathematics knowledge for teaching I want to support in my course. The first, formal way that I usually introduce teachers to these different kinds of mathematics and mathematical knowledge is in the syllabus as follows:

> In this course we will explore classical, community, and critical mathematical knowledge (Gutstein, 2006, 2007) and what this means to us as early childhood mathematics teachers. Classical mathematical knowledge refers to formal kinds of mathematical knowledge that children often gain and use in school. This form of knowledge is very similar to how the National Council of

Teachers of Mathematics (NCTM, 2000) and others define teaching mathematics for, and learning mathematics with, understanding, as well as the kind of knowledge that supports students' achievement on traditional measures of success (e.g., state assessments). In early childhood mathematics classrooms, using and building on children's informal mathematical thinking is an important aspect of developing classical mathematical knowledge. Community mathematical knowledge is the kind of mathematical knowledge that children use and develop by participating in various home and community practices outside of school. This knowledge is also important to understand and connect to the school curriculum. Critical mathematical knowledge refers to the mathematical knowledge that students need and use to understand, analyze, and critique issues of social (in)justice in their communities and society.

In some ways, teachers see classical mathematics as the "real mathematics" with community and critical mathematics as "extra" or an "add-on" despite my attempts to foreground all of them, showing how they are distinct but related (see Koestler, 2010, 2012). Regardless, all are framed with the importance of using a child-centered pedagogy, where teachers take seriously children's thinking, strengths, and interests when considering their approaches to teaching mathematics.

When I first began as a university-based teacher educator, I framed classical mathematics as the kinds of mathematics teaching and learning that I advocated as "reform-based" or "standards-based." In recent years I have moved purposely away from these terms, in part because of the changing notion of what *reform* means, who the "reformers" are, and what set of standards are currently being used. Although I still do include attention on various sets of standards (e.g., local and state standards, the Common Core Standards), I often use the NCTM's (2000) *Principles and Standards for School Mathematics* as a beginning point because I believe these standards are well-written, well-organized, easy for teachers to understand and use, and they were the standards that I used originally to guide my own work as a teacher.

Cognitively Guided Instruction, the body of work produced by Carpenter, Fennema, Franke, Levi, and Empson (e.g., 1999) and their colleagues, has been a huge influence on my child-centered pedagogical approach as an elementary school teacher, a mathematics coach, and as a teacher educator. This work describing how all children come to school with informal knowledge that supports them in solving different kinds of mathematics problems without having to be told how, and how teachers can use this to inform their instruction, transformed the way I thought about teaching and learning mathematics. As a teacher, I was able to see how all of the students at my school had important real-world experiences, even though those experiences varied widely because of their diverse backgrounds. I also realized the need to connect to these experiences and students' strengths (and the importance of it) with what I was trying to accomplish in the classroom. As I became a more knowledgeable and skillful CGI teacher, I was convinced of these ideas both intellectually and practically because of the success I had with all kinds of culturally, linguistically, and economically diverse children with whom I worked.

THE DILEMMA

As the opening statements reveal, some prospective and practicing teachers believe that the students in front of them, *their students*, are unable to engage in high-quality mathematics learning (and often these deficit and destructive discourses are about children who have been historically marginalized in schools), or at least they do not initially envision how it can happen. This sometimes occurs despite teachers understanding and acknowledging the power of child-centered pedagogies and knowing the evidence that all children bring strengths to mathematics classrooms, in part because it is difficult work to do and teachers may not have the professional development and support to do so. I am interested in challenging the idea that child-centered pedagogy can only work for certain children.

However, this also presents a dilemma for me. I have experienced a lot of tension when some teachers in my courses have voiced perspectives like those above that run counter to the idea and power of child-centered pedagogy for *all* children. On one hand, I want to promote a welcoming classroom community where teachers feel safe, and are open and honest about their practical theories of teaching, learning, and children. Only by doing so do I believe they can effectively broaden and refine these understandings. But on the other hand, I also want to challenge, critique, and disrupt deficit notions when voiced. To do this without teachers feeling shut down or judged or silenced is often difficult.

REFLECTING ON THIS DILEMMA

Although there are multiple ways to productively challenge and disrupt this kind of deficit framing of students, my dilemma centers on how to do so in a way that still communicates respect for teachers' experiences and perspectives (and does not convey a sense of judgment and assessment). The contexts in which I work with certain groups of teachers (e.g., prospective teachers in an undergraduate course, practicing teachers in a graduate-level course, professional-development settings that include both prospective teachers and mentor teachers) also can affect the way that I might respond. Following are three different ways that I have attempted to address this dilemma.

Setting Norms

I work carefully and purposefully to build trust and a strong community when working with teachers. I believe that being explicit about norms and expectations in any learning community is important, especially when there may be status differences (e.g., teacher and student, university-based teacher educators and school-based teacher educators). I often start by suggesting norms for classroom interactions, and sometimes these norms are quite broad, such as *be imperfect or unsure* or *be sure to ask questions when unsure or when you have a different view,* and sometimes they are more specific like *articulate connections or disconnections between experiences and course literature.* (Although I sometimes start by suggesting norms like these, I always ask teachers to suggest some too.) I also try to make a point that although it is important to analyze and critique statements and ideas, we must be careful not to critique the individuals themselves.

Providing "Relevant" Resources

Another way that I have attempted to challenge deficit notions of certain kinds of students is by being very purposeful and thoughtful about the kinds of resources I use in class. For example, I try to find examples of work that have been done in different contexts, especially those contexts that are similar to what teachers are currently experiencing. For example, I previously used a set of videos that did an excellent job at emphasizing student thinking and mathematically productive classrooms, however they mostly featured children who some might assume to be "typical" or "advanced" White students. I now show videos highlighting students with various racial, ethnic, and linguistic backgrounds, and those who may struggle in different ways, so that teachers can more easily envision what it might look like in their distinct context or with simply "different" kinds of kids.

Modeling or Providing "Powerful" Mathematics Lessons

Finally, another strategy that I have used with success is taking prospective teachers on a "field trip" to model powerful lessons in real classrooms (or providing the lessons for the prospective teachers to try out themselves). By powerful lessons, I mean well-developed, mathematically-important lessons that I know from experience work well with just about any group of students, specifically highlight students' unique and powerful thinking no matter what students' prior experiences have been, and are interesting and engaging to students. In these kinds of lessons, teachers can observe their students possibly responding in ways that they may have never noticed otherwise when teaching traditional, teacher-led lessons. In other words, these lessons provide concrete opportunities to see their students' mathematical thinking and strengths. All that said, despite its importance, many university-based teacher educators do not often have the time and space to spend large amounts of time out in the field due to the demands of teaching, research, and other institutional responsibilities.

FINAL THOUGHTS

Although I have just briefly described ways that I attempt to challenge and disrupt deficit thinking, I still struggle with those in-the-moment instances where a teacher makes a disparaging remark about a student or a group of students that I feel a responsibility to challenge. It is difficult because I want teachers to feel safe and comfortable in making their beliefs and perspectives public, but at the same time I feel a need to disrupt deficit discourses that teachers may bring. For example, if a mentor teacher expresses a negative stereotype about their students in front of a prospective teacher, I worry I might undermine our already tenuous relationship by pointing out how her perspective might be damaging and false. In situations like this, I might ask the prospective how this connects or conflicts with what we are learning about in class, or I might try to ground my response in my own experiences as a teacher, but I always try to do it with much concern and care.

In closing, while teachers sometimes have negative notions of their students and make comments about their students not being able do the kinds of things we are advocating, I believe that all teachers truly do want their students to succeed and to be powerful mathematics thinkers. I believe that our role as mathematics teacher educators is to continually push teachers to believe their students are capable problem solvers and support them in being able to do so.

REFERENCES

Ball, D. L., & Cohen, D. K. (1999). Developing practice, developing practitioners: Toward a practice-based theory of professional practice. In L. Darling-Hammond & G. Sykes (Eds.), *Teaching as the learning profession: Handbook of policy and practice* (pp. 3–32). San Francisco, CA: Jossey-Bass.

Carpenter, T. P., Fennema, E., Franke, M. L., Levi, L., & Empson, S. B. (1999). *Children's mathematics: Cognitively guided instruction.* Portsmouth, NH: Heinemann.

Darling-Hammond, L. (2013). *Powerful teacher education: Lessons from exemplary programs.* San Francisco, CA: Jossey-Bass.

Felton, M. D., & Koestler, C. (2012). "Questions and answers *can* mean something": Supporting critical reflection in mathematics education. In R. Flessner, G. R. Miller, K. M. Patrizio, & J. R. Horwitz, (Eds.), *Agency through teacher education: Reflection, community, and learning* (pp. 25–35). Lanham, MD: Rowman & Littlefield Education.

Felton, M. D., & Koestler, C. (2015). "Math is all around us and...we can use it to help us": Teacher agency in mathematics education through critical reflection. *The New Educator, 11*(4), 260–276.

Gutstein, E. (2006). *Reading and writing the world with mathematics: Toward a pedagogy of social justice.* New York, NY: Routledge.

Gutstein, E. (2007). Connecting community, critical, and classical knowledge in teaching mathematics for social justice. In B. Sriraman (Ed.), *International perspectives on social justice in mathematics education* (pp. 109–118). Monograph 1of The Montana Enthusiast. Missoula, MT: The University of Montana and the Montana Council of Teachers of Mathematics.

Handal, G., & Lauvas, P. (1987). *Promoting reflective teaching.* Milton Keynes, England: Open University Press.

Koestler, C. (2010). *(Re)Envisioning mathematics education: Examining equity and social justice in an elementary mathematics methods course* (Unpublished doctoral dissertation). University of Wisconsin, Madison, WI.

Koestler, C. (2012). Beyond apples, puppy dogs, and ice cream: Preparing prospective K-8 teachers to teach mathematics for equity and social justice. In A. A. Wager & D. W. Stinson (Eds.), *Teaching mathematics for social justice: Conversations with educators* (pp. 81–97). Reston, VA: National Council of Teachers of Mathematics.

Lortie, D. C. (1975). *Schoolteacher: A sociological study.* Chicago, IL: University of Chicago Press.

National Council of Teachers of Mathematics. (2000). *Standards for school mathematics.* Reston, VA: Author.

WEAKENING DEFICIT PERSPECTIVES WITH COLLECTIVE AGENCY

A Commentary on Koestler's Case

Higinio Dominguez
Michigan State University

In my response to this case, I will emphasize the idea—not new but relevant to these dilemmas—of doing things with words (Austin, 1962) and recruiting others to enact a kind of agency that is socially distributed and shared, something that is "frequently a property of dyads and other small groups rather than individuals" (Wertsch, Tulviste, & Hagstrom, 1993, p. 337). We all do things with words, which means we use words to convince others to do things for us (or for somebody else). The practicing teacher or teacher candidate who tells us that students cannot problem solve or that they come to school without any out-of-school experiences on which to build mathematical knowledge is asking us to do something with what we hear—for example, agree with me, do not ask me to design a problem-solving lesson, give me tools to teach these students. As mathematics teacher educators, we can use

Cases for Mathematics Teacher Educators, pages 121–124
Copyright © 2016 by Information Age Publishing

words to weaken these views. Doing things with words requires being critical of how we use words with others and how others use words with us. For example, in the university classes that I teach, I emphasize how we will be critical of the words we use in our talk with elementary students, as well as the words we put in print in assignments, because these words that populate the mathematics we teach are not politically neutral (Felton, 2010).

MY POSITIONALITY

Before I frame my response, I would like to say who I am and what I do. I am a nondominant mathematics teacher educator and mathematics education researcher. I have been working with both practicing teachers and teacher candidates in Texas, Illinois, and Michigan. My research focuses on creating research partnerships with teachers of nondominant students to support the development of learning environments in which teachers and students notice each other's mathematical ways of knowing. My teaching focuses on promoting the exploration of teacher candidates' mathematics, elementary students' mathematics, and ways to reconcile these two ways of knowing mathematics. In my teaching and research work, I continuously encounter challenges similar to that described in the case I am responding to. Like Koestler, the author of this case, I have never had an easy time hearing or seeing deficit perspectives while remaining detached. I have learned, however, that those who hold these perspectives are themselves oppressed by these views (Freire, 1993).

Reacting to deficit views in like terms can only make things worse. I can easily find words in the moment when I experience anger, offense, and dismay, only to precipitate even worse reactions. Alternatively, I can search for words in my sense of humanity, care, and liberation and use them to weaken these deficit perspectives. I prefer to think of *weakening* these perspectives instead of *challenging* or *disrupting* them, because I want to achieve specific things with words. Weakening is a very specific result that can be achieved by assembling words strategically so as to persuade others to judge the validity of claims. Challenging or disrupting, on the other hand, are more immediate reactions that can obfuscate a desired end result. Something challenged can produce a multitude of outcomes—some undesirable. Similarly, something disrupted can reengage in time. Whereas something weakened can open possibilities for engaging different perspectives and greater hope for change.

INTERPRETING AND RESPONDING TO THE DILEMMA

As a way of illustrating my familiarity with the case that Koestler presents, I want to briefly describe a recent case that involved a teacher expressing deficit views on her students. I chose this case because I believe it illustrates how I responded with the idea of doing things with words and enlisting others to do things with words as suggested by the idea of collective agency. In my interpretation of Koestler's case, *challenging* and *disrupting* are two words or actions that, as I explained earlier, suggest an emotionally charged response rather than a strategically envisioned outcome. I believe there is a difference between envisioning an outcome—such as weakening a deficit perspective—and using words toward that end while letting teachers make sense of those words (are these words challenges or are they something else?) and thinking about a more immediate reaction as suggested by a challenge or disruption. Koestler's case also suggests the outcome that I am thinking about when she expresses a desire to create environments in which teachers can express their views safely and openly. The case is difficult precisely because it requires us to move above immediate reactions and toward a word-mediated outcome.

My response to this case, as articulated here, obviously hides exactly that aspect of this difficult case; that is, my immediate feelings upon hearing these views. Here's the case. Recently, during a debriefing session with practicing teachers in a professional-development project, one teacher claimed that the students she teaches—children from Central America who cross the border to join a migrant worker parent in the United States—do not know anything about time, which makes it very difficult for her to teach the concept of measuring time.

To weaken this deficit perspective, I turned her claim into a question for the rest of the teachers to consider: Do kids not know anything about time? Notice how a question is a way of asking others to do something, in this case, to reconsider the claim that is on the table. This question caused teachers to move in their seats. Some looked up as if thinking about examples familiar to them; others looked at each other, as if looking for a common ground from where to begin; some looked at the teacher who had made the original claim. Sensing this collective agency promoted by my question, I asked a more focused question: So what have you heard kids say in relation to time? A flow of ideas emanated from *all* teachers, including the one who had made the deficit claim. They, in other words, were collectively gathering evidence that weakened the deficit perspective just heard. To weaken the perspective even more, but only after having heard how the teacher was becoming part of the collective agency being developed among the group, I came back to the teacher and asked her if she had children and if she thought her own children knew something about time. She positioned her children as knowing a lot about time.

Doing things with words in this and similar cases works like this: First, my goal is to weaken a deficit perspective. To begin, I enlist the whole group to help me achieve this weakening, as I recognize that I alone cannot do it.

This is what I did in the example I provided, and I have applied a similar approach among the teacher candidates that I teach. This can be done by turning a claim into a question, then eliciting answers to that question, and finally inviting the deficit claimer to personalize the question. This personalization is key, because deficit views are usually constructed around others, not around ourselves. However, as van Maneen (2004) reminded us, "the other is none other than ourselves" (p. 438).

Finally, as a mathematics teacher educator designing syllabi and learning experiences for teacher candidates, I prefer to remain suspicious of words that I use to describe the teaching approach I want my students to understand and practice. For example, I prefer not to use terminology such as *student-centered pedagogy* because I suspect that these words have already made students do something that may not align with what I want them to do. For instance, students may interpret student-centered approaches as certain students requiring a different pedagogy, or that student-centered pedagogy is not for every student. Being critical about how we use words with others, and what we do with the words of others—particularly when those words are used for framing deficit perspectives—is, in my own experiences with similar cases, an effective way of weakening such perspectives.

REFERENCES

Austin, J. (1962). *How to do things with words.* Cambridge, MA: Harvard University Press.

Felton, M. D. (2010). Is math politically neutral? *Teaching Children Mathematics, 17*(2), 60–63.

Freire, P. (1993). *Pedagogy of the oppressed.* New York, NY: Penguin Books.

van Maanen, J. (2004). An end to innocence: The ethnography of ethnography. In S. Hesse-Biber & P. Leavy (Eds.), *Approaches to qualitative research: A reader on theory and practice* (pp. 427–446). New York, NY: Oxford University Press.

Wertsch, J. V., Tulviste, P., & Hagstrom, F. (1993). A sociocultural approach to agency. In E. A. Forman, N. Minick, & C. A. Stone (Eds.), *Contexts for learning: Sociocultural dynamics in children's development* (pp. 336–356). New York, NY: Oxford University Press.

COMMENTARY 2

BUILDING PARTNERSHIPS TO CHALLENGE AND DISRUPT DEFICIT VIEWS OF STUDENTS AND COMMUNITIES

A Commentary on Koestler's Case

Elham Kazemi
University of Washington

MY POSITIONALITY

I am a mathematics teacher educator at the University of Washington (UW), where I have taught and conducted research since 1999. I studied with Megan Franke at UCLA, and my training in mathematics education was primarily in Cognitively Guided Instruction. When I first came to UW, my course was designed wholly around learning about CGI frameworks for understanding children's thinking. Each week, my students read about and studied student work or video cases of children's mathematical learning. I wanted to share with them my passion for learning about children's thinking that I had developed while at UCLA. In those first years as an assistant professor, my

Cases for Mathematics Teacher Educators, pages 125–129
Copyright © 2016 by Information Age Publishing
All rights of reproduction in any form reserved.

students would end the course feeling much more positive about their own mathematical identities, and I would commonly hear, "I loved your course. Math Rocks! I learned a lot about children's thinking, but I didn't learn how to teach." These were humbling words to hear. I cared deeply about my students' experiences and took to heart what they were trying to tell me. Many conversations, experiments, and experiences later, I met Magdalene Lampert at a conference in 2004 and along with Megan Franke and others began a radical transformation of our math methods courses (see Kazemi, Ghousseini, Cunard, & Turrou, 2015; Lampert et al., 2013).

As an immigrant to the U.S. from Iran, moving to Arizona when I was 11, I have deep memories of feeling like an outsider and wanting to cover up and pass for an American-born citizen to feel safe and protected from discrimination. My own experiences have shaped how I think about schooling and society. I think schools should be places where children and their families are invested, where they are known, and where they feel a strong sense of belonging. My own commitments to equity and social justice stem from the view that education is central in our ability to nurture and empower diverse communities in the perpetual struggle for liberation from the fracturing effects of racism, sexism, and classism.

INTERPRETING AND RESPONDING TO THE DILEMMA

Koestler's dilemma, as I understand it, is figuring out how to respond to our teachers when our efforts to encourage our teachers' practical expertise gives way to the sharing of perspectives that convey or reflect deficit discourses about students, their families, or their communities. Koestler writes:

> On one hand, I want to promote a welcoming classroom community where teachers feel safe and are open and honest about their practical theories of teaching, learning, and children. Only by doing so do I believe they can effectively broaden and refine these understandings. But on the other hand, I also want to challenge, critique, and disrupt deficit notions when voiced. To do this without teachers feeling shut down or judged or silenced is often difficult.

Koestler describes three ways that she has tried to manage these challenges, all of which resonated with me.

1. Setting norms that anticipate these challenges. Koestler offers two general norms she promotes: *be imperfect or unsure; be sure to ask questions when unsure or when you have a different view;* or more specific *articulate connections or disconnects between experiences and course literature.* I too try to pay attention to the way norms are stated and cultivated. I have never thought to make it normative to express

disconnections, but I can see the value. It makes visible that connections and disconnections will be experienced and that our class can be a place to surface both of them. Checkpoints along the way during a class, 3 weeks in, 6 weeks in, and at the end are ways that I try to keep these norms and my students' experiences of them alive and in the forefront. Having to respond and discuss prompts such as, *"I used to think... now I think..."* can underscore that our perceptions are always changing with the experiences that we have. I have decided in the last 3 years to offer some specific norms about the way we use language in class. For example, *"Use specific language to describe what you see students doing, rather than labeling students. Avoid labels such as 'low' and 'high.'"* My experience with this kind of statement has been that it helps us have a collective conversation about the power of our language and our actions in building positive learning environments. A close colleague of mine, Elizabeth Dutro, has also taught me to introduce at the outset of class that it is important for us to notice, pause, and analyze our language or our actions when we think we are conveying deficit views or positioning students and families in unfavorable ways. Having a routine and modeling it ourselves might be a way of navigating those uncomfortable moments. "Let's pause here to analyze our language. What might we be conveying when we say, '*these low kids...*'"

2. Providing "relevant resources" including video resources. Koestler nicely describes the importance of having video resources that are of high quality and reflect the richly diverse linguistic and cultural populations that our teachers serve. I agree that such resources are important. I want to expand on this idea by underscoring the importance of collecting locally relevant written and video cases in our partner schools that reflect the families and communities we serve. Although not always as polished, I think we can slowly, over time capture and use stories and cases that depict the kinds of teaching practices and relationships that we are hoping to nurture.

3. Modeling or providing powerful athematic lessons. The final idea that Koestler shares is taking prospective teachers on "field trips" to model powerful lessons in real classrooms (or providing lessons for the prospective teachers to try out themselves). I could write volumes about this. I have found it invaluable to find ways—however small—to teach alongside our students. I have been teaching at the University of Washington since 1999, and before that I worked with Megan Franke at UCLA for 5 years. There are many teachers and community members who can be our mentors and partners as teacher educators. Moving my math methods course to a partner school has allowed us to develop close relationships with

social-justice-oriented teachers, to come to know particular classes of students and to plan and teach with particular children in mind. I do know that these types of experiences are not silver bullets.

Because teaching and learning is always situated within the context of larger societal structures and hierarchies, I have been challenged recently to find ways of mediating our teachers' experiences as they make sense of ideas in their own field placement. Each year in our program, I try to take on something new, something that puts me in relation to my students in the field and that challenges me to see up close how our teachers are working in their varied field placements. This year for example, our math methods instructional team cancelled several on-campus sessions in our second quarter to instead visit our partner schools. Once there, we planned a lesson with all of the prospective teachers and the university supervisor assigned to that school (anywhere from 2 to 6 people), including their mentor teachers when possible, discussed their classroom context, and then visited one or two classrooms to try out that lesson together. Just like in methods, we gave each other permission to pause during instruction to ask each other questions or make suggestions about next steps to take in the lesson (affectionately called "teacher time outs," see Gibbons, Hintz, Kazemi, & Hartmann, 2016). These visits were a next step in pushing on perennial questions of teaching: how to make use of adopted texts and still elicit student thinking, how to work on explanations, how to design a lesson that builds on students' funds of knowledge.

REFLECTING AND LEARNING TOGETHER

Recently I had the chance to listen to Jamy Stillman talk about the work of teacher educators (Anderson & Stillman, 2013; Stillman & Anderson, 2015). I also reread a paper by Peter Murrell (1994) about the experiences of African American boys in discussion-intensive classrooms. Stillman and Anderson argued that teacher educators often lament the quality of student teaching experiences, and they make a provocative observation that the end goal is not to find the perfect placement or to continually shift concerns about teacher education to the quality of field placement. Instead, they argued that it is our job, no matter the quality of the field placement, to play a meditational role in helping our teachers make sense of the complex contexts in which they work. Murrell reminded us that how students read an academic situation, whether designed to position them competently or expose their vulnerabilities, has everything to do with how they respond. I find it invigorating to think about the dual realities of positioning our teachers as competent at the same time that we want to enable them to be vulnerable with one another. Koestler's case helps us engage in an important dialogue about how we create critically reflective communities within and beyond our teacher education courses.

REFERENCES

Anderson, L., & Stillman, J. (2013). Making learning the object: Using cultural historical activity theory to analyze and organize student teaching in urban high-needs schools. *Teachers College Record, 115*(3), 1–36.

Gibbons, L., Hintz, A., Kazemi, E., & Hartmann, L. (2016). *Teacher time out: Educators learning together in and through practice.* Manuscript submitted for publication.

Kazemi, E., Ghousseini, H., Cunard, A., & Turrou, A. C. (2015). Getting inside rehearsals: Insights from teacher educators to support work on complex practice. *Journal of Teacher Education, 67,* 18–31.

Lampert, M., Franke, M., Kazemi, E., Ghousseini, H., Turrou, A. C., Beasley, H., Cunard, A., & Crowe, K. (2013). Keeping it complex: Using rehearsals to support novice teacher learning of ambitious teaching in elementary mathematics. *Journal of Teacher Education, 64,* 226–243.

Murrell, P. C. (1994). In search of responsive teaching for African-American males: An investigation of students' experiences of middle school mathematics curriculum. *The Journal of Negro Education, 63,* 556–569.

Stillman, J., & Anderson, L. (2015). Minding the mediation: Examining one teacher educator's mediation of two preservice teachers' learning in context(s). *Urban Education, 51,* 683–713.

COMMENTARY 3

CREATING INVITATIONS TO DISRUPT DEFICIT DISCOURSES

A Commentary on Koestler's Case

Amy Noelle Parks
Michigan State University

MY POSITIONALITY

There is much in Courtney Koestler's case that I find myself identifying with, which perhaps is not surprising, given that I am also a White, middle-class woman who has spent much of her adult life working on mathematics education in early childhood contexts, both as a classroom teacher and as a university professor at large state universities in the South and the Midwest. Like Koestler, I too have had many moments in the classroom where I have been confronted by a comment about children or families that I found both surprising and disquieting.

In talking about these moments with other educators, I have put together a repertoire of strategies that I pull on in moments like this, although inevitably some I use more than others. For example, I have a colleague who

Cases for Mathematics Teacher Educators, pages 131–134

in these moments routinely asks: "Why do you think it's okay to talk about the children in your class in that way?" There is much I admire in that question: the way it explicitly confronts the damage being done, the way it forces the speaker to engage, and the way it surfaces unspoken assumptions.

However, as much as I admire it, the truth is I have never been able to make that question come out of my mouth. Perhaps, it is because I am too much immersed in the "culture of niceness" for which scholars have critiqued those who teach in early childhood and elementary classrooms (Hoo, 2004; McIntyre, 1997). Perhaps it is because I can still remember when "saving" the students in my own elementary classrooms was all I knew about social justice, and so I am too quick to empathize with my own university students now. Whatever the reason, I have sought ways of interrupting deficit discourses that feel more like invitations than interrogations. My desire to interact in this way is very similar, I think, to Koestler's desire to "challenge, critique and disrupt deficit notions when voiced . . . without teachers feeling shut down or judged or silenced."

INTERPRETING AND RESPONDING TO THE DILEMMA

I have used versions of each of Koestler's strategies—setting norms, providing relevant resources, and modeling powerful mathematics lessons—in my own courses and have found them productive. Beginning the course by explicitly stating the norm that we will be talking about students, families, and ourselves in strength-based ways lets my students know that I expect them to be careful about their language. In addition, it creates a nurturing atmosphere for talking about children and works to create a space where the many young women in my courses notice how often they tend to speak negatively about their own mathematical competence, their teaching, and even their bodies.

Similarly, using videos from my own years as an elementary teacher, where I used Cognitively Guided Instruction in schools that served ethnically diverse children from mostly low-income families, serves two purposes. The videos challenge the idea that this kind of teaching works with only middle-class, White kids from the Midwest, and repositions me as a practitioner as well as an academic. In fact, the stories I tell from my time in the classroom become some of my most powerful and persuasive resources.

In the interest of promoting conversation, I will also add to Koestler's strategies with a couple of my own.

Trusting My Students

When I started teaching in university classrooms, I had a great desire to leap in immediately to correct both mathematical misconceptions and problematic ideas about students and families. I have come to recognize that in both cases if I can slow down the conversation, my students can often get back on track on their own. For example, in my last methods course after an assignment that required my students to describe the children in their classrooms and their families, many students referred to families that "cared about school" as opposed to families that "did not care about school." To help them focus on this language, I wrote both categories on the board and asked them to operationalize them by suggesting behaviors that would indicate caring or not caring.

After adding things like, "not coming to conferences," "not correcting homework," and "not sending in supplies," to the "not caring" list, one of my students interrupted, saying that maybe there were other explanations for these behaviors other than "not caring." Multiple students then went on to suggest reasons that caring parents might fail to check homework or come to a conference. Other students told stories about parents whom they knew cared who had done various things on the "not caring" list for one reason or another. The result was an understanding that "not caring" was only one of many possible interpretations for the behaviors listed, and probably not a very productive one. The fact that the challenge came from another student made the conversation more powerful; however, it's unlikely it would have occurred unless I focused the class's attention on the issue.

Focusing on Practices Rather Than Beliefs

Within mathematics education and teacher education research more broadly, there has been a lot of emphasis on documenting and changing preservice teachers' beliefs (e.g., Achinstein & Barret, 2004; Brown, 2004). In the early years of my work as a mathematics methods instructor, I spent a lot of time on beliefs in my classroom, choosing readings and activities that were designed to challenge my students' beliefs. However, more recently I stopped paying attention to my students' beliefs in favor of attending to their pedagogical practices. This has been based on a conviction that success in diverse classrooms with powerful mathematics practices will do more to foster the changes in belief that I desire than any exercises I might do within the methods course (as well as the equally certain conviction that a lack of success teaching in real classrooms will immediately challenge any changes in beliefs about students or mathematics that I might have been able to create in the course). To this end, each semester I choose two to three pedagogical practices upon which to focus. In choosing focal practices, I look for pedagogies that are relatively easy for beginning teachers to learn, that move easily across contexts, and that support significant mathematical thinking on the part of all students.

To date, I have had the most success by highlighting Cognitively Guided Instruction (Carpenter, Fennema, Franke, Levi, & Empson, 1999), Number Talks (Parrish, 2014), and Complex Instruction (Featherstone, Crespo, Jilk, Oslund, Parks, & Wood, 2011), all of which provide beginning teachers with detailed pedagogical moves they can make in the classroom and which support teaching that is centered on both children and mathematics. My students are much more willing to believe in child-centered, mathematically powerful, equity-oriented pedagogies when they have had some success with them. Calling on this strategy means I must sacrifice some other topics and activities in my courses in order to provide my students with enough scaffolding to ensure they are successful. However, I've come to believe that fostering my own students' success is the most important tool in my repertoire in terms of challenging my students' beliefs about what their children are capable of doing in the mathematics classroom.

REFERENCES

Achinstein, B., & Barret, A. (2004). (Re)framing classroom contexts: How new teachers and mentors view diverse learners and challenges of practice. *Teachers College Record, 106,* 716–746.

Brown, E. L. (2004). What precipitates change in cultural diversity awareness during a multicultural course: The message or the method? *Journal of Teacher Education, 55,* 325–340.

Carpenter, T. P., Fennema, E., Franke, M. L., Levi, L., & Empson, S. B. (1999). *Children's mathematics: Cognitively guided instruction.* Portsmouth, NH: Heinemann.

Featherstone, H., Crespo, S., Jilk, L., Oslund, J., Parks, A. N., & Wood, M. B. (2011). *Smarter together! Groups, status, and complex instruction in the elementary mathematics classroom.* Reston, VA: National Council of Teachers of Mathematics.

Hoo, S. S. (2004). We change the world by doing nothing. *Teacher Education Quarterly, 31*(1), 199–211.

McIntyre, A. (1997). *Making meaning of Whiteness: Exploring racial identity with White teachers.* Albany, NY: State University of New York Press.

Parrish, S. (2014). *Number talks: Helping children build mental math and computation strategies, grades K–5.* Sausalito, CA: Scholastic Math Solutions.

CHAPTER 7

CASE X

Opportunities for America's Youth

Kimberly Melgar and Dan Battey
Rutgers University

INTRODUCTION TO THE DILEMMA

Students enter our institution's elementary teacher preparation program and take their only mathematics methods course in the fall of their senior year. The following year, they complete their student teaching as part of a Master's degree in education. Prior to the program, students major in subject matter outside of education and do not need to take mathematics-specific content courses. The goals for the methods course are for the preservice teachers to

- develop knowledge of the mathematics in the elementary grades,
- learn knowledge about the development of children's mathematical thinking,
- build instruction based on the development of students' mathematical thinking,
- reflect on their instructional practices, and
- understand how equity and access inside and outside of the mathematics classroom impact various learners.

Cases for Mathematics Teacher Educators, pages 135–142
Copyright © 2016 by Information Age Publishing
All rights of reproduction in any form reserved.

135

The final goal is integrated into the class using three approaches: (a) understanding relational aspects of the classroom that can impact learning for different groups of learners, (b) examining ways to connect to students' community and cultural knowledge in authentic ways, and (c) focusing explicitly on social and political topics that impact student success and failure in mathematics. Some of the students have taken a multicultural education course prior to this and some have not. However, students analyze classroom videos for issues of equity, read about funds of knowledge, design lessons that draw on community knowledge, and mathematize the social world to look at equity issues institutionally. This case particularly builds on the third approach.

As future educators, it is important to think about how various student circumstances can lead to different schooling experiences. The preservice teachers first watch the 4-minute video "A Tale of Two Schools" (Al Jazeera, 2009), which compares students in two neighboring yet vastly different school contexts and provides a way to make notions of fairness and equity concrete. "Issues That Impact Success Among America's Youth" is a project created by the first author that explores this idea by helping preservice teachers think about and analyze the impact a student's experiences can have on their ability to be successful in mathematics and life (see Figure 7.1). In this project,

---

Why are there more men than women working on Wall Street? Why are there millions of children in Africa starving and without clean drinking water? Why are more minority men incarcerated than white men? Why do students who come from wealthy households have a greater chance of graduating from high school than students from low-income households?

Many educators and researches struggle to answer these social justice questions everyday. Social justice is a concept concerned with equal justice, not just in the courts, but also in all aspects of society. This concept demands that people regardless of race, gender, and socioeconomic status have equal rights and opportunities. Everyone, from the poorest person in the South Bronx to the wealthiest person in Staten Island deserves an even playing field. However, this is not always the case.

The purpose of this project is to explore the following questions:

1. What do the words "just" or "fair" mean, and what defines equal?
2. Who should be responsible for making sure society is a just and fair place?
3. How do we implement policies regarding social justice?

Your job is to investigate these questions. Working with a partner, you will be given a scenario about the life of a person. Brainstorm the different variables in that person's life and how those variables contributed to where they are in their life now. Write an argument depending on the relationship or correlation you find between the variables. Think about whether or not the outcome of your person's life would be different if they were given different opportunities.

Figure 7.1 Student task description.

pairs of preservice teachers are given two narratives describing individuals and asked to determine if there is a correlation between the students' life experiences and expected outcomes related to a various social justice issues (see Figure 7.2). The social justice issues embedded in the narratives include

---

Scenario 1

My name is Kimberly. I grew up in New Jersey with my mother, father and younger brother. Growing up my parents didn't make too much money, but we lived a pretty good life. Each of my parents made about $25,000 dollars a year or $50,000 combined. My parents immigrated to the U.S. from Central America and neither parent went to college although my mom would later on go back to school and become a Spanish teacher. During high school I was involved in a ton of activities from track and field to student council. I really enjoyed getting involved in my school especially since it was the only thing I was allowed to do. At the time violence and other crimes were on the rise in my neighborhood. As strict as they were, my parents were always involved and active in my life supporting my love for extracurricular activities. Staying in sports and other clubs gave me something positive to do. After high school I went to Syracuse University where I studied Psychology and Biology and then got my masters from Brooklyn College in Education. As a teacher now I make more than what my parents used to make together!

Scenario 2

My name is John. I have a twin sister named Linda and we grew up with both of our parents. My mother is Caucasian and my dad is Mexican. When I was 7 years old my parents got divorced and I moved to Long Island to live with my mother. I loved going to school. Each of the private schools that I went to helped to prepare me for college. I remember being in classrooms where there may have been at most 13 other students. My teachers always made learning fun. I took the SAT on the same day as my sister Linda. I thought the test was pretty easy. There were some things that I didn't know, but overall I felt confident and I scored so well that I received many college scholarships. I am going to graduate this year and I look forward to getting my first real job.

*My father's salary is $35,000 per year. My mother's salary is $80,000 per year.

Scenario 3

My name is Linda. I have a twin brother named John and we grew up with both of our parents. My mother is Caucasian and my dad is Mexican. When I was 7 years old my parents got divorced and I moved to Staten Island to live with my father. I had a great childhood but the schools that I went to were not so good. I remember never feeling safe in school even though we had metal detectors. One of my 10th grade teachers came to school the first month and then never came back. We had 15 different substitute teachers for math that year. I liked school but never felt challenged. My dad encouraged me to go to college so I took the SAT. I was so discouraged when I got my score. I am embarrassed to say that I got an 800. As a result I decided to go to a community college. When I got there I was so far behind that I had to take remedial classes before I could actually take my first college course. I am fighting to make ends meet and I am not sure how I am going to pay for school next year, when I will no longer receive financial aid.

*My father's salary is $35,000 per year. My mother's salary is $80,000 per year.

Figure 7.2 Scenario examples.

student dropout rates, incarceration rates, levels of education, salaries, high school graduation rates by race, single-parent households, poverty level by neighborhood, and so forth. After examining the narratives, all of the students are given the same series of graphs and tables (e.g., Figure 7.3) that, to

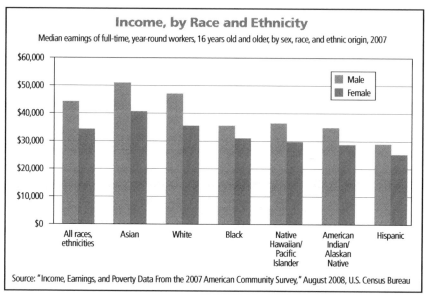

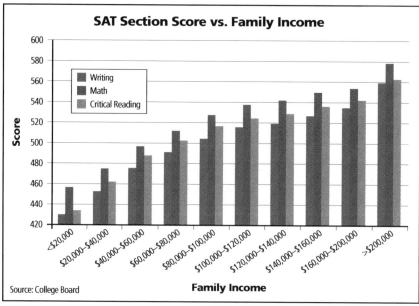

Figure 7.3 Examples of data given to students.

varying extents, relate to the narratives. The students choose which graphs they will examine in relation to their narratives.

About 85% of students in the program are European American women, from middle- to upper-class New Jersey families, while a few students are African American and Latino and some are first-generation college students that are by far the minority in terms of elementary preservice teachers. This activity asks the preservice teachers to examine social issues that can seem very foreign to them, or may not have directly impacted them throughout their own life experiences. They are given sets of data with which to reason and justify a claim. Part of the goal of the project is also to support preservice teachers in creating project-based curricula in a way that also builds algebraic thinking and data-based reasoning. The narrative component of the activity gives them a sense that this is not just about statistics, but about people and how these issues affect the students in or around their own communities. For this reason, the various narratives used are "local," meaning that they are cases of real people living in the densely populated New York and New Jersey area.

THE DILEMMA

After the conclusion of the activity, the teacher educator and preservice teachers debrief the activity, highlighting ways in which this activity can be implemented in their own classrooms and concerns they have about the activity. Figure 7.4 shows the responses of three preservice teachers during

---

THE DILEMMA

Preservice Teacher #1
"It would be better to do this activity in a diverse setting. I don't think you can do it in a school with all White students. I think it could just deepen their stereotypes since the data confirms them, even if some of the stories don't. I'd also be worried if there were one or two students from a particular racial or low SES group that they would be singled out. I guess you could make all of the narratives positive for White students so they couldn't build more stereotypes from the activity."

Preservice Teacher #2
"It might be students' first opportunity to think about these issues. They could be naïve around issues of race and SES to students. I guess one option would be to leave out the race of students in the stories and focus on data analysis and interpretation."

Preservice Teacher #3
"I'm worried that parents would react badly to doing the activity. Many of them don't talk to their children about these issues and might not want teachers discussing issues of race and socioeconomics. If students bring the issues home, parents might not like that."

Figure 7.4 The dilemma: Preservice teachers' reaction and concerns with the activity.

the activity debrief. We examined the preservice teachers' work to understand the perspectives on race and socioeconomic status (SES) that they were bringing to the activity.

The responses from preservice teachers highlight feelings teachers have about how to address the relationship of race and SES as it relates to students' backgrounds and their trajectories to success. The preservice teachers show perspectives that discussing race is applicable in settings with non-Whites, avoidance of discussing race, and worries about parental backlash to discussing race. In this sense, the preservice teachers are struggling with whether to make race explicit and how to make it explicit. The dilemma then is how to support preservice teachers in negotiating issues of race in the classroom, when they are currently novices in thinking through these issues themselves. The comments above illustrate three main ideas teacher educators need to address to support preservice teachers in navigating obstacles surrounding the implementation of social-justice activities in mathematics classrooms: privilege, color consciousness, and dealing with outside influences.

COMMENTARY ON THE DILEMMA

Authors' Reflections

Challenging Privilege

In the first response in Figure 7.4, we can see that one of the concerns preservice teachers have about this activity is that it cannot work in predominantly White settings. The concern here is not only that the activity can reinforce stereotypical views of particular groups, but also about challenging White privilege. Preservice teachers in the process of questioning their own privileges may not be ready to challenge their own students' ideas of privilege. This activity alone is not enough to prepare preservice teachers to engage in these types of discussions with students. Therefore, it is important that these kinds of activities not be stand-alone in a mathematics methods class but rather continue throughout the class and throughout the program. As teachers grapple with this activity, it is important for them to think about the value in addressing preconceptions head on and challenging their students' assumptions. Preservice teachers need to reflect on how to help students address their privilege, how to guide students to question the policies that lead to the drastic educational disparities in racial and social classes, and how to use activities like this to break down stereotypes. And much of this work needs to be done while the preservice students are coming to an understanding of their own privilege. From the standpoint of teacher educators, it is important to ask ourselves two questions: (1) to what extent are preservice teachers prepared to address issues of race and SES as they move into the classroom? and (2) how can we better prepare teachers to be knowledgeable, adept, and ready to support children in challenging their privilege as well?

Developing Color Consciousness

For preservice teachers who are still deepening their understanding of race and SES, they sometimes think that children do not have much insight about these social issues. Having critical conversations with young students can seem scary, and one option that is sometimes raised is to remove topics that seem controversial, such as race and SES. This, however, makes the activity decontextualized and colorblind, defeating the purpose of the activity in raising students' critical consciousness through mathematics. Positioning students as naïve about race and SES is not an excuse to avoid teaching about it. Everyone is naïve about these issues at some point in their lives, but everyone needs opportunities to engage in dialogue that can develop deeper understandings about them. This activity can instead serve as a way for teachers to elicit students' understandings of race and SES. From this perspective, teachers need to challenge students' notions of race and SES as well as develop a more robust understanding of race as a social construct.

This requires teachers to develop understandings that build more color-conscious stances. Ullucci and Battey (2011) detailed four critical understandings to develop color consciousness:

1. Challenging neutrality by racializing whiteness.
2. Validating the experiences and perspectives of the oppressed.
3. Naming racist educational practices and developing Culturally Responsive Teaching.
4. Challenging neutrality in policy and seeing institutional racism.

In the article, these scholars offered a number of activities to support preservice teachers in gaining each of these understandings. The activity presented here intentionally uses narratives based on real people that challenge commonly held stereotypes, combatting the concern that this activity design may reinforce deficit narratives. The narratives are a way to support teachers in broadening their experience with stories of oppressed groups. The project then can help foster conversations about race and SES through data in efforts to address societal issues while, at the same time, considering the mathematical concepts of variance and outliers in the lived experiences of different student groups. Additionally, the activity examines structural aspects of society that impact educational success. In this way, the activity develops the second and fourth understandings from Ullucci and Battey (2011).

These four understandings can serve as a framework for the types of activities we use in building color consciousness with preservice teachers. In thinking about developing each of these understandings, it is important that teacher educators model how to acknowledge and racialize whiteness, what conversations about race and SES look like with students, and how to learn to identify institutional racism. Our experience is that much more time needs to be allotted to these issues in teacher education and in ways that support preservice teachers in having critical conversations with young students. Additionally, preservice teachers should have opportunities not only to discuss culturally responsive teaching strategies, but to practice implementing these strategies in the mathematics classroom while actively naming race and SES in the instruction.

Dealing With Outside Influences

One of the many challenges preservice teachers will face is the delicate balance of communicating with parents and the impact that parents can have on their teaching. One of the concerns expressed by preservice teachers about this kind of culturally relevant mathematical activity is the possibility of a negative reaction from parents. The concerns here are that parents may not want students to partake in these kinds of conversations and that there could be backlash. As educators, we are not merely imparting our students with knowledge about mathematics, but knowledge about the world in order to become contributing citizens to society. This involves knowing, acknowledging, and addressing race as the landscape of America continues to change. We encourage our preservice teachers to keep open lines of communication with parents and to keep them informed about student work, but this necessitates sustained engagement with parents and communities.

Although we encourage mathematics teacher educators to work with parents and administrators, which is a long-term solution for many new teachers who are placed in all types of schools, many of which are not dedicated to social justice. We are realistic in realizing that it may not be appropriate for a first-year teacher to enact social justice in their classroom or school. Teachers must understand their students, the school, and the community before deciding what to teach. Having sustained relationships with parents means talking to them more than when their child is a behavioral problem. It means participating in the community, advocating for children, and welcoming parents into the classroom, rather than viewing parents as adversarial. Only with sustained engagement in the community can teachers allay concerns that they are pushing a specific ideology.

REFERENCES

Al Jazeera (2009). A tale of two schools [Video file]. Retrieved from https://www.youtube.com/watch?v=5xdfVAPvv9A

Ullucci, K., & Battey, D. (2011). Exposing color blindness/grounding color consciousness: Challenges for teacher education. *Urban Education, 46*, 1195–1225.

COMMENTARY 1

THE DELICATE BALANCE OF A THREE-LEGGED STOOL

A Commentary on Melgar and Battey's Case

Erika C. Bullock
University of Wisconsin–Madison

MY POSITIONALITY

I am a Black woman and secondary mathematics teacher educator (MTE) at an urban-serving university in Memphis, Tennessee. Memphis's population is approximately 63% Black, 29% White, and 6% Latina/o. Based on 2013 data (Delavega, 2014), the city's poverty rate is 27.7% (compared to 15.8% nationally), with 33.5% of Blacks (27.6% nationally), 47.0% of Latinas/os (24.8% nationally), and 9.8% of Whites (11.1% nationally) living in poverty. Child poverty is at 45.7% (22.2% nationally). On the basis of these statistics, Memphis has the highest overall and child poverty rates "among Metropolitan Statistical Areas (MSAs) with populations greater than 1,000,000" (p. 2). Why is this information relevant to a discussion about mathematics methods courses? Because both place and social location matter in (mathematics) teacher education (Gutiérrez, 2012; Milner, 2007; Rousseau Anderson, 2014).

In urban environments like Memphis, effective teacher education must attend to content knowledge, pedagogical content knowledge, and racial/cultural knowledge with fidelity (Howard & Milner, 2013). I acknowledge that equity as a construct encompasses more than just race, but I use race as the basis for my discussion here. Extant research addressing equity in (mathematics) teacher education has documented the cultural mismatch that often exists between preservice teachers (PSTs)—who are most often White, middle-class women—and the students they will likely teach (Hampton, Peng, & Ann, 2008). This mismatch renders racial/cultural knowledge particularly difficult to grasp and places a burden on MTEs to approach developing this knowledge with intention. For these reasons, the dilemma that Melgar and Battey pose in this case is significant related to the process of building racial/cultural knowledge toward teacher mathematics with equity in mind: how mathematics teacher educators who are sensitive to issues of equity support and create opportunities for PSTs to acknowledge their dysconscious racism (King, 1991) and to develop a vocabulary for addressing these issues with their future students.

INTERPRETING THE DILEMMA—RETURNING TO MEMPHIS TO SUPPLEMENT THE CASE

Like the context in the commentary, PSTs in the secondary mathematics licensure program in which I teach take one methods course, so time is at a premium. Additionally, these PSTs are predominately White, come from suburban and rural locations, and have limited exposure to diversity, particularly with respect to race. Using the statistics referenced earlier provides a means for students to understand the city's demographics, the disparities that exist among groups, and the dire situation in which we find ourselves with respect to the rest of the country. Memphis as a place is another actor in the teacher education process as PSTs must understand the city and consider how its demography affects the schools and classrooms in which they will likely work. But what happens next? How do we move away from using the statistics to gaze upon those who have been marginalized through racialization toward the material pursuit of racial competence (Milner, 2010)?

Oh, that these questions elicited a simple response! The reality is that there is not a set of prefabricated solutions that can be offered because there is not a single path to equity (Gutiérrez, 2012). The MTE's work is to seek ways to confront inequalities built upon systemic racism within the context in which she or he works. That is, the MTE is responsible for complicating the ideas of mathematics, mathematic teaching, and mathematics learning with full acknowledgement of the ways in which these three ideas have been racialized (Martin, 2013). In so doing, MTEs must create opportunities for themselves and their students (i.e., PSTs) to grapple with the evidence of racism in their own lives.

Howard and Milner (2013) have offered a framework for urban teacher education that supports building racial/cultural competence through direct engagement with race and racism in a way that is useful for mathematics teacher education. However, they discouraged teacher educators from assuming a myopic focus on such issues that shortchanges content knowledge and pedagogical content knowledge. Within Howard and Milner's framework, teacher educators balance the development of content knowledge, pedagogical content knowledge, and racial/cultural knowledge with equal attention. I argue that this tripartite approach should not be limited to urban teacher education, as the focal elements are essential to quality teacher education without qualification. Therefore, like a three-legged stool, teacher education can only be effective if it attends to these three components equally.

RESPONDING TO THE DILEMMA—BALANCING ACTS

In settings like the one presented in this case and my own, Howard and Milner's (2013) framework is instructive to the MTE as she or he seeks to do equity work with PSTs. Maintaining the balance of the three-legged stool means that activities and assignments in methods courses should attend to the three components of successful teacher education equally. In my own teaching, I have chosen to avoid activities that focus singularly on any one of the three focal elements. Rather, I choose to address building racial/cultural competence by mathematizing those issues in a way that allows students to access both content and pedagogical content knowledge in the process. Of course, it is not always possible to attend to each equally, but maintaining an intentional focus on this balance has allowed me to use the limited time in my methods course wisely.

Resources such as *Rethinking Mathematics: Teaching Social Justice by the Numbers* (Gutstein & Peterson, 2013) and the Radical Math website (www.radicalmath.org) provide a foundation upon which MTEs can build to address content knowledge, pedagogical content knowledge, and racial/cultural competence. However, it is important to supplement these activities particularly related to issues of race and culture in ways that acknowledge the complexity of these issues and resist attempts to use the mathematics to overshadow the larger social significance of the task content. Planning these types of activities for PSTs is also a pedagogical opportunity. If the goal is for PSTs to eventually teach their own students in ways that embrace principles of social justice and culturally relevant pedagogy (Ladson-Billings, 2011), then introducing them to the process of designing quality mathematics lessons that attend to race and culture can be instructive for their practice.

Another balancing act necessary in the development of racial/cultural competence is connected to the ways in which MTEs choose to use activities that address issues of race and culture. Absent careful planning, activities that rely on stereotypes—as a whole or in part—can often backfire by reinforcing those stereotypes. For example, the data describing poverty in Memphis cited earlier in this commentary can support powerful mathematical investigations of race and poverty. As I mentioned, the majority of PSTs with whom I work are middle-class White women from suburban or rural areas. Absent careful planning and monitoring, these PSTs could leave investigations using this poverty data with evidence supporting stereotypes about poor Black and Brown children. Therefore, planning activities for PSTs requires that the MTE not only use and contextualize such data, but also that we directly address the inequalities undergirding the data and the stereotypes. This thorough planning and execution will prepare PSTs not only to provide equitable experiences for their students, but also to encourage their future students to use mathematics to critically examine the world around them.

REFERENCES

Delavega, E. (2014). *2013 Memphis poverty fact sheet*. Memphis, TN: University of Memphis School of Urban Affairs and Public Policy. Retrieved from http://www.memphis.edu/socialwork/docs/2014povertyfactsheet.pdf

Gutiérrez, R. (2012). Context matters: How should we conceptualize equity in mathematics education? In B. Herbel-Eisenmann, J. Choppin, D. Wagner, & D. Primm (Eds.), *Equity in discourse for mathematics education: Theories, practices, and policies* (pp. 17–33). Dordrecht, The Netherlands: Springer.

Gutstein, E., & Peterson, B. (2013). *Rethinking mathematics: Teaching social justice by the numbers* (2nd ed.). Milwaukee, WI: Rethinking Schools.

Hampton, B., Peng, L., & Ann, J. (2008). Pre-service teachers' perceptions of urban schools. *Urban Review, 40,* 268–295.

Howard, T. C., & Milner, H. R., IV. (2013). Teacher preparation for urban schools. In H. R. Milner IV & K. Lomotey (Eds.), *Handbook of urban education* (pp. 199–216). New York, NY: Routledge.

King, J. E. (1991). Dysconscious racism: Ideology, identity, and the miseducation of teachers. *The Journal of Negro Education, 60,* 133–146.

Ladson-Billings, G. J. (2011). "Yes, but how do we do it?" Practicing culturally relevant pedagogy. In J. Landsman & C. Lewis (Eds.), *White teachers/diverse classrooms: Creating inclusive schools, building on students' diversity, and providing true educational equity* (pp. 33–46). Sterling, VA: Stylus.

Martin, D. B. (2013). Race, racial projects, and mathematics education. *Journal for Research in Mathematics Education, 44,* 316–333.

Milner, H. R. (2007). Race, narrative inquiry, and self-study in curriculum and teacher education. *Education and Urban Society, 39,* 584–609.

Milner, H. R. (2010). Reflection, racial competence, and critical pedagogy: How do we prepare pre-service teachers to pose tough questions? *Race Ethnicity and Education, 6,* 193–208.

Rousseau Anderson, C. (2014). Place matters: Mathematics education reform in urban schools. *Journal of Urban Mathematics Education, 7*(1), 7–18.

VALIDATING AND CONTEXTUALIZING PRESERVICE TEACHERS' RESISTANCE TO SOCIAL JUSTICE PEDAGOGY IN MATHEMATICS

A Commentary on Melgar and Battey's Case

Niral Shah
Michigan State University

MY POSITIONALITY

The issues and dilemmas raised by this case resonate with me as both a researcher and a teacher educator. As a researcher, I study processes of racialization in mathematics education, as well as broader issues of equity across STEM domains. As a mathematics teacher educator at a large public

Cases for Mathematics Teacher Educators, pages 149–153
Copyright © 2016 by Information Age Publishing
All rights of reproduction in any form reserved.

university, I am also committed to preparing teachers with the knowledge and dispositions needed for teaching mathematics in robust and equitable ways. Personally, I view myself as occupying a contradictory positionality with respect to race. On the one hand, as a person of color in the U.S. context, I have been the target of explicit and implicit racial discrimination throughout my life. On the other hand, as someone who is considered a member of a "model minority," I have also benefited from certain racial privileges. This positionality informs how I approach my work and understand this case.

INTERPRETING THE DILEMMA

The case describes a classroom activity designed to foster empathy for people from historically marginalized groups, particularly with respect to race and class. The authors report that engaging preservice teachers in this activity elicited concerns and resistance. Similar reactions from preservice teachers to discussions about social justice have been well documented in the literature (Bonilla-Silva & Forman, 2000). What is less clear, though, is how teacher educators should make sense of and respond to these reactions. In my view, the preservice teachers' statements point to two larger questions: (1) Which audiences need to be engaged in social justice-oriented pedagogy, and (2) what are the pragmatic challenges associated with implementing this kind of teaching, especially in mostly White, class-privileged contexts?

The first question regarding audience is implicit in the comments made by the first two preservice teachers. Preservice teacher #1 expresses concern that the activity would be better to do in a "diverse setting," rather than in an all-White setting. Similarly, preservice teacher #2 proposes that it might be better to modify the task by deleting race from the activity, because students may be "naïve" about such issues. Embedded in both teachers' concerns is an important question: *Who* needs an activity like this? Given that they remain the dominant racial group in the U.S., I contend that Whites are a group that needs such an activity. Denying White students the opportunity to learn about and reflect on issues of social justice would only reproduce historical patterns of inequity.

It is also problematic to assume that students—especially White students—are "naïve" about issues of race. As Leonardo (2009) and other race scholars have argued, Whites actually have considerable knowledge about race and racism. However, this does not mean that Whites and people of color share the same perspective on race. In that sense, preservice teacher #2 raises a legitimate concern that there may be a tension between the perspectives certain students bring to the classroom and the goals of the activity. And yet, this does not mean that removing all explicit mentions of race from the activity is an effective way of responding to that tension. As the authors of the case note, doing so would dilute one of the original goals of the activity.

Besides the question of audience, a second theme in the preservice teachers' comments concerns the pragmatic challenges associated with implementing this kind of activity in schools. Preservice teacher #1's concern that the activity might inadvertently perpetuate socially unjust ideologies by reinforcing students' preexisting stereotypes is legitimate and resonates with findings in the literature (see Esmonde, 2014). The teacher also pointed out that the activity might spotlight students of color, which could be harmful to those students in a majority White classroom context. This issue resonates with me, as the racial demographic of my teacher education courses also tends to be mostly White with relatively fewer students of color. Finally, preservice teacher #3 calls attention to potential discomfort and resistance from parents.

On the one hand, none of these pragmatic challenges justify avoiding or diluting social-justice-oriented pedagogy. On the other hand, the pedagogical and political complexity involved in doing this kind of work cannot be underestimated. What are some ways to think about these challenges, and how can mathematics teacher educators support preservice teachers in grappling with them?

RESPONDING TO PRESERVICE TEACHERS' CONCERNS— CONSIDERATIONS FOR MATHEMATICS TEACHER EDUCATORS

If preservice teachers had raised these concerns in a course of mine, my first response would be to validate their concerns. One reason to do this is that, simply put, their concerns *are* valid. Implementing a social justice pedagogy in historically conservative settings like many U.S. schools requires teachers to navigate a complex terrain of cultural, emotional, and political issues. Another reason to validate preservice teachers' concerns has to do with what we know about how people learn (Bransford, Brown, & Cocking, 1999). The learning process can only start from learners' prior knowledge and experiences. Regardless of whether they are learning about multiplying fractions or racial justice, when learners encounter new ideas that do not immediately sync with how they previously understood that topic, there is likely to be anxiety, uncertainty, and resistance. From that perspective, the reactions by the preservice teachers in this case are entirely reasonable. And further, their resistance can be understood as an important and necessary part of the learning process, in that it shows they are attempting to grapple with the possibilities and risks associated with teaching in certain ways (Garrett & Segall, 2013). It is crucial that we honor where preservice teachers are starting from

152 ■ N. SHAH

on issues of social justice, as well as the considerable institutional and political forces they may face when they enter the classroom.

That said, there are several concrete actions teacher educators can take to facilitate our students' trajectories toward enacting more socially just pedagogies. One dimension of this work involves having preservice teachers think about how local classroom decisions relate to what happens more broadly in society. For example, if teachers are considering modifying an activity to take out explicit mentions of controversial topics, a teacher educator might engage them in thinking through the potential costs of that decision to society writ large. How might avoiding controversial topics affect how students see the world and the people they become? A similar reflection process could be useful if a teacher wants to avoid using such activities with White students—what impact would this local pedagogical decision have on how those same White students vote as *adults* later in life, especially on policies that affect people of color?

The dilemmas posed by this case also underscore the need to think about how such activities fit within the broader structure and mission of a teacher education program. I agree with the authors of the case that activities of this kind are less effective when they are "stand-alone" experiences. How can themes of equity and social justice be woven into all of the coursework and fieldwork that preservice teachers complete across the multiple years of a teacher education program? What opportunities are there for coordination across subject areas? For example, in addition to a mathematics methods course, the activity described in this case could readily be used in a social studies methods course. If we can engage preservice teachers in making such connections during their training, it might encourage them to build similar interdisciplinary collaborations when they enter the classroom full-time.

As a final point, it is important to remember that new teachers are entering a complex system consisting of multiple stakeholders, including parents, administrators, corporations, and policymakers. Further, that system has considerable historical inertia. This means that any move—however small—to change that system will be difficult. If sustainable change is possible, teachers must proactively try to include those stakeholders in their efforts to implement socially just pedagogies. Or at the very least, teachers must attempt to communicate the rationale behind a particular teaching approach and solicit feedback from those groups. Teacher educators can support this work by building into their courses opportunities for preservice teachers to practice the kinds of interactions and scenarios they might encounter. Doing so would acknowledge that teaching for social justice is about more than designing a high-quality curriculum, but that it also requires navigating the politics that come with implementing that curriculum in mainstream schools.

REFERENCES

Bonilla-Silva, E., & Forman, T. A. (2000). I am not a racist but . . . : Mapping White college students' racial ideology. *Discourse & Society, 11,* 50–85.

Bransford, J. D., Brown, A. L., & Cocking, R. R. (1999). *How people learn: Brain, mind, experience, and school.* Washington, DC: National Academy Press.

Esmonde, I. (2014). "Nobody's rich and nobody's poor . . . it sounds good, but it's actually not": Affluent students learning mathematics and social justice. *Journal of the Learning Sciences, 23,* 348–391.

Garrett, H. J., & Segall, A. (2013). (Re)considerations of ignorance and resistance in teacher education. *Journal of Teacher Education, 64,* 294–304.

Leonardo, Z. (2009). *Race, whiteness, and education.* New York, NY: Routledge.

CONCEPTIONS OF EQUITY AND THEIR IMPACT ON STUDENTS' OPPORTUNITIES TO LEARN MATHEMATICS

A Commentary on Melgar and Battey's Case

Marilyn Strutchens
Auburn University

MY POSITIONALITY

As a middle-class, African American mathematics teacher educator born and reared in a rural working-class home and currently employed at a predominantly White university, I understand the scenarios on the basis of my lived experiences and the reactions of the prospective teachers as a result of my work with secondary and elementary prospective and inservice teachers. In my work with teachers, one of my main goals is to help them to understand what equity is. I usually do this through helping them to think about conceptions of equity. I have found Secada (2003) and Bartell and

Cases for Mathematics Teacher Educators, pages 155–159

Meyer (2008) to be helpful in this regard. These articles help teachers to think about their own beliefs about equity as they examine others' conceptions, and the impact those conceptions have on students' mathematical success or failure. Below are a few conceptions of equity, which I use from different sources:

> Lipman's (2004) concept of equity includes "the equitable distribution of material and human resources, intellectually challenging curricula, educational experiences that build on students' cultures, languages, home experiences, and identities; and pedagogies that prepare students to engage in critical thought and democratic participation in society." (p. 3)
>
> Equity is extended from a unidirectional exchange—as primarily benefitting growth of students and student groups that have historically been denied equal access, opportunity and outcomes in mathematics to a reciprocal approach. We contend that as a field we need to think of diversity as a resource for the learning of mathematics for all students. (Civil, 2007)
>
> Equity means "being unable to predict mathematics achievement and participation based solely upon student characteristics such as race, class, ethnicity, sex, beliefs, and proficiency in the dominant language." (Gutiérrez, 2007, p. 41)

I use these conceptions of equity to launch conversations about students' opportunities to learn; the importance of establishing bidirectional relationships with students' families; and the impact teachers', administrators', students', and others' beliefs and stereotypes have on students' engagement and achievement in mathematics. We also discuss a myriad of ways to make mathematics class more equitable, such as pedagogical strategies ascribed by the National Council of Teachers of Mathematics Standards documents (1989, 1991, 1995, 2000, 2014), culturally relevant pedagogy (Ladson-Billings, 1995), social-justice lessons (Gutstein, 2003), using multicultural literature as a context for mathematical problem solving (Strutchens, 2002), and others. I also try to help teachers see the similarities and differences in these different approaches to equitable pedagogy (Banks & Banks, 1995). Banks and Banks (1995) defined equitable pedagogy as "teaching strategies and classroom environments that help students from diverse racial, ethnic, and cultural groups attain the knowledge, skills, and attitudes needed to function effectively within, and help create and perpetuate, a just, humane, and democratic society" (p. 152).

INTERPRETING THE DILEMMA

One might say that I have an eclectic stance when it comes to how one can achieve an equitable mathematics classroom. Although I think that it is important to see students as being a part of a particular cultural group and acknowledge particular customs or traits that might impact the students on the basis of their backgrounds, I also think that it is important to acknowledge and support students as children in a particular age group who have some intersecting interests and developmental needs. Thus, it is important to use critical education theory and critical race theory as lenses to examine educational situations. *Critical education theory* as defined by Weiler (1988) is a critical view of the society, asserting that society is both exploitive and oppressive but also capable of being changed. *Critical race theory* (CRT) within the context of education is a framework or set of basic insights, perspectives, methods, and pedagogy that seeks to identify, analyze, and transform those structural and cultural aspects of education that maintain subordinate and dominant racial positions in and out of the classroom (Solórzano & Yosso, 2002, p. 25). These theories overlap in that they both deal with societal issues that are oppressive but can be changed. However, critical race theory focuses on issues that are akin to particular racial groups, which need to be explored and discussed by teachers so that they understand what their students may have experienced or might experience as being a part of a particular racial group. I find like the authors of this case study that stories about people's lived experiences help teachers to think about their roles in lessening the oppressive nature of society and helping students to have more options.

There are several connections that can be made between "Case X: Opportunities for America's Youth" and several of the conceptions of equity mentioned above, critical education theory, and critical race theory.

1. It appears to me that the prospective teachers were looking at the information and statistics from a deficit viewpoint rather than a critical educational view: society is oppressive but can be changed (Weiler, 1988).

2. Moreover, the prospective teachers were only considering a surface-level relationship with parents. Helping them understand and embrace the idea of creating a bidirectional relationship (Civil, 2007) among the teacher and the parents around students' mathematics education could change their perspective and help them to see that they could actually contact the parents before they implement a lesson that has elements of social class and race/ethnicity to seek input from the parents. Parents would then know what the lesson entails ahead of time and may be able to provide some important insights to the teachers.

3. Finally, the educator is faced with the challenge of covering a broad spectrum of equity issues in one course. The goal may need to be to help the teachers to see themselves as change agents through the social-justice activity that was created with looking at the different affordances of school environments and their impact on students' futures.

RESPONDING TO THE DILEMMA

Overall, I think that the intentions of the mathematics educators in the case are laudable. However, I wonder if separating the assignment into two distinct activities would have helped them to accomplish their goals better:

First, I would have used The Tale of Two Schools Activity with the video, scenarios, and statistics as a social justice lesson for the prospective teachers. This activity would have been used to really help them to see the importance of students' opportunities to learn, school environments, and the impacts that these factors have on students' identities as learners and doers of mathematics and other subjects. I would ask them to think about the inequities that exist in education and the consequences. I would have a frank discussion with them about the differences in the two schools in the video and the scenarios. Moreover, I would help them to think about the differences that they could make for students who attend a school with inadequate resources and the importance of expectations. What kinds of grants are available to teachers? What partnerships could teachers develop with parents? Businesses? The objective would be to move them towards actions. Then, after moving them through these phases, I would ask them to develop a social justice or culturally responsive lesson that is developmentally appropriate for their respective grade levels.

Possible Takeaways:

- Given the time constraints and the opportunities mathematics teacher educators have with prospective teachers in programs, we should provide them with opportunities to observe and experience equitable pedagogical strategies and constructs that will enable them to help all of their students to reach their full mathematics potential. Moreover exposing prospective teachers to stories of people who have overcome obstacles related to their race or social economic status may enable them to see their role as change agents as needed and influential.
- Moreover, there should be a balance in methods courses between ensuring that prospective teachers develop a sense of agency about other people or social-political situations and students developing

positive mathematics identities within the classroom regardless of their race/ethnicity, socioeconomic status, or facility with the English language.

REFERENCES

Banks, C. A., & Banks, J. A. (1995). Equity pedagogy: An essential component of multicultural education. *Theory Into Practice, 34,* 152–158.

Bartell, T. G., & Meyer, M. R. (2008). Addressing the equity principle in the mathematics classroom. *Mathematics Teacher, 101,* 604–608.

Civil, M. (2007). Building on community knowledge: An avenue to equity in mathematics education. In N. Nasir & P. Cobb (Eds.), *Improving access to mathematics: Diversity and equity in the classroom* (pp. 105–117). New York, NY: Teachers College Press.

Gutiérrez, R. (2007, October). *Context matters: Equity, success, and the future of mathematics education.* Paper presented at the annual meeting of the North American Chapter of the International Group for the Psychology of Mathematics Education, Reno, NV. Retrieved from http://www.allacademic.com/meta/p228831_index.html

Gutstein, E. (2003). Teaching and learning mathematics for social justice in an urban, Latino school. *Journal for Research in Mathematics Education, 34,* 37–73.

Ladson-Billings, G. (1995). Toward a theory of culturally relevant pedagogy. *American Educational Research Journal, 32,* 465–491.

Lipman, P. (2004, April). *Regionalization of urban education: The political economy and racial politics of Chicago-metro region schools.* Paper presented at the annual meeting of the American Educational Research Association, San Diego, CA.

National Council of Teachers of Mathematics. (1989). *Curriculum and evaluation standards for school mathematics.* Reston, VA: Author.

National Council of Teachers of Mathematics. (1991). *Professional standards for teaching mathematics.* Reston, VA: Author.

National Council of Teachers of Mathematics. (1995). *Assessment standards for school mathematics.* Reston, VA: Author.

National Council of Teachers of Mathematics. (2000). *Principles and standards for school mathematics.* Reston, VA: Author.

National Council of Teachers of Mathematics (2014). *Principles to actions: Ensuring mathematical success for all.* Reston, VA: Author.

Secada, W. G. (2003). *Conceptions of equity in teaching science.* Unpublished manuscript.

Solórzano, D., & Yosso, T. (2002). Critical race methodology: Counter-storytelling as an analytical framework. *Qualitative Inquiry, 8,* 23–44.

Strutchens, M. (2002). Multicultural literature as a context for mathematical problem solving: Children and parents learning together. *Teaching Children Mathematics, 8,* 448–454.

Weiler, K. (1988). *Women teaching for change: Gender, class and power. Critical studies in education.* New York, NY: Bergin & Garvey.

CHAPTER 8

HEARING MATHEMATICAL COMPETENCE EXPRESSED IN EMERGENT LANGUAGE

Judit Moschkovich
University of California Santa Cruz

INTRODUCING THE DILEMMA

I teach a course for undergraduates interested in mathematics education. The majority are mathematics majors in their junior or senior years interested in teaching (some may pursue a credential after their undergraduate degree). Some may have tutored or observed a K–12 classroom. They are not technically prospective teachers (they are not in a credential program), nor are they student teachers (they are not student teaching in a classroom). However, I will refer to these undergraduates as "prospective teachers" because I think the issues they face are similar to those that prospective teachers face, when they have little experience in classrooms or with children.

This is the first course these prospective teachers have ever taken focusing on mathematics education.[1] The course provides an introduction to principles and practices for mathematics education. We examine how empirical

Cases for Mathematics Teacher Educators, pages 161–170
Copyright © 2016 by Information Age Publishing

research on learning and teaching mathematics provides principles for mathematics instruction, lesson planning, curriculum design, assessment, and policy. The course also provides an introduction to national standards by the National Council of Teachers of Mathematics (NCTM) and the Common Core State Standards (CCSS) for K–12 mathematics instruction, exemplary mathematics curricula, and equity in mathematics classrooms. Coursework includes readings and written assignments. In-class work includes lectures, small-group discussions of readings and analyzing interview or classroom data (videotapes, transcripts, interviews, written work).

The course addresses the following questions: How can empirical research on learning and teaching mathematics provide principles for mathematics education? What are different views of learning and teaching mathematics? What are mathematical proficiency and conceptual understanding? How can conceptual understanding be assessed? What are principles for effective mathematics teaching for conceptual understanding? What are issues of equity in mathematics classrooms?

THE DILEMMA

I have found that many prospective teachers hold the following three problematic beliefs about mathematics learners who are bilingual or learning English:

1. English learners cannot participate in mathematical discussions.
2. Everyday or home languages are obstacles to doing and learning mathematics.
3. Mathematical discourse requires formal vocabulary to express mathematical ideas.

These beliefs are reflected in the repeated comments I have heard in my 20 years working in mathematics education, comments such as

These students cannot speak English or Spanish well. They don't have ANY language.

Mathematical discussions are fine for native English speakers, but we cannot expect ELs to talk about math. They need to learn English first.

Mathematics vocabulary is precise; we cannot talk about mathematics using everyday language.

I heard two students talking in a classroom in Spanish—what if the teacher cannot understand them?

My dilemma is that, because of what I know about conceptual change, I know that telling prospective teachers that they are wrong is not likely to have an impact on their misunderstandings about English learners and mathematics. For example, I know that just telling them that research shows that these beliefs are not the case will not be sufficient for them to rethink these beliefs. This is a dilemma for me for many reasons: I cannot let prospective teachers state these beliefs as if they were facts, and I do not want them to go into classrooms with English learners assuming that these beliefs are true. These beliefs are not necessarily or principally the result of racist attitudes. In some cases, these beliefs arise from sincere intentions to help and support English Learners; sometimes students who are English learners themselves may hold these beliefs.

My dilemma stems from my commitments to constructivist assumptions about learning, even when the learning is not mathematical but changing one's beliefs about a student population. This dilemma has been with me for many years. When I first started doing research on the topic of language and mathematics learning, I was surprised that many naïve beliefs about the relationship between language and mathematics had seeped into the scholarly literature. I now see that many of these lay beliefs are, in fact, intuitive conceptions about what language is, how we learn it, and how it relates to mathematical thinking. Other beliefs are more likely the result of both intuitive conceptions about language (e.g., code switching reflects that the speaker is experiencing a difficulty) as well as sociopolitical attitudes about language (e.g., we should not mix up two languages; it reflects a deficiency). As an English learner myself and as an advocate for children from nondominant communities, I struggled with my initial reactions (anger, frustration, and indignation) to these naïve beliefs.

As my work with both preservice and inservice teachers progressed, I realized that if I really assume a constructivist stance and I really see conceptual change as the learner's own reconstruction of ideas, then I have to consider how to respond to these naïve beliefs in other ways, and I have to develop opportunities for prospective teachers to deconstruct and reconstruct these ideas for themselves. An interesting byproduct of this approach has been what I might call *cognitive compassion*. I see myself now remaining much calmer when I hear evidence of these beliefs. I respond with more compassion because I frame the beliefs as naïve or intuitive ideas that require attention, not correction, in much the same manner that intuitive conceptions in science (e.g., that the distance from the earth to the sun causes the seasons) require attention and reconstruction.

My research has been devoted to documenting the resources that learners, especially English learners, use to make sense of mathematics. I use examples from my research with teachers and PSTs as they reconsider their beliefs, especially deficit models of learners.

MY RESPONSE TO THE DILEMMA

To address this dilemma, I use a conceptual-change perspective to address these beliefs. I have designed activities for participants to examine these beliefs themselves though discussions grounded in short classroom vignettes. Activities involve a video, a transcript, a set of discussion questions, and my own analysis of the vignette. I developed this approach for my work with PSTs and teachers, building on my participation in the Chèche Konnen teacher workshops. The Chèche Konnen approach to working with teachers is much broader and longer term than any principles for working with one video or classroom vignette (see Rosebery & Warren, 1998, for examples). However, I distilled my experiences in the Chèche Konnen teacher workshops into a few principles to structure a lesson or workshop (1–2 hours) using video.

These activities can be used in a variety of ways depending on the particular context. When working with prospective teachers, I generally use the following format. First, I show the video several times in class. Prospective teachers then work in small groups reading the transcript aloud and discussing their responses to the vignette and reflection questions based on the "Guidelines for Discussion" (see Figure 8.1). Then, either in class or as homework, I ask the prospective teachers to read my analysis of the vignette and respond to additional questions. My analysis of the vignettes describes my own responses to the discussion questions and highlights hearing the mathematical ideas in student contributions, uncovering the mathematical competence students show (not their errors or deficits), and seeing how the teacher supports mathematical discussion.

One vignette I use shows students in a third-grade bilingual classroom in an urban school, identified as Limited English Proficient, communicating mathematically using multiple resources (see Figure 8.2).[2] After viewing, reading, and discussing this vignette I then have prospective teachers read and respond to my analysis of this vignette. In my analysis of this vignette I highlight the following issues:

1. *Student participation:* The vignette shows that ELLs can and do participate in discussions as they grapple with important mathematical content. Students were discussing definitions for quadrilaterals, describing the concept of parallelism, and using mathematical practices (making claims, generalizing, imagining, hypothesizing, and predicting what will happen to two lines segments if they are extended indefinitely). To communicate about these mathematical ideas, students used everyday expressions, objects, gestures, and other students' utterances as resources.

GUIDELINES FOR DISCUSSION

Looking at video and reading transcripts
- Read the transcript aloud at least once (If you are working in a group, assign parts to the people in your group).
- Assume responsibility for making sense of what students are saying
- Take a few minutes to think about what each student says, how she/ he says it, to whom, how it's taken up by others, etc.
- You can make some notes on the transcript about questions you have, points you want to make, lines that you find interesting.

Discussing and exploring a video/transcript
- Keep the conversation grounded in the transcript/video, talk about these children rather than children in general.
- Use the transcript to ground your comments, questions, or claims.
- Stay focused on the discussion questions.
- Assume that students are making sense and that there is knowledge and expertise in what they are saying.

These guidelines are adapted from work with the Chèche Konnen Center, TERC, Cambridge, MA.

DISCUSSION QUESTIONS

Focus on the students:
    1. How are students communicating mathematically?
    2. What mathematical ideas are students talking about?
    3. What resources do students use to communicate mathematically?
Focus on the teaching:
    4. How is the teacher supporting student participation in a mathematical discussion?

Figure 8.1 Guidelines for discussing vignettes.

2. *Mathematical practices:* Julian was participating in two mathematical practices: abstracting and generalizing. He described an abstract property of parallel lines—that they do not meet—and made a generalization by saying that parallel lines will never meet.

3. *Language resources:* Julian used his first language, gestures, and objects (running his fingers along parallel sides of a paper rectangle), and everyday language, the colloquial expressions "go higher" and "get together" rather than the formal terms *extended* or *meet*.

4. *Teacher's support of mathematical discussion:* The teacher moved past Julian's uses of the word *parallela* and focused on the mathematical content of Julian's contribution. He did not correct Julian's English, but asked questions probing for meaning. This teacher did *not* focus directly on vocabulary but instead on mathematical ideas and

---

VIGNETTE: "THEY NEVER GET TOGETHER"

The lesson excerpt presented here (from Moschkovich, 1999) comes from a third-grade bilingual classroom in an urban California school. In this classroom, there were 33 students identified as English Learners. In general, this teacher introduced students to topics in Spanish and then later conducted lessons in English. The students had been working on a unit on two-dimensional geometric figures. For several weeks, instruction had included vocabulary such as radius, diameter, congruent, hypotenuse, and the names of different quadrilaterals in both Spanish and English. Students had been talking about shapes, and the teacher had asked them to point to, touch, and identify different shapes. The teacher identified this lesson as an ESL mathematics lesson in which students would be using English in the context of folding and cutting to make tangram pieces.

Excerpt 1:

(1) *Teacher:* Today we are going to have a very special lesson in which you're really gonna have to listen. You're going to put on your best, best listening ears because I'm only going to speak in English. Nothing else. Only English. Let's see how much we remembered from Monday. Hold up your rectangles high as you can. [Students hold up rectangles] Good, now. Who can describe a rectangle? Eric, can you describe it? Can you tell me about it?

(2) *Eric:* A rectangle has [pause] two [pause] short sides, and two [pause] long sides.

(3) *Teacher:* Two short sides and two long sides. Can somebody tell me something else about this rectangle, if somebody didn't know what it looked like, what, what [pause] how would you say it?

(4) *Julian:* Paralela [holding up a rectangle, voice trails off].

(5) *Teacher:* It's parallel. Very interesting word. Parallel. Wow! Pretty interesting word, isn't it? Parallel. Can you describe what that is?

(6) *Julian:* Never get together. They never get together [runs his finger over the top side of the rectangle].

(7) *Teacher:* What never gets together?

(8) *Julian:* The paralela [pause] they [pause] when they go, they go higher [runs two fingers parallel to each other first along the top and base of the rectangle and then continues along those lines], they never get together.

(9) *Antonio:* Yeah!

(10) *Teacher:* Very interesting. The rectangle then has sides that will never meet. Those sides will be parallel. Good work. Excellent work.

Figure 8.2 Sample vignette: "They never get together."

arguments as he interpreted, clarified, and rephrased what students were saying. This teacher provided opportunities for mathematical discussion by moving past student vocabulary or grammar errors, listening to students, and trying to understand and focus on the mathematics in what students said and did. Some of the teacher strategies that supported students' participation in mathematical arguments in this example are: using gestures and objects for clarification, building on what students said (Lines 4 and 5), asking for clarification (Line 7), and rephrasing student statements using more formal language (Lines 8 and 10).

I ask prospective teachers to read my analysis for homework and answer the following questions:

1. What surprises you about the analysis?
2. Are there differences between the analysis and the ideas or issues that you discussed?
3. Does this analysis open up any new perspectives for you?

SUMMARY

If teachers perceive formal mathematical vocabulary as the only linguistic resource, there is little room for addressing these students' mathematical ideas, building on them, and connecting these ideas to the discipline. Instead, mathematics instruction for ELs should focus on mathematical content and provide opportunities for students to participate in mathematical practices.

This case shows how to structure a discussion that focuses on hearing mathematical competencies in student contributions. Teachers can move towards recognizing the multiple language resources that students use to express mathematical ideas. Students may show evidence of competencies in mathematical practices by using everyday language as a resource. Teachers need to first support students as they display these competencies (orally or in writing) and then, later and over extended time periods, in learning to communicate in more formal mathematical language.

The vignette highlights that hearing the mathematical content, uncovering the mathematical practices, and seeing the language resources in student contributions are complex tasks. Two crucial questions for uncovering students' mathematical competencies when these are expressed through language are:

1. What is the mathematical content of a student's contribution?
2. What mathematical practices (describing patterns, abstracting, generalizing, etc.) are evident in student contributions during a discussion?

This case also develops prospective teachers' awareness and acceptance that students may use imperfect, everyday, colloquial, or home languages to express important mathematical ideas, show evidence of competence in mathematical practices, and participate in mathematical discussions.

REFLECTION ON THE DILEMMA

Participants have shared comments that lead me to believe that the structure for working on the vignette made a difference. I have heard prospective teachers say that they had never realized that students could express correct mathematical ideas in everyday language, that they were struck by Julian's persistence to communicate, and that they were convinced that Julian knew something about parallel lines and parallelograms. PSTs have also described how they can see the teacher supporting the mathematical discussion.

Following the process of watching the video, reading the transcript, and focusing on these particular discussion questions (see #1 and #2 above) helps many participants see and hear things that contradict some of their beliefs about English Learners. Since a teacher's strategies for hearing mathematical competence might be different depending on the grade level and the lesson topic, I now include examples from different grade levels and mathematical topics.

However, my concern is that these kinds of discussions are only the beginning, and much more work needs to be done to follow up. For example, the first few times I used this activity, I heard several comments from participants that showed me they were concerned with what they could or should do next if they were teaching this class. I now include a follow-up activity in which I ask participants to design a follow-up lesson that builds on this discussion, starts from students' informal ways of talking, and supports students in developing more formal mathematical ways of talking.

If participants work in classrooms, I would also ask them to follow up with these activities:

- Collect examples of students using home language or everyday language when doing mathematics in a classroom. Focus on the students' language resources. Reflect on the language resources you hear students using during a lesson. Look for instances of students using their home language to communicate mathematically and students using everyday language to communicate mathematically.
- Observe a mathematics lesson in a classroom, watch a video of a lesson online, or read a transcript of a lesson, using the same focus questions. After observing (or watching or reading), then use the set of discussions questions to reflect on how you are learning to see and hear mathematical competence.

ACKNOWLEDGMENTS

The analysis of the lesson excerpt presented here was supported by Grants #REC-9896129 and #ROLE-0096065 from the National Science Foundation. The Math Discourse Project at Arizona State University videotaped this lesson with support by a National Science Foundation grant.

NOTES

1. In California, a teaching credential is a post B.A. degree, so undergraduates cannot work towards a credential or major in education, they can only take education courses or complete a Minor in Education.
2. This vignette is one of several I have used in this manner. An analysis of this vignette was described in Moschkovich (1999), the analysis of a subsequent discussion in the same elementary classroom is described in Moschkovich (2007). Other vignettes are available in Moschkovich (2009, 2011, 2014a, and 2014b) and Zahner and Moschkovich (2011); Moschkovich (2009, 2011) use examples from secondary classrooms.

REFERENCES

Moschkovich, J. N. (1999). Supporting the participation of English language learners in mathematical discussions. *For the Learning of Mathematics, 19*(1), 11–19.

Moschkovich, J. N. (2007). Examining mathematical discourse practices. *For The Learning of Mathematics, 27*(1), 24–30.

Moschkovich, J. N. (2009). How language and graphs support conversation in a bilingual mathematics classroom. In R. Barwell (Ed.), *Multilingualism in mathematics classrooms: Global perspectives* (pp. 78–96). Bristol, England: Multilingual Matters Press.

Moschkovich, J. N. (2011). Supporting mathematical reasoning and sense making for English learners. In M. Strutchens & J. Quander (Eds.), *Focus in high school mathematics: Fostering reasoning and sense making for all students* (pp. 17–36). Reston, VA: National Council of Teachers of Mathematics.

Moschkovich, J. N. (2014a). Building on student language resources during classroom discussions. In M. Civil & E. Turner (Eds.), *The common core state standards in mathematics for English language learners: Grades K–8* (pp. 7–9). Alexandria, VA: TESOL International Association.

Moschkovich, J. N. (2014b). Language resources for communicating mathematically: Treating home and everyday language as resources. In T. Bartell & A. Flores (Eds.), *Embracing resources of children, families, communities and cultures in mathematics learning* (TODOS Research Monograph, Volume 3, pp. 1–12) . San Bernadino, CA: Create Space Independent Publishing Platform.

Rosebery, A. S., & Warren, B. (1998). *Boats, balloons & classroom video: Science teaching as inquiry.* Portsmouth, NH: Heinemann Educational Books.

Zahner, W., & Moschkovich, J. N. (2011). Bilingual students using two languages during peer mathematics discussions: ¿Qué significa? Estudiantes bilingües usando dos idiomas en sus discusiones matemáticas: What does it mean? In K. Tellez, J. Moschkovich, & M. Civil (Eds.), *Latinos/as and mathematics education: Research on learning and teaching in classrooms and communities* (pp. 37–62). Charlotte, NC: Information Age.

COMMENTARY 1

TEACHING PRESERVICE TEACHERS TO SUCCESSFULLY POSITION ENGLISH LEARNERS

A Commentary on Moschkovich's Case

Kathryn B. Chval and Rachel J. Pinnow
University of Missouri–Columbia

OUR POSITIONALITY

During the past 26 years, I (Kathryn) have facilitated professional development for teachers across the experience spectrum. In addition, I have researched teaching practices to understand how to best equip teachers for the complexity of teaching. Throughout those experiences my passion for children, especially English learners (ELs), who have been denied access to learning has grown. As a mathematics teacher educator (MTE), I recognize my own limitations. First, I have to recognize my privilege as well as the fact that I have not experienced what these children encounter. Second, my educational preparation focused on mathematics education and teacher

Cases for Mathematics Teacher Educators, pages 171–175
Copyright © 2016 by Information Age Publishing
All rights of reproduction in any form reserved.

education rather than second-language learning and teaching. As a result, I joined forces with Rachel Pinnow, an applied linguist, to pursue a research agenda that would draw upon the strengths of multiple fields. I (Rachel) have had experiences as a second-language learner that have shaped my understanding of the role of the teacher in providing effective learning opportunities, as well as fostering social interactions vital to joining the classroom as a respected member. Together we began to focus on the role of classroom interaction in fostering ELs' mathematical content knowledge and English language learning.

As I have engaged with prospective teachers (PSTs), I (Kathryn) have recognized the importance of designing tasks that provoke PSTs to share their beliefs as well as tasks that will provide opportunities for them to confront their beliefs and assumptions (e.g., Chval, 2004; Lannin & Chval, 2013). As demonstrated by Chval and Pinnow (2010), PSTs may have limited experiences and resources to draw upon when considering instruction for ELs because they have not had opportunities to develop them yet. For example, some assume they would outsource the work to ESL teachers and translators, isolate ELs to support them (prior to school) rather than structuring entry into classroom communities, and shape instruction based on country of origin (i.e., students from some countries would need remediation and others would need "enrichment"). Moreover, "most people, including PSTs, make assumptions about the knowledge and capabilities of others, depending on how they look, act, and speak" (Vomvoridi-Ivanovic & Chval, 2014, p. 116).

INTERPRETING THE DILEMMA

This case highlights a number of dilemmas. First, MTEs cannot employ a deficit view of PSTs that positions them as lacking knowledge or agency, as we do not want them to in turn employ that perspective with their own students. Second, we want to challenge unproductive beliefs about teaching, learning, mathematics, and children. As MTEs, how do we also address beliefs and assumptions about children from different countries and cultures, use of native language, and the role of English language (e.g., ELs should not be given challenging cognitive or academic tasks until their English language proficiency is near native-like)? How do we change their perceptions of what is helpful for ELs? How do we help PSTs identify their fears (e.g., if students speak in their native language, I won't understand what they say or won't be able to manage them)? Third, designing tasks for these discussions is challenging and takes specialized knowledge that many MTEs have not yet developed.

As the case author notes, many PSTs express the belief that native languages are a hindrance and that an English-only environment is the best learning environment. MTEs can anticipate these beliefs and utilize tasks that will promote critical discussion of them. This can be done through discussions of transcripts, videos, or case studies that provide PSTs with opportunities to consider classroom interactions from ELs' perspectives. For example, in this case, prior to watching the video, ask PSTs to describe a rectangle, giving different constraints. Some can only use spoken English, another can only use gestures, another can only use a physical model, another can only use drawings or representations, and one can use multiple resources. What are the benefits of multiple resources? Why would we constrain children and limit the resources they can draw upon? Next ask them to anticipate how third-grade students would describe a rectangle using English or Spanish, what misconceptions third graders may demonstrate while describing rectangles, and how a third grader would explain "parallel." These types of discussions set the stage for viewing the video because many PSTs want to know if their predictions are "right." In this case, did any of them anticipate the use of gestures to describe "parallela"?

RESPONDING TO THE DILEMMA

MTEs could also use this case to address the role of positioning (Harré & Van Langenhove, 1999). Is Julian positioned as a spectator who is denied participation based on his English proficiency? ELs are often positioned inequitably in peer-to-peer and whole-class discussions, which makes it more difficult for them to gain access to academic debate and discussion (Pinnow & Chval, 2015). Inequitable positioning constrains ELs' access to learning opportunities necessary for developing both advanced mathematical ability and English language competencies (Pappamihiel, 2002; Pinnow & Chval, 2014; Yoon, 2008). Building on PSTs' desire to be effective, will they be willing to admit that the choices they make as teachers may deny access, reduce opportunities, induce fear and anxiety, and facilitate inequities in their classrooms? Not likely. Facilitating awareness about these issues will promote a desire for them to learn how to draw on students' resources as they teach mathematics.

Although there is evidence that the teacher in this case has made a space for ELs to interact equitably and productively, PSTs could benefit from activities designed to expose possible assumptions they might hold *about students* based upon their ethnicity, race, and educational background so that PSTs can develop more critical and reflective pedagogy. MTEs teach in different contexts that should influence the design and implementation of tasks (e.g., elementary versus secondary, languages represented in student populations in local schools, and language backgrounds of PSTs). For example, the PSTs in our classes complete field experiences in a district with 47 different languages among students from a variety of educational settings (e.g., transnational students, immigrants and refugees with limited formal schooling or interrupted schooling, and ELs from high-income private schools). In our context, it is important to display videos and have discussions that emphasize the use of multiple semiotic resources for thinking, strategizing, explaining, and communicating in mathematics classrooms. By drawing upon the multiple modes of communication that ELs utilize in the classroom, teachers can leverage these elements to promote both mathematical sense making and second-language acquisition.

We conclude with another dilemma. The design of tasks related to teaching ELs should not be a collection of isolated activities. MTEs need a framework as well as tools to guide decisions, make connections, and establish a coordinated course of study. Moreover, PSTs need frameworks and tools that they can use in their classrooms to guide their decisions and instruction. The following guidelines have supported our decisions.

- Problematize the objective so that PSTs develop self-awareness of their beliefs, assumptions, and fears and want to learn to "solve that problem" (e.g., I don't want to position my students to be silent spectators who do not have access to mathematics; I don't want to be afraid if ELs speak in their native languages).
- Provide simulations where PSTs consider the perspective of ELs (e.g., If I were an EL, I would not want my teacher and peers to think that I wasn't capable of learning mathematics).
- Set the stage prior to using images of effective instruction with ELs in order to build the resources that PSTs can draw upon when they teach.
- Discuss practices that are not "helpful" for ELs.
- Draw on PSTs' resources and desires in order to encourage them to pursue professional development related to EL instruction throughout their careers. They cannot learn everything necessary in one semester; therefore, it is essential to encourage them to pursue knowledge throughout their careers.

REFERENCES

Chval, K. B. (2004). Making the complexities of teaching visible for prospective teachers. *Teaching Children Mathematics, 11,* 91–96.

Chval, K. B., & Pinnow, R. (2010). Preservice teachers' assumptions about Latino/a English language learners. *Journal of Teaching for Excellence and Equity in Mathematics, 2*(1), 6–12.

Harré, R., & Van Langenhove, L. (1999). *Positioning theory: Moral contexts of intentional action*. Oxford, England: Blackwell.

Lannin, J. K., & Chval, K. B. (2013). Challenge beginning teacher beliefs: Use these specific strategies to confront assumptions about the teaching and learning of mathematics. *Teaching Children Mathematics, 19,* 508–515.

Pappamihiel, N. E. (2002). English as a second language students and English language anxiety: Issues in the mainstream classroom. *Research in the Teaching of English, 36,* 327–355.

Pinnow, R. J., & Chval, K. B. (2014). Positioning ELLs to develop academic, communicative, and social competencies in mathematics. In M. Civil & E. Turner (Eds), *Common core state standards in mathematics for English language learners: Grades K–8* (pp. 21–33). Alexandria, VA: TESOL Press.

Pinnow, R., & Chval, K. (2015). "How much you wanna bet?": Examining the role of positioning in the development of L2 learner interactional competencies in the content classroom. *Linguistics and Education, 30,* 1–11.

Vomvoridi-Ivanovic, E., & Chval, K. B. (2014). Challenging beliefs and developing knowledge in relation to teaching English language learners: Examples from mathematics teacher education. In B. Cruz, C. Ellerbrock, A. Vasquez, & E. Howes (Eds.), *Talking diversity with teachers and teacher educators: Exercises and critical conversations across the curriculum* (pp. 115–130). New York, NY: Teachers College Press.

Yoon, B. (2008). Uninvited guests: The influence of teachers' roles and pedagogies on the positioning of English language learners in the regular classroom. *American Educational Research Journal, 45,* 495–522.

PREPARING OUR NEW TEACHERS (AND OURSELVES) TO "HEAR MATHEMATICAL COMPETENCE"

A Commentary on Moschkovich's Case

Crystal Kalinec-Craig
University of Texas at San Antonio

BACKGROUND AND POSITIONALITY

As a White female from a middle-class background, I taught middle and high school mathematics classes for 6 years in the United States and overseas in Germany before I began my studies in 2007 for mathematics teacher education. As a new teacher, I assumed that mathematics was a universal language and assumed that one's culture, race, gender, ethnicity, class, and native language neither would nor should play a role in how anyone learned mathematics. Furthermore, I encouraged my students to articulate their mathematical thinking with precise terminology, pronunciations, and justifications that were aligned with my own expectations. As a doctoral

Cases for Mathematics Teacher Educators, pages 177–181

student in southern Arizona, I had a number of transformative experiences that challenged and reframed what I thought I knew about teaching, learning, and doing mathematics. During those 5 years, I learned more about students who came from a background different than my own, many of whom were undocumented immigrants or nonnative English speakers in a primarily Latin@[1] neighborhood. Over time I noticed a dramatic shift in my thinking. When learning, teaching, and doing mathematics, I recognized that one's culture, race, gender, ethnicity, and class matters. *A lot.* And more importantly, I learned how crucial it was for a teacher to honor a child's native language while teaching mathematics. Now as a mathematics teacher educator at a Hispanic-Serving Institution in Texas, one of my primary goals is to help my prospective elementary teachers develop a sense of "hearing mathematical competence" in their practice.

INTERPRETING THE DILEMMA

Moschkovich's case is one that resonates loudly with me because I once professed similar beliefs and misconceptions as a young teacher, and I also hear these same misconceptions from my prospective elementary teachers each semester. Therefore, I situate my interpretation as not only a K–12 teacher, but also a teacher educator in higher education and will briefly discuss the broader contexts that frame Moschkovich's case.

Helping educators develop sensitivity for diverse ways students communicate their mathematical thinking is a complex and urgent issue. A layer of complexity exists, considering that many of today's teachers have few firsthand experiences as second-language learners and need specific and explicit support to incorporate linguistic strategies in their mathematics instruction (Hollins & Guzman, 2005). Teachers who use sentence stems (or partially completed sentences), gestures, visuals aids, and cognates can help their students to learn mathematics more easily in a second language by leveraging their existing knowledge situated in the native language (Wiest, 2008). And although some teachers may believe that they will "never teach English Language Learners," these linguistic strategies are ones that can benefit *all* students during mathematics instruction, especially for those who are not yet fluent in the primary language of instruction.

There is also a sense of urgency pressing on the preparation of new teachers. Given the recent influx of immigrants to the United States, teacher preparation programs must address the issue of helping new teachers to elicit and validate the mathematical thinking of nonnative English speakers. The Rio Grande Valley section of the U.S. Customs and Border Protection agency has recorded more than 6,000 unaccompanied undocumented children from Mexico and Central American countries who have crossed into the U.S. since October 2014 (United States Customs and Border Protection, 2015). Since October of 2013, the Rio Grande Valley section has accounted for nearly 72% of the immigration that includes undocumented workers, families, and unaccompanied children. Our immigrants bring a wealth of experiences and knowledge rooted in their native languages, and as a result, they need well-prepared teachers who will help bridge their mathematical knowledge in their native language with their new language. The issue that Moschkovich poses is one that should not be solved by encouraging our students to limit their focus on English fluency above all else.

RESPONDING TO THE DILEMMA

When I consider the issues posed by Moschkovich in her case, I am reminded of my unique position as a former K–12 teacher and now teacher educator. Each semester, I am prepared for the misconceptions that my new elementary teachers profess when we discuss the intersection of native language and mathematics instruction because I once held those same beliefs and stereotypes as a young teacher. My experiences as a teacher overseas and then as a doctoral student instigated a transformative shift in my thinking and perspective. And as a result, I continue to seek new ways to help my prospective teachers experience a similar shift in their thinking and consider three questions each semester.

The first question I consider is in what ways can I foster a safe space for my prospective teachers to discuss and reframe their misconceptions and stereotypes about the role of language while teaching and learning mathematics? Similarly to Moschkovich, I too find it counterproductive to simply inform new teachers that their beliefs may be misinformed. Therefore, after we collectively discuss our prior educational experiences and what they believe to be elements of quality teaching, I explicitly emphasize that my role is to support them to develop a new and more critical lens for what they believe to be good teaching. Sometimes our eyes might deceive us into thinking that a particularly strategy is helpful for all students while learning mathematics. It is always a fragile dance to honor the knowledge and experiences of my new teachers while helping them to form new visions of teaching that promote equity for all students they may teach in the future.

Secondly, I consider what kind of experiences can help my prospective teachers to learn about effective linguistic strategies for teaching mathematics? Throughout the semester, we read scholarly articles about teaching mathematics for nonnative English speakers and the strategies that can support all students to communicate their thinking. But this is sometimes not enough. Because the majority of my prospective teachers are native English speakers, I stage an experience where they solve a mathematics task in a language that many of them are typically not familiar with, German. During the task, I use and clearly emphasize each of the strategies posed in the articles we have read while the prospective teachers solve an addition problem about the total age of my pets. I use pictures, hand gestures, and sentence stems to help my new teachers solve the task. This semester, when I debriefed them on their experience of solving the mathematics task posed in German, I heard one of the prospective teachers tell another colleague at her table, "I didn't know German, but Dr. Craig helped us because she used the strategies that we just read about in all those articles. I guess it really does work." Which leads me to ask my final question: How can I help my new teachers sustain the practices that they learn in class?

I presume many teacher educators like myself are constantly searching for new ways to help their prospective teachers sustain the learning long after they graduate from our programs. Although I recognize my own limitations as a mathematics teacher educator for one methods course, I also believe that it is my role to help my prospective teachers learn how to teach mathematics while also developing a critical lens on their instructional decision-making. My prospective teachers should constantly ask themselves "Is this strategy effectively helping ALL of my students to be successful in mathematics, and how do I know this?" This question similarly drives my own practices in the classroom.

CONCLUDING THOUGHTS AND TAKEAWAYS

Moschkovich's case highlights both a challenge and a goal for mathematics teacher educators. Recognizing the wide variety of experiences that our new teachers bring to their teacher preparation program also means we must be prepared for when their visions of teaching might initially include misconceptions and unintentional biases. I am an example of a teacher who shifted (and is still shifting) her misconceptions and biases about teaching in a way that will eventually lead to a critical, reflective lens about the intentions and implications of my practice. In the spirit of Lisa Delpit (1988), our teachers need to do more than *listen* for correct mathematical thinking; our teachers need to be ready to *hear* the mathematical richness and competence that is situated within our students' native languages.

NOTE

1. I utilize the word Latin@ to honor those who do not identify as male or female (Gutierrez, 2012).

REFERENCES

Delpit, L. D. (1988). The silenced dialogue: Power and pedagogy in educating other people's children. *Harvard Educational Review, 58*, 280–298.

Gutierrez, R. (2012). Embracing Nepantla: Rethinking "knowledge" and its use in mathematics teaching. *Journal of Research in Mathematics Education, 1*, 29–56. doi: 10.4471/redimat.2012.02

Hollins, E., & Guzman, M. T. (2005). Research on preparing teachers for diverse populations. In M. Conchran-Smith & K. Zeichner (Eds.), *Studying teacher education: The report of the AERA panel on research and teacher education* (pp. 477–548). Mahwah, NJ: Erlbaum.

United States Customs and Border Protection. (2015). *Southwest border unaccompanied alien children.* Retrieved from http://www.cbp.gov/newsroom/stats/southwest-border-unaccompanied-children

Wiest, L. R. (2008). Problem-solving support for English language learners. *Teaching Children Mathematics, 14*, 479–484.

POSITIONING, STATUS, AND POWER: FRAMING THE PARTICIPATION OF EL STUDENTS IN MATHEMATICS DISCUSSIONS FOR PROSPECTIVE TEACHERS

A Commentary on Moschkovich's Case

Maria del Rosario Zavala
San Francisco State University

MY POSITIONALITY

I am an American-born Latina, the first of my Peruvian family to be born in the United States. I grew up mostly in White and Asian neighborhoods as my father moved our family across the U.S. ever in search of his American dream. My identity as a Latina was constructed over time as I grappled with how little I looked like my best friends, how much I loved to use Spanish as

Cases for Mathematics Teacher Educators, pages 183–187
Copyright © 2016 by Information Age Publishing
All rights of reproduction in any form reserved.

a way to connect with people, and how I seemed to be able to use my gifts for math and my linguistic abilities to tutor the Mexican kids at a local parochial school when I was a teenager. About 15 years ago I *was* that student in their senior year in the mathematics department at UCSC, taking Judit's class, working in a high school in Watsonville through a program at UCSC designed to give undergraduate mathematics majors opportunities to work in classrooms in preparation for teaching. My education ultimately took me down a different path away from high school mathematics and towards elementary school teaching, where I enjoyed a few years as a literacy and mathematics support teacher before moving into teacher education.

Many ideas in Judit's chapter resonate with me. I wasn't the student saying that ELs cannot participate in mathematics discussions, because I knew if we switched the language to Spanish many more kids participated. However I do remember classmates raising these issues and me not being sure how to challenge those statements. I believe that this was in part due to the socialization my classmates and I had about what were "proper" ways of doing mathematics. Imagine a room full of juniors and seniors finishing their degrees in mathematics, men and women, Whites and Latin@s, people who thought they had mathematics figured out because they could write a proof about why there are an infinite number of prime numbers—we held particular beliefs about what it meant to participate meaningfully in mathematics. These beliefs were mostly about individual achievement. No one knew or cared to talk about the myths of the achievement motivation narrative when it comes to EL or other marginalized communities in mathematics—we didn't have enough experience drawing connections between institutional racism and educational disparities.

Coming from such a paradigm it's not hard to see how we made assumptions about the abilities of EL students. Certainly, these assumptions border on (or cross the border completely into) racism because they are hard to separate from the assumptions about the learners, those brown kids that we assume have little formal schooling. Latino critical scholars remind us of how overlapping aspects of identity, such as gender, race, immigration status, languages spoken, all contribute to how Latin@s define their experiences of subordination and power in the United States (Yosso, Villalpando, Delgado Bernal, & Solórzano, 2001). These *intersectionalities* are important for us to attend to as teachers who are trying to understand our students as whole human beings. So I argue that when my classmates made general statements that put down EL students because of their linguistic abilities, they put them down as people—that's the intersectionality of language prejudice and racism.

INTERPRETING THE DILEMMA

Now as a scholar in mathematics education I research issues of power and participation in mathematics education with Latinos, and so I tend to see every issue through a lens of who has power and who does not. My reading of the dilemma is that of both a need to address underlying prejudice about EL students and a need to address what counts as legitimate participation in mathematics, with a group of students who I am guessing are largely White and good at math. I see Judit's response as very important work in which she is trying to help the students develop new lenses on EL learners. This is a huge task, and the tools she uses are important. I know that the time we had to discuss and analyze video cases in her class began to give me vocabulary for understanding issues of racism in mathematics classrooms. Now in my current job as a teacher educator, I know the power of video to impact preservice teachers' ideas of what is good teaching and what is possible. But I also know how hard it is to challenge deficit perspectives without prospective teachers evoking ideas of exceptionalism, like the idea that students on video must be different from the "real" students the preservice teachers actually work with.

RESPONDING TO THE DILEMMA

Most importantly, I see the need for prospective teachers to unpack their own multiple forms of privilege. If I don't help my prospective teachers interrogate how deficit assumptions are linked to a belief system that is not based on reality, it doesn't matter how many videos of Latino EL students talking about mathematics they see. But I think this kind of reflection is bigger than a mathematics class. I think it's an issue for the institution. However within a particular class we could still support prospective teachers to reflect on why they are going into the profession, what beliefs they have about what makes children successful in mathematics, and what they truly believe that they as a teacher can impact.

Beyond the need for self-reflection on privilege, I also find it important to hold prospective teachers accountable for positioning EL students as important mathematical thinkers. For example, one of my preservice teacher cohorts recently completed 4 days of planning, teaching, and reflecting on lessons taught in a local elementary school to fifth graders in Spanish, in a school that is about 50-50 White and Latin@. In preparation for this, they read and discussed an article on positioning EL students as *agentive problem solvers*—authors of mathematical ideas (Turner, Dominguez, Maldonado, & Empson, 2013). This article allowed us to develop the language of *positioning*, as in "what can we do to *position* our students as important mathematical thinkers?" Adopting the language of positioning was useful to both remind prospective teachers that students' mathematical identities are not static but rather malleable, and emphasize the teacher's ability to influence student mathematical identity, making them accountable for how the students are seen by others as capable mathematicians. Each day they were asked to reflect on what they did to position the Latin@ students in their small group as good mathematical thinkers. This drew their attention to the importance of creating opportunities for EL students to participate in significant ways in discussions and also to their own power to reframe whether a student is indeed really lacking in the language needed to participate, or whether it is their own responsibility to listen to what the student says and help interpret what is mathematically important about their contribution.

Closely related to positioning is the idea of status (Featherstone, Crespo, Jilk, Oslund, Parks, & Wood, 2011). I introduce the idea of status to my preservice teachers to help connect observable behavior to power relationships in the classroom, and also once again to reinforce the idea that students' identities as mathematicians are not fixed. With many of my preservice-teacher methods classes I use video watching through the lens of status and power to analyze what a teacher can do to support the successful involvement of traditionally marginalized students in classroom discussions. We watch Ball's interaction with Mamadou (from http://deepblue.lib.umich.edu/handle/2027.42/78024), and I facilitate discussions about how the teacher valuing the understanding that the student did show is more important than the wrong answer, and also discussion of what the importance of positioning this particular student, a young African American male, may mean given the historical achievement of this population in mathematics.

To me, the only way we get prospective teachers to stop making deficit statements about EL students is to help them make other kinds of statements, which involves helping them interrogate what counts as legitimate mathematics, helping them acknowledge their privilege, and helping them ask relevant questions instead of making sweeping generalizations. I think that means we have to give them tools that connect their observations to assumptions of what counts as participating in mathematics, and also allows us to challenge those assumptions and beliefs—tools like framing participation through status and positioning. We also need to hold them accountable to involving EL students in mathematics discussions so that ideas of exceptionalism can be challenged. Videos and modeling analysis, like Judit does, are an important tool to do this work. And then to go deeper we must find ways to draw out the assumptions and support prospective teachers to challenge them with their own observable evidence using a framework that helps them see how EL students do not have to be positioned as lower status students, but rather that they could be positioned as some of the most important mathematical thinkers in the class.

REFERENCES

Featherstone, H., Crespo, S., Jilk, L., Oslund, J. A., Parks, A. N., & Wood, M. B. (2011). *Smarter together!: Collaboration and equity in the elementary math classroom.* Reston, VA: National Council of Teachers of Mathematics.

Turner, E., Dominguez, H., Maldonado, L., & Empson, S. (2013). English learners' participation in mathematical discussion: Shifting positions and dynamic identities. *Journal for Research in Mathematics Education, 44,* 199–235.

Yosso, T., Villalpaldo, O., Delgado Bernal, D., & Solórzano, D. G. (2001). *Critical race theory in Chicana/o education.* National Association for Chicana and Chicano Studies Annual Conference. Paper 9. Retrieved from http://scholarworks.sjsu.edu/naccs/2001/Proceedings/

CHAPTER 9

TRACKING IN A LOCAL MIDDLE SCHOOL

Do You See What I See?

Dorothy Y. White
University of Georgia

As a mathematics educator who is dedicated to issues of equity and culture in mathematics classrooms, I am aware of structural inequities present in classrooms and schools. My years of experience help me notice the messages teachers send about their views of mathematics, teaching, and students. In this case, I share my dilemma of helping a class of secondary mathematics preservice teachers (PSTs) notice the inequitable practices in tracked mathematics classrooms. In particular, I struggled to help PSTs examine how the different ways two teachers organized their classrooms, assigned mathematical tasks, and allocated class time and space resulted in differential learning opportunities for students across race and perceived academic ability. At the end of the course, I do not think the PSTs saw what I saw: the tracking of Black and special education students, the consequences of tracking for those students' mathematics learning, and the benefits of tracking for White and gifted education students.

Cases for Mathematics Teacher Educators, pages 189–196
Copyright © 2016 by Information Age Publishing
All rights of reproduction in any form reserved.

CONTEXT

Recently, I taught three semesters of a secondary mathematics methods course with a 10-week field component at a local middle school, *Math Middle* (pseudonym). *Math Middle* enrolled approximately 575 students in Grades 6–8 with a student population of 53% Black, 30% White, 9% Hispanic, and 60% free or reduced lunch. The Black student population was close to the overall proportion in the district; however, the school's proportion of White students was higher than the district average of 19%, and the Hispanic population was lower than the district average of 22%.

The methods course is the second of three mathematics pedagogy courses in our teacher education program and focuses on assessment and equity. During the course, I assigned chapters from several texts, including *Mathematics Assessment: A Practical Handbook for Grades 6–8* (Bush, Leinwand, & Beck, 2000), *Mathematics for Every Student: Responding to Diversity in Grades 6–8* (Malloy & Ellis, 2008), *More Good Questions: Great Ways to Differentiate Secondary Mathematics Instruction* (Small & Lin, 2010), and *Mathematics and Teaching: Reflective Teaching and the Social Conditions of Schooling* (Crockett, 2008).

Teaching the course in a middle school during the middle 10 weeks of the semester provided PSTs with opportunities to learn about students' mathematical thinking as they worked with small groups of 3–4 students. There were two mathematics teachers and a special education teacher at each grade level. PSTs worked in all six mathematics classrooms, spending 3 weeks at each grade level and rotating in each class in a given grade. They worked with students on mathematics tasks designed by the teacher. My Teaching Assistant and I observed and modeled ways to engage the middle-school students in mathematics tasks (Philipp et al., 2007). Each week, the PSTs wrote an activity report describing the mathematics tasks they worked on with students; what they learned about students' thinking; what they learned about teaching a small group of students, including their successes and challenges; and what they would change in the next session and why. The case I describe is based on my experiences teaching one semester of the methods course at *Math Middle*. In this semester, there were three incidents that prompted my dilemma: (a) look and learn activity, (b) working in the 7th-grade classrooms, and (c) the final activity report. Although the incidents happened in one semester, I had the same sinking feeling I have had each time I have taught this course, that my PSTs "*couldn't see what I saw.*"

THE CASE

Look and Learn Activity

On our first classroom visit, I had my PSTs engage in a *Look and Learn Observation* activity. In this activity, PSTs work in pairs to observe how mathematics is taught and the ways students participate in lessons. Each pair consists of a macro- and a micro-observer who observe the same classroom for 15 minutes, recording what is happening in the class in 5-minute intervals. The macro-observer records the teacher's actions and the classroom context. The micro-observer focuses on a small group of no more than four students and records the students' actions.

One pair, a White female and White male, shared their observations. After the female PST talked about the class from a macro-perspective (in which she described the classroom, students, and math tasks in general), the male PST shared what he learned from observing a group of three boys. He started by saying that he didn't know the boys' names so I told him to use pseudonyms. He said, "Ok, I'll call the three guys, "*Lil Man*," "*Ese*," and "*Bro*." He and the class laughed but I was stunned. I thought he would use actual names for the boys instead of stereotypical terms. My first response was, "Are you talking about the group with the Black boy, the Latino boy, and the White boy?" He said "Yes" and started to look down. I said "Ok" and let him continue sharing his observations. Once all the pairs had shared, I revisited the issue of race. I talked about the importance of not being colorblind to students' races because pretending that all students are the same ignores and devalues students as individuals with their own cultural backgrounds. I knew the PSTs would work with a variety of students in the school and wanted them to feel comfortable talking about students and seeing their identities. I did not explicitly mention why I found the PST's stereotypical comments offensive because I knew the PSTs might be uncomfortable talking about race around me. I am the only African American professor they would encounter in the teacher education program, and I did not want them to become guarded with their observations. I thought my brief lecture would create a space for them to talk about students' race rather than pretend not to notice.

Working in 7th-Grade Classrooms

During the first 3 weeks at *Math Middle*, the PSTs worked with students in the two 6th-grade classrooms. I focused my attention on teaching PSTs how to ask probing questions to solicit and assess students' mathematical thinking and how to differentiate tasks to accommodate students' ideas. In preparation for my PSTs to transition to the next grade, I briefly observed the two 7th-grade mathematics classrooms to get a sense of how they were organized and the type of instruction present in the classrooms. When I sat in on the classes, I noticed several stark differences. The first thing I noticed was the difference in the number and racial compositions of the students in the two classes. One class (Ms. Bond's) had 14 students and was about 80%

White and 20% Black; the second class (Ms. Miller's) had about 30 students, with 90% Black and 10% White students. Ms. Bond's room had the desks arranged in clusters of 4 so each student sat across from someone. In contrast, the larger class, Ms. Miller's room, had students' desks paired and arranged in rows. There were two teachers in the larger class and one teacher and a student teacher in Ms. Bond's.

The most noticeable difference between the classes was the noise level. In Ms. Bond's room, students talked and worked together so the class was not quiet, but not so loud that the students could not hear Ms. Bond when she called their attention. However, Ms. Miller's class was louder because as some students called out answers to the warm-up problems, other groups of students had side conversations and were not paying attention to the lesson. As a result, Ms. Miller and Ms. Hawkins (the special education teacher) spoke even louder to be heard. When I asked the teachers about their classes, I learned Ms. Bond taught 7th-grade content for advanced 6th-grade students, and Ms. Miller's room was an inclusion class with special education and general education students.

After leaving these classrooms, I immediately thought of Oakes's work (2008) on tracking students in different mathematics classrooms primarily by race and Kunjufu's (2005) research on the overrepresentation of African American male students in special education classes. I was concerned about the way PSTs would view students from diverse backgrounds while working in these classes.

For the 3 weeks the PSTs worked in the two 7th-grade classes, I noticed Ms. Bond's class was engaging. Although teacher-led, the students engaged in problem-solving activities and used manipulatives. She began the lesson with Powerpoint slides she designed followed by many real-world examples on the Smartboard to motivate her students. Ms. Miller was lively and started each class with 4–5 warm-up problems posted on the board. Students were expected to copy and solve the problems on their own and be prepared to share their solutions. As students worked, Ms. Miller went around the class and told each student that she loved them. She expected them to say it back to her and when they didn't, she waited before going to the next student. She spent about 10 minutes every class with this routine and then went over the warm-up problems. If students solved the problem correctly, she praised them or in some cases, gave them a treat. After the warm-up, Ms. Miller gave students a worksheet of problems to solve, but due to the arrangement of the desks, she could never get around to the back of the class. The seven PSTs were assigned to groups of students and were challenged to find a space to work with them given the desk arrangements. Many of the students did not hear the directions and began to talk and disengage. As the noise level increased, Ms. Miller turned off the lights "to calm everyone down." As a result, the room only had the natural light from the windows, making it hard to see the worksheet problems. Most of the PSTs were frustrated in the class and couldn't wait to rotate to another class.

After 3 weeks, we transitioned to 8th grade, and PSTs worked in two classrooms. Unlike the 7th grade, these classes had similar demographics and class configurations, with the main difference being the teachers' pedagogical styles. I was happy to move on but was still disturbed by what I had seen in Ms. Miller's class.

Final Activity Report

When we returned to campus, I addressed equity issues and differentiation in the last 3 weeks of classes. As a class, we read and discussed a case by Crockett (2008) on "Race and Teacher Expectations," and I taught lessons on culturally relevant pedagogy, teaching for social justice, and ways to accommodate special education students without lowering the cognitive demand of tasks. At the end of the class, PSTs wrote a final reflection paper comparing and contrasting any two classrooms at *Math Middle*. In particular, they had to reflect on the ways the teachers managed time, space, and learning resources for their students and encouraged active and equitable engagement of all students. They were also asked to identify aspects of the learning environment they would like to emulate in their future classrooms. As a class we talked about the different classes at *Math Middle*, and I answered any questions they had about the assignment.

All but 2 of the 14 PSTs compared Ms. Miller's class to that of another teacher. Ms. Miller was described by the PSTs as energetic, caring, and engaging (something they wanted to emulate). They thought her telling each student she loved them was a good way to establish a relationship with them. Interestingly, they found the desk arrangement as a way all students could see the board, but some also mentioned that it limited the teachers' ability to get around to each student and did not support students' working together. They also liked how she accepted all students' answers and praised them after they answered questions.

Ms. Miller's classroom environment was most often described as chaotic, rushed, and with no classroom norms. PSTs mentioned that the worksheet tasks were not group worthy and why students were off task. However, half of the PSTs blamed the students for Ms. Miller's classroom practices because they talked too much about nonmath things, did not pay attention, lacked interest or motivation, and had behavior problems with each other. In contrast, students from Ms. Bond's room were described as motivated, smart, calm, not too noisy, and worked well together.

I was very surprised that PSTs did not mention anything about the students' special education status. PSTs felt Ms. Miller's students needed more attention, but they could not see that students needed accommodations based on their particular learning needs. For example, no PST mentioned students needing instructional accommodation, such as a calculator. They spoke of the students in broad terms of what they lacked. This surprised me because while at *Math Middle* two special education teachers had taught PSTs lessons: one on different types of learning challenges and a coteaching model and the other on reading and planning based on students' IEPs. Instead the PSTs seemed to lack an understanding that some students learn differently and do not benefit from traditional instruction. They did not seem to make the connection.

I was encouraged when I read a reflection from a graduate student who sat in on the course when she wrote,

> One barrier to teaching and learning that seems hard to overcome is the "hidden" tracking that seems more prevalent today than when I first started teaching ten years ago. Witnessing class sizes of 30, where most students have special needs, seems to prohibit engaging forms of teaching and learning. On the other hand, accelerated classes of 14 seem inequitable when occurring at the same time. Will these students with special needs be supported to accelerate? Will they have opportunities for rich discourse? Will their teacher tap into their strengths?

Unfortunately, her thoughts and concerns were not echoed in the PSTs' reflections.

REFLECTIONS ON THE DILEMMA

I am not naïve to think that PSTs will see classrooms the way I do in one semester. It was obvious to me that the classes were racially segregated, but I am not sure if PSTs could see what I saw. This experience continues to bother me because I do not think I helped move PSTs to think about the repercussions of tracking on students' mathematical experiences. I wanted them to notice that the classroom environments, tasks, and teacher expectations were very different across the two classes. I wanted them to wonder and notice that the students in the inclusive classroom were predominately African American whereas the students in the other class were predominately White. I wanted them to notice that Ms. Miller spent too much time telling the students that she loved them instead of using the time to teach them mathematics.

Reflecting on the semester, I wish I had assigned and referenced the case on race and teacher expectations earlier in the semester and included a reading on tracking as the PSTs worked in the school. I wish I had talked with the PSTs about the long-term effects on students' learning when they experience mathematics in classrooms like Ms. Miller's. I wish I could have connected what they noticed to help them think more deeply about issues of race, class, and special education. In the end, I just did not know how to make the connections in a way that used the shared experiences we witnessed in Ms. Miller's class and the general ideas of tracking. Was I too busy focusing on the PSTs learning how to teach mathematics to call their attention to these important issues that affect the teaching of mathematics? Were PSTs uncomfortable talking about race with me because I am an African American? Even though I tried to open a space for them to talk about these issues, I feel I missed an opportunity, and I know that unless the issue of tracking and other inequities in the classroom are addressed, PSTs will never make real changes in the classroom.

FINAL THOUGHTS

In writing this case, I hesitated to position myself as an African American woman because in my experience, colleagues erroneously assume that my Blackness bestows upon me some special skills to see inequities in schools and the world. As a result, they often use my Blackness as an excuse to ignore equity issues in their courses because they weren't born with my "special skills." What they fail to understand is that I am able to see tracking and other forms of school inequities because I *learned* they exist. Before then I, along with many Black teachers and graduate students, didn't think schools would systematically treat students this way. I have worked with several African American teachers who do not notice tracking in their schools and classrooms. Moreover, they often do not see how they employ inequitable practices.

My Blackness helped me understand tracking as a racialized phenomenon because it explained why so many Black and Hispanic students were not enrolled in gifted programs. I have known many gifted Black and Hispanic students, and learning about tracking and other forms of inequity in schooling answered a lot of questions for me. My colleagues' assumptions minimize all my years of graduate study, working in classrooms, and commitment to improving the mathematics education of all students. I hope colleagues learn that a lens on equity is a learned skill that benefits us all.

REFERENCES

Bush, W. S., Leinwand, S., & Beck, P. (2000). *Mathematics assessment: A practical handbook for grades 6–8.* Reston, VA: National Council of Teachers of Mathematics.

Crockett, M. D. (2008). *Mathematics and teaching: Reflective teaching and the social conditions of schooling.* New York, NY: Routledge.

Kunjufu, J. (2005). *Keeping black boys out of special education.* Chicago, IL: African American Images.

Malloy, C. E., & Ellis, M. W. (2008). *Mathematics for every student: Responding to diversity in grades 6–8.* Reston, VA: National Council of Teachers of Mathematics.

Oakes, J. (2008). Keeping track: Structuring equality and inequality in an era of accountability. *Teachers College Record, 110,* 700–712.

Philipp, R. A., Ambrose, R., Lamb, L. L. C., Sowder, J. T., Schappelle, B. P., Sowder, L., Thanheiser, E., & Chauvot, J. (2007). Effects of early field experiences on the mathematical content knowledge and beliefs of prospective elementary school teachers: An experimental study. *Journal for Research in Mathematics Education, 38,* 433–476.

Small, M., & Lin, A. (2010). *More good questions: Great ways to differentiate secondary mathematics instruction.* New York, NY: Teachers College Press.

COMMENTARY 1

UNPACKING EXPECTATIONS AND LENSES IN MATHEMATICS CLASSROOM OBSERVATIONS

A Commentary on White's Case

Lynette DeAun Guzman
Michigan State University

MY POSITIONALITY

Throughout my K–12 schooling experiences, I was enrolled in gifted programs, and I distinctly remember how this tracking impacted how I saw myself. For instance, I was quite aware of negative stereotypes about Latin@ and Black students in mathematics; however, I would distance myself from these narratives because I thought I was *different* than *those people*. I had a very different consciousness than I do now. I did not see structural inequities in classrooms and schools. Instead, I viewed meritocracy as key to the game and my success. My views were challenged, though, during my interactions in postsecondary education.

Cases for Mathematics Teacher Educators, pages 197–201
Copyright © 2016 by Information Age Publishing
All rights of reproduction in any form reserved.

As an undergraduate student, I regularly volunteered as a tutor and mentor for mathematics outreach programs within the context of schools with large populations of low-income families of color. I worked in classrooms labeled as remedial mathematics and quickly observed differences in how teachers structured those classroom spaces compared to well-funded schools I visited. Thus, I started questioning what I saw happening in mathematics classroom spaces, especially with respect to how students saw themselves as people who do (or not do) mathematics. These experiences led me to graduate studies with a focus on equity in mathematics education.

I am a Latina mathematics teacher educator at a predominantly White institution. My experiences over the past 3 years include teaching mathematics content courses for prospective elementary teachers, shadowing as a field instructor for secondary mathematics interns, and providing support to graduate teaching assistants in developing their practice in both mathematics and nonmathematics contexts. Reflecting upon my earlier experiences highlights Dr. White's point about how she *learned* to see structural inequities in schools. Although it is important to recognize that people make meaning of their own experiences in the world, our reflections through these journeys may be very different.

INTERPRETING THE DILEMMA

In this case, I see a dilemma in the differing expectations and lenses for secondary mathematics preservice teachers (PSTs) and their teacher educator during methods course experiences. More specifically, this dilemma involves differences in foundational beliefs and philosophies for what teaching and learning mathematics looks like. Additionally, this dilemma involves how educators develop lenses to see structural inequities in classrooms and schools. This particular case contrasts two particular 7th-grade mathematics classrooms at Math Middle School; however, I would imagine that this dilemma is likely to arise in observing other grade levels, schools, and districts across the United States. In my response, I begin with rethinking an interaction after the first classroom observation and then primarily attend to suggestions for activities during the 3-week observation period in 7th-grade classrooms.

First, I was struck by one particular interaction Dr. White described during the *Look and Learn Activity* when a PST referred to middle school students as "Lil Man," "Ese," and "Bro." Although I agree that the comments needed to be addressed, I might have responded to that interaction differently. Reading the exchange, it seemed to me that there was already a low level of comfort in explicitly pointing out race when describing middle-school students. I did not see the comments as colorblind; instead, I saw the comments as racialized. Inasmuch as I have not yet experienced the dilemma that is presented in this case, the suggestions I share next stem from reflecting on my own experiences in learning how to see structural inequities in mathematics classrooms. Additionally, I draw on conversations and suggestions from other teacher educators who face similar challenges in their work.

RESPONDING TO THE DILEMMA

When I am working with novice teachers, I have to constantly remind myself that what is obvious to me is not necessarily obvious to other people. We have evidence in mathematics teacher education that supports this statement from literature on professional noticing across differing levels of expertise (e.g., Jacobs, Lamb, & Philipp, 2010). Therefore, one strategy for addressing the dilemma presented in this case would be to use an activity called "Willing to be Disturbed." This reflective activity involves identifying and critically examining your own interactions with people that illuminate implicit aspects of your knowing and has been used successfully in my doctoral courses as a way of influencing how I regularly reevaluate my own assumptions and beliefs about the world. I have not fully incorporated "Willing to be Disturbed" into my own courses, but I think it would be a useful exercise in learning to teach mathematics.

In addition to individual reflection activities, I would suggest inviting a guest who also observed the classrooms to join our initial discussions. Considering the power relations between a teacher educator and PSTs due to course grading and evaluation, I see it useful to bring in someone such as a doctoral student, who may be a former teacher developing expertise in education, into discussions about observations in field sites. For example, Dr. White commented on a reflection she read from a graduate student observing the methods course that seemed to highlight the concerns Dr. White saw in the two 7th-grade classrooms. Inviting this graduate student to participate in discussions might provide another nonthreatening perspective for PSTs to consider and engage with while unpacking their own noticing. Bringing in guest speakers is one strategy that Dunn, Dotson, Ford, and Roberts (2014) used in response to PST resistance in their multicultural courses.

After continued observations and journaling, I would suggest bringing in readings related to topics of discussion and themes across individual reflections to further conversations about our noticing in the classrooms. Some specific examples that may be useful for this particular dilemma include Wickett (1997), which touches on investigating bias in the classroom from a teacher's perspective; Lambert and Stylianou (2013), which focuses on providing cognitively demanding tasks for all students; and Martin (2009), which highlights how racial identities matter in mathematics teaching and learning. Each of these articles lends a perspective for PSTs to consider and engage with while unpacking their own noticing. Additionally, these examples of readings relate to implicit bias in mathematics classroom spaces, which may serve as a possible entry point to conversations about structural inequities in classrooms and schools.

Finally, I would suggest focusing on shifting discussions toward more critical reflections across the multiple weeks of observations and learning in the classrooms. An important observation from Dr. White's case is her statement that "most of the PSTs were frustrated in [Ms. Miller's] class," and I would want to open those feelings of frustration in a critical discussion. Why are PSTs frustrated in the class? What aspects of the classroom space, activities, and interactions relate to PSTs' frustration? How do we consider multiple perspectives and alternative choices that help us frame our interpretations differently? Unpacking what "makes us frustrated" in Ms. Miller's class foregrounds PSTs actively voicing and thoughtfully dissecting their expectations and lenses for teaching and learning mathematics.

CLOSING THOUGHTS AND TAKEAWAYS

In closing reflections as a teacher educator, I would use the dilemma in this case to further think about how I might respond differently in initial interactions with my students when critically examining our own observations in mathematics classrooms. In my experiences both as a participant and a facilitator, lectures seem to shut down discussions, rather than open up a space to critically examine what we think, do, and say. In other words, telling people what they *should* do is not enough for our efforts in promoting equity in mathematics education. I am still learning how to productively navigate these conversations but also acknowledge that this is challenging work in mathematics teacher education. In closing, I want to echo calls for ongoing dialogue and collaborations among teacher educators, at all levels of experience, who challenge structural inequities (Dunn et al., 2014). We will all benefit in learning from and supporting each other through our efforts to improve mathematics education.

REFERENCES

Dunn, A. H., Dotson, E. K., Ford, J. C., & Roberts, M. A. (2014). "You won't believe what they said in class today": Professors' reflections on student resistance in multicultural education courses. *Multicultural Perspectives, 16,* 93–98.

Jacobs, V., Lamb, L., & Philipp, R. (2010). Professional noticing of children's mathematical thinking. *Journal for Research in Mathematics Education, 41,* 169–202.

Lambert, R., & Stylianou, D. A. (2013). Posing cognitively demanding tasks to all students. *Mathematics Teaching in the Middle School, 18,* 500–506.

Martin, D. B. (2009). Does race matter? *Teaching Children Mathematics, 16,* 134–139.
Wickett, M. (1997). Uncovering bias in the classroom: A personal journey. In J. Trentacosta & M. J. Kenney (Eds.), *Multicultural and gender equity in the mathematics classroom* (pp. 102–106). Reston, VA: National Council of Teachers of Mathematics.

COMMENTARY 2

SEEING ISN'T ALWAYS BELIEVING: RECOGNIZING RACE DYSCONCIOUSNESS IN THE PRESERVICE TEACHER CONTEXT

A Commentary on White's Case

Danny Bernard Martin
University of Illinois at Chicago

I would like to begin by thanking Dr. White for her willingness to share these episodes. I also want thank her for making visible, through her own ability to see, conditions of schooling that limit opportunities for some students while simultaneously providing enhanced opportunities for others. I also commend her for sharing examples of a phenomenon that continues to receive attention in teacher preparation practice and research: the difficulties that some White students have in recognizing racial disparities and in-school inequities and failing to reflect on and interrogate those disparities even when they are made visible.

Cases for Mathematics Teacher Educators, pages 203–207
Copyright © 2016 by Information Age Publishing
All rights of reproduction in any form reserved.

MY POSITIONALITY

The issues raised by the episodes are very familiar to me. Like Dr. White, I am a mathematics educator who is committed to attending to issues of equity and social justice in my research and my teaching. Specifically, my research over the last 25 years has focused on understanding the salience of race and identity in Black learners' mathematical experiences, taking into account sociohistorical forces, structural forces, community forces, school forces, and individual agency (e.g., Leonard & Martin, 2013; Martin, 2000, 2009). I have taught mathematics for the past 28 years to students of all age ranges. For the past 12 years, I have worked at a large urban public university in the city of Chicago. I primarily teach mathematics content courses for elementary teachers. Occasionally, I teach elementary mathematics methods. In each of these strands of my teaching, I try to incorporate ideas and content from the other.

My perspectives on teaching, learning, and issues of equity are informed by my readings and interpretations of culture-practice theory, cultural-ecological theory, critical theories of race, and racial identity development theory. In terms of my own identity, I consider myself a critical Black scholar. Actualizing this identity entails challenging anti-Black racism and discourses that construct Black children as mathematically illiterate and intellectually inferior to children in other social categories. This identity also brings with it a responsibility to challenge and change practices that limit Black children's access to high-quality mathematics learning opportunities.

I will respond to the episodes in the order that they were presented, then offer some overall commentary on the larger issues of differential opportunities for students and critical reflection by preservice teachers.

INTERPRETING AND RESPONDING TO THE DILEMMA

Look and Learn Activity

The idea of pairing students and having one focus on microlevel interactions and the other focus on macrolevel issues in the classroom is quite interesting. It provides students with different grain sizes and levels of analysis to assess what they think is happening in the classroom. In this episode, we are drawn to a microlevel observation that appears to be colored by stereotypes and racial knowledge appropriated from the macrolevel context of race in the United States. To some degree, I shared the reaction of Dr. White when she noted that she was stunned when the White male preservice student referred to the three boys in his observation as "Lil Man," Ese," and "Bro." These labels were used to index individual Black, Latino, and White boys, respectively. My shared reaction is tempered by the fact that literature on racial attitudes and White preservice teacher education shows that many White students have appropriated negative or stereotypical views of non-White group members (e.g., Bonilla-Silva, 2006; Picower, 2009). So, although I may have been stunned, I cannot say that I was completely surprised. I also share Dr. White's inclination to revisit the issue of race. I think it is important to interrupt moments such as these so that all students can learn from them.

However, this is also the point in the episode where my views differ from those of Dr. White. First, whereas Dr. White indicates she revisited the issue of race because she wanted to talk "about the importance of not being colorblind to student's race," I would argue that the White male student was, in fact, very cognizant of race. It takes a certain kind of racial knowledge and audacity to make the racially-coded name associations that were made by the male student. Moreover, I believe that he was not being colorblind, and that he was not treating the students the same, unless we accept that he was stereotyping all three boys with equal ease. It would not be a large leap to assume that the student was already familiar with a variety of stereotypes associated with the chosen names. That is why he chose them. Therefore, he was, in my view, very much attending to issues of race. Like Dr. White, I too, would have revisited the issue of race. However, my own choice would have been to point out, for educative purposes, the ways in which the student apparently *did* see race.

Dr. White also notes that she "wanted them to feel comfortable talking about students and seeing their identities" and that she believed "the PSTs might be uncomfortable talking about race around me." Here, I contend that the White male preservice teacher had already assumed a state of comfort, to the degree that he was willing to invoke the racially-coded names in the presence of his Black professor and the rest of his peers. Striving for even higher levels of comfort could, in fact, allow preservice students to further entrench themselves in their beliefs. The time for debriefing could have been a perfect opportunity to be even more explicit about race and, perhaps, raise several counternarratives (e.g., del Rosario Zavala, 2014; Varley-Gutiérrez, Willey, & Khisty, 2011; Terry, 2011) about Black, Latina/o, and even White students.

Working in 7th-Grade Classrooms

The realities of these two classrooms are not new to me. They are emblematic of many contexts where Black and special education students are segregated within school contexts and given access to less rich learning opportunities. Preservice teachers need to grapple with these realities and understand how they come to be as a result of structural and institutional forces (Milner, 2010; Milner & Laughter, 2013). Preservice teachers also need to understand how, even in the context of these inequities, they must continue to love and respect, rather than dehumanize and depersonalize, their students, as demonstrated by Ms. Miller. That love should not be rooted in deficit views or not be tied to lowered expectations or lowered demands for higher level thinking (Aguirre, Mayfield-Ingram, & Martin, 2013). Dr. White mentions the frustration of the preservice teachers with Ms. Miller's class. I wonder about the ways she could have unpacked that frustration in the moment and to what degree students could have been encouraged to consider structural and institutional forces (e.g., Lipman, 1998) as contributing to the conditions in Ms. Miller's class.

Final Activity Report

Dr. White does note that on her return to class, she engaged in a number of activities to illuminate disparities and issues of equity. This contrasts with the fact that when students were asked to reflect on their classroom experiences, they blamed students for creating the "chaotic" atmosphere in Ms. Miller's classroom. I wonder if preservice teachers were aware of or made to be aware of their color-coded characterizations of the students in Ms. Miller's class versus those in Ms. Bond's class. How would they respond if they were asked to teach their own class composed of students like those in Ms. Miller's classroom? If I interpret Dr. White's reflection correctly, it seems that her efforts did not gain much traction among these students. Of course, learning to teach is a developmental process. These young adult preservice teachers will grow and mature into their professional identities as teachers. But, teacher educators should not ignore the reality that many preservice teachers, despite the best efforts in teacher education, struggle to get past their deep-seated beliefs, especially about children whose schools have not served them well.

FINAL REFLECTION

The episodes in this case raise several questions for me. For example, I harken back to a question raised in one of my papers (Martin, 2007). In that paper, I asked, "Who should teach mathematics to African American children?" The answer to this question is not a simple one. But, in preparing preservice teachers to serve Black and other children, I argue for deepening preservice students' racial knowledge, in addition to their content knowledge. Although I understand that deepening one's knowledge is a ongoing, developmental process, there is an urgency that comes with knowing that disproportionately large numbers of Black, Latin@, indigenous, and poor students need teachers who are able to view their full humanity. This urgency is in tension with the waiting game of having preservice teachers getting to a point of being "comfortable" teaching these students. How long will the quest for quality education for Black, Latin@, indigenous, and poor students be contingent on the comfort levels and sensibilities of a mostly White teaching pipeline? Dr. White has given us much to think about as a result of these episodes.

REFERENCES

Aguirre, J., Mayfield-Ingram, K., & Martin, D. (2013). *The impact of identity in K–8 mathematics learning and teaching: Rethinking equity-based practices.* Reston, VA: National Council of Teachers of Mathematics.

Bonilla-Silva, E. (2006). *Racism without racists: Color-blind racism and the persistence of racial inequality in the United States.* Lanham, MD: Rowman & Littlefield.

del Rosario Zavala, M. (2014). Latina/o youth's perspectives on race, language, and learning mathematics. *Journal of Urban Mathematics Education, 7*(1), 55–82.

Leonard, J., & Martin, D. B. (Eds.). (2013). *The brilliance of Black children in mathematics: Beyond the numbers and toward new discourse.* Charlotte, NC: Information Age.

Lipman, P. (1998). *Race, class and power in school restructuring.* Albany, NY: State University of New York Press.

Martin, D. B. (2000). *Mathematics success and failure among African American youth: The roles of sociohistorical context, community forces, school influence, and individual agency.* Mahwah, NJ: Erlbaum.

Martin, D. B. (2007). Beyond missionaries or cannibals: Who should teach mathematics to African American children? *The High School Journal, 91*(1), 6–28.

Martin, D. B. (Ed.). (2009). *Mathematics teaching, learning, and liberation in the lives of Black children.* London, England: Routledge.

Milner, H. R. (2010). *Start where you are but don't stay there: Understanding diversity, opportunity gaps, and teaching in today's classrooms.* Cambridge, MA: Harvard Education Press.

Milner, H. R., & Laughter, J. C. (2013). But good intentions are not enough: Preparing teachers to center race and poverty. *The Urban Review, 47*, 341–363.

Picower, B. (2009). The unexamined whiteness of teaching: How white teachers maintain and enact dominant racial ideologies. *Race Ethnicity and Education, 12*, 197–215.

Terry, C. L. (2011). Mathematical counterstory and African American male students: Urban mathematics education from a critical race theory perspective. *Journal of Urban Mathematics Education, 4*(1), 23–49.

Varley-Gutiérrez, M. V., Willey, C., & Khisty, L. L. (2011). (In)equitable schooling and mathematics of marginalized students: Through the voices of urban Latinas/os. *Journal of Urban Mathematics Education, 4*(2), 26–43.

COMMENTARY 3

IDENTITY, CONTEXT, AND CONVERSATIONS ABOUT RACISM

A Commentary on White's Case

Joy Oslund
Michigan State University

Education does more than impart knowledge; it changes who the learner is, how she sees the world, and how she can relate to others. In the short story, "Of the Coming of John," W. E. B. Du Bois (1903/1994) described a young Black man who goes away to school and returns to his hometown newly aware of the evils of segregation. The following excerpt shows his family coming to grips with changes in John's demeanor:

> "John," she said, "does it make every one—unhappy when they
> study and learn lots of things?
> He paused and smiled. "I am afraid it does," he said.
> "And, John, are you glad you studied?"
> "Yes," came the answer, slowly but positively.
> —W. E. B. Du Bois (1903/1994, p. 149)

Cases for Mathematics Teacher Educators, pages 209–213
Copyright © 2016 by Information Age Publishing
All rights of reproduction in any form reserved.

This tragedy ends with John being driven from his teaching job in the town's Black school. The mayor ended John's relationship with the students for fear he would alert them to the injustices he learned about at school.

As mathematics teacher educators, many of us have "[studied] and [learned] lots of things"—some of which are ugly and sad, and often not considered topics for "polite" conversation. I suspect most of us have experienced moments in our teaching that are not unlike the one described in the case, when our students reveal stereotypes and we are faced with a decision about our response, or when we struggle—and often fail—to help our students see educational contexts and systemic injustices in the ways we have learned to see them. This case prompts us to think about our students' identities, our own identities, and the teacher education (TE) context.

PSTs come to us with diverse identities, experiences, and uneven willingness to talk about racism. We want to respect these but build on them, acknowledging that PSTs are at an early stage in their professional learning trajectories. We cannot expect them to be where we are in terms of their professional development, and yet we strive to help them interrogate their perspectives and better understand the systemic inequalities that affect the schools in which they will soon be teachers. This type of reflection challenges PSTs' identities and may make them unhappy. As we are struggling to engage PSTs in this work, they struggle to preserve their own identities and relationships (Rex, 2011). It takes skill to push our students to interrogate their perspectives while simultaneously building the trust required for further learning conversations.

Issues of PSTs' developing identities are particularly complicated when engaging PSTs in K–12 classrooms. One known difficulty is PSTs' sense of familiarity and competing purposes of K–12 schools and TE programs. While we try to help PSTs see systemic racism, they often focus on what to do (or not do) as teachers (Feiman-Nemser & Buchmann, 1985). In addition to convincing students that they can talk about race in our classes, we must also convince them to trust us enough to fully engage in tasks that they do not see as having practical application.

Our own intersecting identities affect the possible responses we might consider. As a White, female mathematics teacher educator, I am privileged to speak to my mostly White, mostly female elementary PSTs about racism without worrying they will perceive me as "the angry Black woman"—a prevailing stereotype. This does not mean PSTs will be willing to talk with me about markers of privilege and oppression. PSTs often resent what they perceive as the liberal leaning of the university and professors pushing social justice in their courses (Kitchen, 2012). Any of us may respond in ways that promote or impede the building of trust that allows PSTs to have difficult conversations with us.

The TE context complicates the issue. The fact that Dr. White is the only African American professor these students will have in their teacher education program was a consideration in her response to the stereotypical pseudonyms and her relationship to the White teacher educators who comprise much of this book's audience. The small numbers of teachers and teacher educators of color (Ladson-Billings, 2005) is more evidence of the same systemic racism and tracking Dr. White saw in the middle-school classrooms. Additionally, as teacher educators, we are guests in the K–12 schools where our PSTs study. In this case, the PSTs saw the students as the problem in an inclusion classroom, whereas the teacher educator saw issues in the teaching. We want to point out elements of a teacher's practice that we do not want PSTs to emulate. However, we don't want to build reputations as guests who enter the school then criticize or blame teachers. Teachers and students in both classrooms are situated within a larger situation in which injustice pervades. Our goal is to help students to see beyond the particular actions of students and teachers to a critique of the larger system.

RESPONDING TO THE DILEMMA

How and when do we effectively point out PSTs' prejudices and still create and maintain spaces where they feel free to talk about issues of injustice? How do we teach about things that have the potential to make our students unhappy while promoting the trust needed to continue teaching them? As a White person dedicated to helping others understand systemic racism, my first tendency is often to immediately point out stereotypes. However, in my experience, this sometimes shuts down the conversation, as Dr. White feared. When I've done this, PSTs have sometimes perceived this as my intolerance of others' opinions. Indeed it is—I do not want to tolerate racism, sexism, or ablism in my classes. Yet, I have wondered about this move's effectiveness. Has it prevented further conversations that could have been more effective? Has it made PSTs feel like they can't wrestle with important ideas in my classroom? And, if enough students complain of my intolerance often enough, will it affect my annual evaluations and my ability to continue teaching? If so, will it have been worth it?

Narrative pedagogy offers a possible way forward. Engaging PSTs in storytelling and critical study of narratives can help learners *re-story* their understandings of situations (Goodson & Gill, 2014). Asking PSTs to create and analyze their own narratives of events in the field placement, as well as events in their own schooling histories, could help the PSTs defamiliarize the classroom context and their assumptions about students, teaching, and learning (Juzwik, 2011). This is difficult identity work that promotes changes in understanding and coming to grips with former understandings and experiences in ways that are notably humanizing (Goodson & Gill, 2014).

In our own TE classrooms, Sandra Crespo and I have used narrative for these purposes (Oslund & Crespo, 2014). We provide PSTs with sets of classroom photographs to analyze, then ask them to narrate the story of the lesson shown and arrange the photographs to illustrate their story. As PSTs narrate stories, they talk about aspects of the context that they imagine playing into teachers' decisions, prompting them to verbalize their ideas about race and class. For example, some PSTs have suggested that the teacher is concerned about including students learning English, warranting this suggestion by noting physical characteristics of the students or classroom. Others have noted the condition of the furniture and materials pictured and made statements about the school's location as urban or suburban. These ideas become an object of inquiry, as we ask PSTs how they came to their conclusions, what assumptions they are making, and what other interpretations would be possible. The activity allows students to critique ideas without criticizing a particular teacher they know.

Perhaps the biggest takeaways from this case are (1) we should commit to discussions of racism in our classes, including planning ahead during syllabus design, knowing it will be messy and we will not always know what to do, and (2) we should be gracious with ourselves and our students as we continue to seek to do the uncertain work of learning about systemic injustices together.

REFERENCES

Du Bois, W. E. B. (1994). *The souls of Black folk.* New York, NY: Dover. (Original work published 1903)

Feiman-Nemser, S., & Buchmann, M. (1985). Pitfalls of experience in teacher education. *Teachers College Record, 87,* 53–65.

Goodson, I., & Gill, S. (2014). *Critical narrative as pedagogy.* New York, NY: Bloomsbury Academic.

Juzwik, M. (2011). Exploring cultural complexity in teacher education through interactional and critical study of classroom narratives. In L. A. Rex & M. Juzwik (Eds.), *Narrative discourse analysis for teacher educators: Managing cultural differences in classrooms* (pp. 105–130). New York, NY: Hampton Press.

Kitchen, R. S. (2012). Closing remarks. In L. J. Jacobsen, J. Mistele, & B. Sriraman (Eds.), *Mathematics teacher education in the public interest: Equity and social* justice (pp. 253–266). Charlotte, NC: Information Age.

Ladson-Billings, G. (2005). Is the team all right? Diversity and teacher education. *Journal of Teacher Education, 56,* 229–234.

Oslund, J., & Crespo, S. (2014). Classroom photographs: Reframing what and how we notice. *Teaching Children Mathematics, 20,* 564–572.

Rex, L. (2011). Introduction: Narrative discourse analysis for teacher educators: Considering participation, difference and ethics. In L. A. Rex & M. Juzwik (Eds.), *Narrative discourse analysis for teacher educators: Managing cultural differences in classrooms* (pp. 1–29). Cresskill, NJ: Hampton Press.

PART II

CONVERSATIONS ABOUT INEQUITIES IN MATHEMATICS CONTENT COURSES

Mathew D. Felton-Koestler
Ohio University

Marta Civil
The University of Arizona

The idea that mathematics is culture-free and neutral has been challenged by many (Bishop, 1988; Gutstein & Peterson, 2013; Powell & Frankenstein, 1997). Although discussions about this idea and the related themes of equity and social justice in mathematics education may take place in the context of mathematics methods courses or professional development experiences, exploring issues of equity and social justice seems to be less common in mathematics content courses. Based on our own informal conversations with colleagues in departments of mathematics as well as our own experiences, a common explanation centers on the idea that there are certain expectations of the kind of content to be included in mathematics courses, which may make it difficult or impossible to organically incorporate other

Cases for Mathematics Teacher Educators, pages 215–218
Copyright © 2016 by Information Age Publishing
All rights of reproduction in any form reserved.

perspectives. Also, the students themselves may expect a mathematics content course to be about "just the mathematics." However, especially because of this last reason, we feel it is critical that content courses for prospective teachers include spaces for conversations about inequities. We think it is important that students see that mathematics is not neutral, is not culture-free, and is not value-free (Bishop, 1988). The cases and commentaries in this section of the book provide examples of mathematics teacher educators grappling with these issues.

Civil's case, "'This Is Nice but They Need to Learn to Do Things the U.S. Way': Reactions to Different Algorithms," considers prospective teachers' reactions to algorithms used in non-U.S. countries, and in particular whether the common sentiment quoted in her title—that ultimately students need to learn the U.S. algorithms—masks a feeling that the U.S. method is somehow superior. D'Ambrosio's commentary highlights the emotional damage this can do to students whose mathematics is not valued, whereas Philipp and Murrays frame this as a specific case of the need for teachers to learn to value and notice *children's* mathematics.

In "Using Mathematics to Investigate Social and Political Issues: The Case of 'Illegal Immigration,'" Felton-Koestler considers how to support his prospective teachers in maintaining ownership over investigations into social issues of their choosing, while still pushing back against problematic portrayals of nondominant groups (undocumented workers in this case). In response, Brown addresses the need to discuss sensitive issues with teachers and prospective teachers and to support them in looking at the different sides of a situation, whereas Terry highlights the need to balance prospective teachers' ownership of ideas with the need to ensure that prospective teachers understand how to mathematize these sensitive issues. Celedón-Pattichis emphasizes providing multiple opportunities for prospective teachers to engage in using mathematics to analyze sociopolitical issues.

Mistele and Jacobsen's case, "Searching for Cohesion in a Mathematics Course for Social Analysis," centers on the dilemma of how to organize and design a course that successfully blends social issues and mathematical content. They found that when organized around social issues the mathematics often became mechanical and peripheral to exploring the social issue, but when organizing around mathematical concepts the social issues became contrived. The commentaries offer several perspectives on dealing with this tension: Pierson Bishop suggests easing the expectation that *all* of the mathematics be integrated with social issues or the expectation that the course comprehensively cover the mathematical content; Fernandes recommends prioritizing the mathematics and selectively integrating well-suited social issues into this content; and Zahner suggests exploring a smaller number of social issues with multiple forms of mathematics over time.

In "Not Called to Action (or Called Upon to Act): Can Social Justice Contexts Have a Lasting Impact on Preservice Teachers?" Simic-Muller explores how effective she is in reaching her social justice goals if her prospective teachers are engaged by and concerned about the social injustices, like sweatshop labor, they explore but never take action designed to remedy these injustices. In response, Gutstein emphasizes that action is different for different people and that beginning to explore these issues may be an appropriate entry point for these prospective teachers. Quander suggests encouraging smaller forms of action, such as writing a letter or engaging in social media. Finally, Powell highlights the importance of the prospective teachers having ownership over the topic if they are going to be asked to act.

Finally, in "Who Counts as a Mathematician?" Strickland considers not only who counts as a mathematician, but what it means to come from an underrepresented group. When teaching a history-of-mathematics course, she asked her students to write a paper on an underrepresented mathematician, which led to the question of who counts, both as a mathematician and as underrepresented. The commentaries provide suggestions for how to further support students in unpacking the meanings of *mathematician* and *underrepresented*: De Araujo suggests doing this through more explicit conversations of inclusion and exclusion criteria for these terms, Shockey through an exploration of what culture is, and Eli through additional assignments that provide space for exploring boundary cases.

Several themes can be seen across the cases. Civil's and Strickland's cases are ultimately concerned with who counts—with who is valued—in mathematics. In both cases this can be seen as a question of whose mathematics is included as part of the mathematical cannon. In Civil's case the question is of whose algorithms are taught and honored in classrooms. In Strickland's case the question is of whose contributions are recognized by dominant tellings of mathematical history.

Mistele and Jacobsen's and Felton-Koestler's cases both highlight the difficulties in balancing mathematical content with investigations of social issues. In Mistele and Jacobsen this is of central concern to the design of the course. In Felton-Koestler's context, a frequent struggle is how to encourage substantive uses of mathematics without taking over his prospective teachers' projects.

Strickland's, Felton-Koestler's, and Simic-Muller's cases examine the tension that may exist between the instructor's equity goals and the prospective teachers' views and dispositions. Is someone incarcerated for murder an underrepresented mathematician (Strickland)? When does teaching cross over from social justice to dictating one's perspective (Felton-Koestler)? Does it count as social justice if no one acts (Simic-Muller)?

As you read each case and the commentaries that follow, we invite you to consider the following questions:

- How does the context of a content course contribute to this case? How might it change in another setting?
- In what ways are the mathematical content and issues of equity integrated?
- How do these cases and commentaries help you address the potential tension between "teaching content" and addressing issues of power, privilege, and oppression?

REFERENCES

Bishop, A. J. (1988). Mathematics education in its cultural context. *Educational Studies in Mathematics, 19*, 179–191.

Gutstein, E., & Peterson, B. (2013). Introduction. In E. Gutstein & B. Peterson (Eds.), *Rethinking mathematics: Teaching social justice by the numbers* (2nd ed., pp. 1–6). Milwaukee, WI: Rethinking Schools.

Powell, A. B., & Frankenstein, M. (Eds.). (1997). *Ethnomathematics: Challenging Eurocentrism in mathematics education*. Albany: State University of New York Press.

CHAPTER 10

"THIS IS NICE BUT THEY NEED TO LEARN TO DO THINGS THE U.S. WAY"

Reactions to Different Algorithms

Marta Civil
The University of Arizona

CONTEXT FOR THE CASE

This case is not based on a specific event but rather builds on several experiences with preservice and practicing teachers mostly in mathematics content courses for elementary teachers. A typical topic in these courses is a discussion of different algorithms for arithmetic operations. As I reflect on the several incidents that moved me to write this case, I realize that probably the first one happened when I showed a group of preservice teachers the way I had been taught how to subtract. Instead of the traditional algorithm that most people in the U.S. learned (the regrouping one), I learned what is often called the "equal addition algorithm." The preservice teachers struggled to make sense of this different algorithm and were trying to interpret it through the one they knew, thus focusing on "borrowing." Their reactions to the algorithm

Cases for Mathematics Teacher Educators, pages 219–225
Copyright © 2016 by Information Age Publishing
219

itself were mixed, from "does it always work?" to wondering why I was showing them this. As I shared with them other algorithms, some of the reactions included Vicky's, "I do believe that you could eventually convince him that learning to carry is easier and leaves less room for error," in reference to an algorithm for subtraction in which a child used negative numbers instead of the traditional regrouping. As I wrote in 1993:

> Why did Vicky believe that carrying was easier? Perhaps it was just that she was used to that method? One of my goals in this course was to get students to look at the contents of school mathematics with a critical attitude. It was not an easy task. It is hard to be critical of that which one does not understand but which seems to be accepted by a large group of people. (Civil, 1993, p. 95)

At the time I was concerned about how preservice teachers may react to their future students bringing in alternative approaches if they themselves did not know how to make sense of them. As I wrote in the excerpt above, I wanted them to look at school mathematics with a critical attitude. However, I was not as aware then, as I am now, of the concept of valorization of knowledge (de Abreu, 1995). The question that I often raise now is "What happens when the students (children) involved belong to groups whose knowledge has historically not been recognized or valued by schools?" Let me illustrate this point in what follows.

THE DILEMMA

Several years later, I asked a group of preservice teachers in a mathematics content course to read an article by Perkins and Flores (2002), "Mathematical Notations and Procedures of Recent Immigrant Students." The article presents some notations and algorithms that are usually taught in Latin American countries and that are different from the ones traditionally taught in schools in the United States. Overall students reacted with a mixture of surprise at there being so many different approaches and some uneasiness as to what it is that they were supposed to do as teachers ("how can we be expected to know all these different ways?"). One student said something that is at the root of my dilemma: "This is nice but they need to learn to do things the U.S. way." Unfortunately I do not remember me saying anything about this statement at the time, so I do not know what prompted this prospective teacher's comment nor do I have any kind of follow up to the comment. In fact, this is part of my dilemma: What could I have said?

Hence, the dilemma I present in this case is, as a mathematics teacher educator, what should I do when preservice teachers or practicing teachers do not seem to value others' ways of doing mathematics, particularly when the "others" come from nondominant communities? How do I respond to "this is nice but they need to learn to do things the U.S. way"?

MY REFLECTION

Over the years since this event happened, I have wondered: Was this preservice teacher expressing a concern for immigrant students in that they would need to adapt to the new system and therefore it would be better for them if they learned the algorithms traditionally taught in U.S. schools? Was it that she did not feel comfortable as a teacher to understand all the different approaches that students may bring with them? Or was it an issue of valorization of knowledge, by which the knowledge that immigrant children bring was not as valuable as the one that they would acquire in the U.S. schools?

More recently, I often share with preservice and practicing teachers a different representation for the division algorithm in Mexican schools (see Figure 10.1). A main reason for sharing this algorithm for division comes my many years of working with Latina/o parents both in Arizona and in North Carolina. When I first start talking to them about mathematics, one of the first things they bring up is the difference in the division algorithm in the U.S. and in countries in Latin America, particularly Mexico (because most of the parents in my work come from that country). Their comments corroborate Perkins and Flores's (2002) observations that in Mexico there is a sense of pride at being able to do the subtraction mentally instead of writing it down like it is done in the traditional U.S. algorithm. As I checked recently with a group of teachers from Mexico, they confirmed that when children first learn to divide, they write all the steps down, but the idea is to get them to the version shown in Figure 10.1, where some of the steps are done in the head.

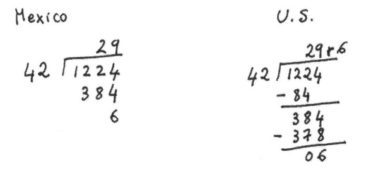

Figure 10.1 Division in Mexico and in the United States.

In my work with preservice and practicing teachers, I often show them the algorithm for division in Mexico and then show them a video where two mothers who were schooled in Mexico shared their reactions to the division algorithm in the U.S. (Civil & Planas, 2010):

Marisol: When I looked at how he [her son] was dividing, he subtracted and subtracted and that he wrote all the equation complete I said, I even said, "what teacher wants to make things complicated. No, son, not that way! This way!" And he learned faster with this procedure [the one from Mexico] . . . but if visually you see such a mess [referring to the U.S. algorithm] then, really, it's the truth, it's the first feeling I had.

Verónica: I tried to do the same with my child with divisions, that he didn't write everything, but he says, "no, no, mom, the teacher is going to think that I did it on the computer." "You don't need to write the subtraction son", I say, "you only put what is left." . . . "No, no, my teacher is going to think that I did it on the computer, I have to do it like that." "Ok, you think that . . . , but I want to teach you how we learned." And I did teach him, but he still uses his method, and that way he feels safe that he is doing his homework as they told him to. The same thing with writing above what they borrow and crossing it out, I tell him, "I remember our homework did not have to have any cross-outs," whereas his does. . . . A bunch of numbers on top.

These two mothers conveyed very passionately what they see as advantages in the way they learned how to divide in Mexico: It is faster, it is cleaner, it is less complicated, and so forth. My point is not to claim that one method is better than another, nor to agree or disagree with these mothers' perceptions. I want teachers and preservice teachers to appreciate the diversity in approaches to arithmetic and to understand that they have at least two kinds of responsibilities towards this diversity. One, they have to know how to make sense of the different algorithms that their students (or their parents) may show them. This is certainly part of a content course for teachers. They cannot dismiss them just because they do not understand them. A second responsibility relates to valorization of knowledge. Assuming the teachers and preservice teachers understand the different algorithms, how do they perceive them? The typical reaction I get when I show them the division approach from Mexico is along the lines of, "if they do not show all the work and they make a mistake it is harder to see where they made the mistake and so it is harder for us to help them." Several questions come to my mind: Is it really about "showing their work"? Although I understand that we want students to show their work, what is an appropriate balance in how much work they show? What work should be shown? As I reflect on my own mathematics education, I was expected to show my work when solving problems to explain my reasoning, but I was not expected to show my work in the division algorithm I learned (which is similar to the one taught in

Mexico). That is, the subtraction part was something that you did in your head, so for me to be asked to show that part of the work, it would have seemed like an odd request. I wonder, are there situations in which we do not expect students to show all the work they could show and yet we are fine with that? My point is that although I am sure that in some cases the teachers' reactions point to a genuine concern, I am often left wondering whether it also masks a feeling of "our method is better than theirs."

Gorgorió and Abreu (2009) described the situation with a student (David) from Ecuador and his mother (Mónica) on the differences between the algorithm for division in Ecuador and in Spain. When the mother met with the teacher, the teacher told her that the method from Ecuador was wrong. As the authors wrote,

> The important issue in the case of David and Mónica is not whether there are or are not differences in the way the division algorithms look, but the reaction of the teacher to this difference. It is unlikely that the teacher really thought that in Ecuador division was taught wrongly. It is more probable that the teacher, unaccustomed to such conversations with parents, did not listen carefully enough to what Mónica was saying and made an unreflective use of social representations.... The cultural nature of mathematics is linked to social representations related to the process of learning mathematics, to who is believed to know and not know mathematics, and to culturally defined correct ways of doing mathematics. (Gorgorió & Abreu, 2009, p. 72)

What are teachers' and preservice teachers' perceptions of who knows and does not know mathematics? Do they believe in a preferred way to do mathematics? Who determines which ways are the preferred ones? What are the consequences for David (in the excerpt above) when hearing that his method (also his mother's method) is "wrong"? Or what about Eliseo, a child of Mexican origin, who attempted to share his mother's way to do an arithmetic operation but did not quite get it, and the teacher said, "Yes, but that's in mama's home. Let's do it the way that we do it in the school" (Civil & Planas, 2010)? What message is she sending Eliseo?

As mathematics teacher educators, what should we be doing to prepare teachers to not only encourage children's home mathematics, but to do so in an authentic way, one that really values the cultural nature of mathematical knowledge? We cannot always rely on having teachers or preservice teachers in the room who will give a counterpoint, as is the case of Caroline, a teacher who viewed the different algorithms that the Latina/o students in her class shared as assets to her instruction:

> The Latino children, if their parents come from Mexico, then they probably did it a different way ... and even the algorithms maybe look a little different. If you're looking at algorithms, they're going to be like "my dad does it this

way" or "my mom does it this way." And so then you're bringing in another way so that they're seeing maybe even a third or a fourth or a fifth way to attack a problem.

POSSIBLE ACTIONS

As I reflect on what I could do differently in my teacher preparation courses, I think that I need to incorporate more culturally based ways of doing mathematics. In content courses we often spend some time discussing different algorithms, but those are presented without a human connection, that is we hardly make an explicit connection to the cultural origin of the algorithms under discussion. Resources are available, for example, the Algorithm Collection Project, http://www.csus.edu/indiv/o/oreyd/ACP.htm_files/Alg. html, by Daniel C. Orey. Articles such as the one by Perkins and Flores (2002) or by Philipp (1996) are also useful resources, both as discussion starters and in terms of the activities suggested. For example, Philipp described an assignment in which preservice teachers are to find out about different algorithms used in their community and explain why they work. Through these readings and activities, I would want to move beyond an awareness that there are different culturally based algorithms into a discussion of valorization of knowledge to address beliefs that preservice teachers and practicing teachers may have about these different algorithms.

We also have human resources that I would like to incorporate more into the courses, such as a panel with parents who attended school in countries other than the U.S. and could come to the content courses to share their approaches and their reflections on mathematics education or having the preservice teachers or practicing teachers interview someone with mathematical experiences different from their own (using, for example, an interview protocol that we developed in our work with parents around "math personal history").[1] My most recent experience was to invite a Mayan teacher to one of the courses with a group of elementary school teachers. The teacher shared his knowledge of the Mayan numeration system and how to add and subtract, going beyond what we usually read in books of "these are the symbols and this is how it works." He shared the cultural significance of each symbol and in so doing he provided a window into his world, his history, his background.

NOTE

1. This interview protocol was developed under Project MAPPS (Math and Parent Partnerships in the Southwest) and CEMELA (Center for the Mathemat-

ics Education of Latinos/as), both funded by the National Science Foundation (NSF) under grants ESI-9901275 and ESI-0424983, respectively.

REFERENCES

Civil, M. (1993). Prospective elementary teachers' thinking about teaching mathematics. *Journal of Mathematical Behavior, 12,* 79–109.

Civil, M., & Planas, N. (2010). Latino/a immigrant parents' voices in mathematics education. In E. Grigorenko & R. Takanishi (Eds.), *Immigration, diversity, and education* (pp. 130–150). New York, NY: Routledge.

de Abreu, G. (1995). Understanding how children experience the relationship between home and school mathematics. *Mind, Culture, and Activity: An International Journal, 2,* 119–142.

Gorgorió, N., & de Abreu, G. (2009). Social representations as mediators of practice in mathematics classrooms with immigrant students. *Educational Studies in Mathematics, 72,* 61–76.

Perkins, I., & Flores, A. (2002). Mathematical notations and procedures of recent immigrant students. *Mathematics Teaching in the Middle School, 7,* 346–351.

Philipp, R. A. (1996). Multicultural mathematics and alternative algorithms. *Teaching Children Mathematics, 3,* 128–133.

WHEN THE "U.S. WAY" IS NOT THE STANDARD!

A Commentary on Civil's Case

Beatriz D'Ambrosio
Miami University

As I began reading the case, I was immediately drawn back to my childhood. Although I was born in Brazil, my family moved to the U.S. when I was 3, so I was schooled in the U.S. through seventh grade. Upon finishing seventh grade, we moved back to Brazil. So, although Brazilian, when in school, I felt like an immigrant. I spoke the language but did not read or write, so I struggled in many classes. I hoped that in science and math classes I would feel more at home. But alas, the algorithms I used did not match those used by my colleagues and teachers. I masked my differences well in those classes, until the day that I was called to the board to solve a problem. All went well until I had to set up and solve a division problem in front of the class. The problem required me to divide two numbers, say 1224 by 42 (using the numbers from this case), and I happily set it up the way we would in the U.S. My colleagues snickered and laughed, my teacher

Cases for Mathematics Teacher Educators, pages 227–231
Copyright © 2016 by Information Age Publishing
All rights of reproduction in any form reserved.

stopped me as I had set up the problem "backwards and upside down." She took the chalk from my hand and "fixed it" producing the following image:

$$1224 \overline{\smash{\big)}\, 42}$$

She left me at the board to finish the problem. I felt humiliated because I had no idea how to proceed. I went home that night feeling ashamed but committed to learn how to fit in. My parents were able to show me the algorithm, which I never really understood until I became a teacher and had to teach it to my students. However, from my parents' instruction, I was able to memorize where all the pieces went, so I would do the problem my way, and put the pieces in the correct place resulting in the following representation:

$$\begin{array}{r|l} 1224 & 42 \\ 384 & 29 \\ 6 & \end{array}$$

I had not recently thought about that episode until the emotions of that day resurged as I read Marta's case.

INTERPRETING THE DILEMMA

I can imagine the case resonating with any preservice teacher who experienced similar episodes of feeling humiliation and shame that I had experienced in my early adolescent years. My humiliation and shame resulted from being different and the anxiety that the difference caused. It resulted from a teacher who did not seize the opportunity to celebrate the difference and explore the rich mathematical opportunity that presented itself, had I had the opportunity to show my colleagues a different algorithm for long division. Had she behaved differently, I might have felt that my knowledge was valued and that my individuality was honored in my classroom community. Instead, I felt silenced and alone.

Of course, my story occurred many years ago, before anyone talked about the mathematics classroom as a community and before we began exploring the power of mathematical discourse for learning (Herbel-Eisenmann & Cirillo, 2009). The idea that my algorithm might have spurred discussion and mathematical conversations requires a vision that can only be construed in the classrooms of today. Yet, even today, how often do we see alternative strategies or solutions valued and celebrated in our mathematics classrooms at the elementary or secondary levels?

RESPONDING TO THE DILEMMA

Over the years I have used the case of Corwin (1989) with my preservice teachers to raise the issues involved in prohibiting a child from solving a problem in his or her unique way. The discussion of this reading results in many emotional testimonials regarding experiences in mathematics classes. Feelings of shame, humiliation, embarrassment, self-silencing, and inadequacies all surface from readings of this type. Students are very articulate in describing the types of relationships they have had with mathematics throughout their personal life histories. I suspect that Marta's case can cause similar responses.

As the case so eloquently illustrates, in our schools, there is a prevailing image of mathematical procedures as immutable and rigid. Fosnot and Dolk (2001) described the research findings of Ann Dowker, who asked 44 mathematicians to solve simple operations, and for the most part, rarely did the mathematicians she interviewed use a standard algorithm to determine the answer to a multiplication or division problem. As preservice teachers read this chapter, they are very surprised to know that, of all people, the mathematicians break away from the use of algorithms if the numbers lend themselves to other, more efficient strategies. They are surprised to learn that there may be something considered, by mathematicians, as more efficient than the algorithms.

I can also see the case generating an opportunity to question how our algorithms came to be. A historical development of algorithms could be explored, and the discussion of how knowledge becomes institutionalized could ensue. The division algorithm as we teach today began to emerge in Europe in 1200 CE when Leonardo de Pisa (also known as Fibonacci), a traveling merchant, published his book *Liber Abaci* and introduced what he had learned about mathematics throughout his travels to the Arab world. In this excerpt, translated from Fibonacci's book published in the year of 1202, he explains to all of the Latin speaking how to divide 10000 by 8 in Figure 10.2.

A more sophisticated version of the division algorithm appeared in the work of Stevin (a Belgian mathematician) in the middle of the 16th century.

The Division of 10000 by 8.

Also if one will wish to divide 10000 by 8, then he puts the 8 beneath the 0 of the first place, and he says $\frac{1}{8}$ of 10 [p29] is 1, and there remains 2; he puts the 1 beneath the 0 in the [fourth] place, [and he puts the 2 above, and he takes $\frac{1}{8}$ of 20 which is 2 and there remains 4; he puts the 2 beneath in the third place] and the 4 above, and he takes $\frac{1}{8}$ of 40 which is 5, which he puts beneath the second place; and the row of places in the quotient is filled up by putting 0 beneath 0 in the first place, as is displayed in this illustration.

```
24
10000
8
1250
```

Figure 10.2 Image from Sigler (2003), p. 54.

Figure 10.3 Image from Chaubert (1998), p. 43.

The image in Figure 10.3 explains, step by step, the algorithm for division used in 16th-century Europe.

The mathematical analysis of the algorithm is a rich source of mathematical problem solving for future students. Furthermore, the exploration of the historical evolution of algorithms can situate the appearance of algorithms as an invention of humanity that has social, cultural roots, outside of the U.S. and even outside of Europe. These taken-for-granted algorithms have a very long history, and they evolve differently throughout the world. The historical development of many other algorithms in different parts of the world has been discussed by Joseph (2010) and can provide a nice basis for development of an understanding of the non-European roots of mathematics.

I can also imagine the power of this case to debunk the myth that the way we "adults in the U.S." do mathematics is somehow better than the algorithms used throughout the world. Although the curriculum privileges certain algorithms, it warrants teachers engaging in the critical analysis of the position of privilege given to certain ways of knowing over others.

By using this case, teachers can be sensitized to the power of using the cultural capital available in a culturally diverse classroom and community. Even in extremely homogeneous communities, the historical development of mathematical topics can contribute to the discussions and curiosities related to the contributions of different cultural groups to the evolution of the knowledge base of a discipline.

I can also imagine the power of unpacking the mathematical reasoning that goes on as people use the algorithms. Much like the Mexican mothers describe in the case, the Brazilian algorithm relies on much mental computation, in particular, relying on an equal addends method of subtraction that is done mentally.

In my use of children's invented algorithms, using real children's work, a similar dilemma occurs. My students ask, when do they learn the real algorithm? They'll say, "I can see the value of their inventions, but when do we show them how to find the answer the way we learned?"

Unfortunately, in today's current political climate, the same teachers who had settled this dilemma in their own minds are now quoting the CCSSM to justify going back to teaching the U.S. standard algorithm. Over several years, they had witnessed the power of children's inventions. The sharing of algorithms resulted in powerful mathematical discourse. Children's ownership of their inventions led to their continued success in using them. Reverting to the U.S. standard algorithm is lamentable. Once again, children will feel inadequate because the U.S. standard algorithm may not make sense to them in quite the same way that their own algorithms did.

The injustice of forcing children to do problems in one particular way can be the highlight of a discussion based on a hypothetical thought experiment about a world that is forced to do subtraction or division using the U.S. standard algorithm!

REFERENCES

Chaubert, J. (Ed.). (1998). *A history of algorithms: From the pebble to the microchip*. New York, NY: Springer-Verlag.

Corwin, R. B. (1989). Multiplication as original sin. *Journal of Mathematical Behavior, 8*, 223–225.

Fosnot, C., & Dolk, M. (2001). *Young mathematicians at work: Constructing multiplication and division*. Portsmouth, NH: Heinemann.

Herbel-Eisenmann, B., & Cirillo, M. (Eds.). (2009). *Promoting purposeful discourse: Teacher research in the mathematics classroom*. Reston, VA: National Council of Teachers of Mathematics.

Joseph, G. G. (2010). *The crest of the peacock: Non-European roots of mathematics*. Princeton, NJ: Princeton University Press.

Sigler, L. (2003). *Fibonacci's* Liber Abaci: *A translation into modern English of Leonardo Pisano's* Book of Calculations. New York, NY: Springer-Verlag.

COMMENTARY 2

NOTICING STUDENT THINKING

A Commentary on Civil's Case

Eileen Murray
Montclair State University

I have been involved in mathematics education as a teacher and teacher educator for close to two decades. I have worked as a high school teacher, a graduate assistant, a middle-school math coach, a professional developer, and a professor of mathematics education. Throughout my experiences, I have reflected upon the ideas in Civil's case many times in many different contexts. Broadly speaking, whose mathematics matters and who gets to decide?

Most recently, I have had the opportunity to teach mathematics content courses for prospective elementary teachers. In these courses, we spend a great deal of time discussing algorithms. While discussing operations, I work to illustrate many different algorithms and provide examples of students' clinical interviews illustrating various ways of thinking. In class, my students and I then discuss the "traditional" algorithm versus alternate algorithms. We consider the shortcomings of different algorithms and how certain ones might make more sense to a particular child. However, I have

Cases for Mathematics Teacher Educators, pages 233–236
Copyright © 2016 by Information Age Publishing
All rights of reproduction in any form reserved.

233

not incorporated more culturally based ways of doing mathematics, nor do I explicitly connect various algorithms to cultural origins. After reading Civil's case, I see the value (and necessity) of adding this component to my courses as a way to challenge prospective teachers' beliefs about mathematics and students.

INTERPRETING THE DILEMMA

At the base of Civil's case, I see a strong connection to the difficulty in appreciating (and noticing) student thinking. I too have had experiences of prospective teachers being confused about a particular algorithm they have never seen. Many times students will ask questions or make comments similar to those Civil experienced in her work. "Why do we need to know this?" "Do I need to memorize all of these ways of thinking?" "Isn't the traditional algorithm easier?" It is at these times that I begin a conversation about the social and cultural nature of mathematics. Almost universally, my students have not considered who decided the traditional algorithm was the way we should all be adding, subtracting, multiplying, or dividing. They have not thought deeply about their own preferred way of doing mathematics. Nor have they considered how they came to that knowledge and preference. Finally, most have not considered that other people may have learned an entirely different way of doing mathematics, and that those ways are also valid.

RESPONDING TO THE DILEMMA

The way that I engage my students in these conversations is to have them consider what their role as a teacher might be. That is, our job as teachers is to listen and attend to our students' thinking. Teachers must learn to respond to children's thinking and to help them build on their own knowledge to make sense of mathematics. In this way, I try to help prospective teachers begin to develop their professional noticing skills (Jacobs, Lamb, & Philipp, 2010). The importance of focusing on children's mathematical thinking lies in the fact that when teachers use instruction that builds on children's mathematical thinking, they provide richer instructional environments and can improve student achievement. Therefore, to develop professional noticing of children's mathematical thinking can help prospective teachers understand how to make sense of children's thinking in ways that highlight core mathematical ideas as well as develop frameworks of children's mathematical understandings (Jacobs et al., 2010). It is through developing these frameworks that teachers can then learn to respond appropriately to children's thinking, which is the most difficult noticing skill for teachers to develop, yet the most valuable in extending student learning (Jacobs & Ambrose, 2008).

However, as stated above, I have not done a good job of including in the conversation the ideas of culturally based ways of doing mathematics or the connection of various algorithms to cultural origins. By removing culture from the conversation, have I inadvertently avoided the fact that "mathematics classrooms are inherently cultural spaces where different forms of knowing and being are being validated" (Nasir, Hand, & Taylor, 2008, p. 206)? The takeaway from this reflection for mathematics teacher educators is that we need to do more than just attend to teachers' professional noticing. We need to work towards helping prospective teachers also learn how to recognize and understand students' cultural backgrounds in order to be prepared to meet the needs of an increasingly diverse student population (Kitchen, 2005).

Mathematics teacher educators can help prospective teachers by attending to their multicultural mathematics dispositions (White, Murray, & Brunaud-Vega, 2012), which are characterized by the three dispositional factors of openness, self-awareness/self-reflectiveness, and commitment to culturally responsive teaching. White and colleagues' work illustrates the need for mathematics teacher educators to provide direction to prospective teachers in understanding "the multiple layers of culture, the cultures surrounding them and those they will create when they teach" (pp. 40–41). This direction is important to disrupt dominant oppressive norms in mathematics classrooms that tend to position students and their thinking in a deficit manner (Chao, Murray, & Gutiérrez, 2014). This direction can also help with Civil's question of what might happen to students' alternative approaches when they belong to groups whose knowledge has historically not been recognized or valued in schools. Specifically, if mathematics teacher educators can attend to and develop both prospective teachers' multicultural mathematics dispositions *and* professional noticing, alternative algorithms for arithmetic operations can be valued and used to enhance instruction, affirm students' identities, and provide all students opportunities to engage in meaningful mathematics learning.

REFERENCES

Chao, T., Murray, E., & Gutiérrez, R. (2014). *NCTM equity pedagogy research clip: What are classroom practices that support equity-based mathematics teaching? A research brief*. Reston, VA: National Council of Teachers of Mathematics.

Jacobs, V. R., & Ambrose, R. C. (2008). Making the most of story problems. *Teaching Children Mathematics, 15*, 260–266.

Jacobs, V. R., Lamb, L. C., & Philipp, R. A. (2010). Professional noticing of children's mathematical thinking. *Journal for Research in Mathematics Education, 41*, 169–202.

Kitchen, R. S. (2005). Making equity and multiculturalism explicit to transform the culture of mathematics education. In A. J. Rodriguez & R. S. Kitchen (Eds.), *Preparing mathematics and science teachers for diverse classrooms: Promising strategies for transformative pedagogy* (pp. 33–60). Mahwah, NJ: Erlbaum.

Nasir, N. S., Hand, V., & Taylor, E. V. (2008). Culture and mathematics in school: Boundaries between "cultural" and "domain" knowledge in the mathematics classroom and beyond. *Review of Research in Education, 32*, 187–240.

White, D. Y., Murray, E., & Brunaud-Vega, V. (2012). Preparing preservice teachers to educate black students: Discovering multicultural mathematics dispositions. *Journal of Urban Mathematics Education, 5*(1), 31–43.

COMMENTARY 3

VALORIZATION OF KNOWLEDGE AS A COMPONENT OF UNDERSTANDING AND BUILDING UPON STUDENTS' THINKING

A Commentary on Civil's Case

Randolph A. Philipp
San Diego State University

As a mathematics educator in the School of Teacher Education at San Diego State University, I generally teach mathematics methods courses and mathematics education graduate courses. I also developed and teach a course on children's mathematical thinking, a laboratory for students enrolled in their first mathematics course for prospective elementary school teachers (PSTs). In the presented case, a student in a mathematics class

Cases for Mathematics Teacher Educators, pages 237–241
Copyright © 2016 by Information Age Publishing
All rights of reproduction in any form reserved.

for PSTs, after reading an article about notations and algorithms taught in Latin American countries and differing from those taught in U.S. schools, said, "This is nice, but they need to learn to do things the U.S. way." The dilemma for Civil is how to respond to this comment, which she interpreted as a failure of her student to "value others' ways of doing mathematics, particularly when the 'others' come from nondominant communities."

INTERPRETING THE DILEMMA

I agree with Civil that this case raises issues of valorization of knowledge, but I think that this valorization is most meaningfully viewed as part of a broader issue about teaching and learning: PSTs are generally unaware of a key principle of teaching—the great pedagogical value in understanding and building upon children's mathematical thinking. Instead, most PSTs approach the teaching and learning enterprise with an image of a teacher as one who disseminates information to children, and from this perspective they see little need to tap into children's preinstructional or informal understandings. As such, valorization of knowledge is a part of a much broader issue: PSTs simply do not recognize the value of listening to their students' mathematical thinking. I believe that learning to value listening to students' mathematical thinking requires unpacking and grappling with the mathematics and with children's engagement with that mathematics. I address these two issues before returning to make final comments about what I might have done in this case.

The Mathematics of Algorithms

The mathematical issue at the heart of the case is algorithms, procedures that can be applied generally. Mathematical algorithms are conventions adopted by societies, much like the U.S. convention of driving on the right-hand side of the street. Thus, not only does considering one convention better than another make little sense, but also the notion that someone else's algorithm is automatically wrong because it is different is ignorance and hubris.

Although algorithms are conventions adopted by societies, they are based upon mathematical principles. For example, the standard U.S. multidigit-subtraction-regrouping algorithm and the equal-addend algorithm, though procedurally and conceptually different, are both based upon principles of place value and the meanings of the operation of subtraction. Furthermore, they share the mathematical conditions that the minuend (top number) must be at least as large as the subtrahend (bottom number), both numbers must be nonnegative whole numbers, and both must be expressed

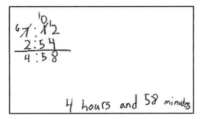

Figure 10.4 One third-grade student's invalid mathematical approach to finding the elapsed time between 2:54 p.m. and 7:12 p.m.

in the same base (which is typically base ten), and one must know how to operate in that base. Recognizing these conditions may seem unimportant, but if one of these conditions is not met, students who approach algorithms without a stance of meaning making find that their algorithm yields incorrect results. For example, most PSTs struggled to analyze the work of a third grader (Figure 10.4) in applying the standard algorithm in a context of elapsed time, to which it does not validly apply.

Also of mathematical note is that the two division algorithms presented in this case, which have identical procedural steps but differ in their representations, are mathematically related in a way that many culturally based algorithms, such as the two subtraction algorithms referred to above, are not. As Civil notes, in Latin American countries, students first learn essentially the U.S. algorithm and only later modify the representation by holding the products to be subtracted in their heads instead of writing them.

Too often, however, students, including adult PSTs, fail to understand the principles underlying algorithms (Philipp, Siegfried, Jacobs, Lamb, Bishop, & Schappelle, 2015). I consider the knowledge that algorithms, as societal conventions, may be viewed only as *a* way, never *the* way, to compute an answer as part of what has been referred to as *specialized* content knowledge (Hill, Sleep, Lewis, & Ball, 2007) for teachers.

Understanding the Students' Thinking

In her reflection, Civil wonders whether the PST was driven by her concern that immigrant students need to adapt to the new system to learn the traditional algorithm, was uncomfortable with the thought of having to learn many different algorithms, or, perhaps, devalued the knowledge of immigrant children. These important questions are designed to get at the PST's thinking, an issue to unpack further. The *Common Core State Standards* in mathematics [CCSSM] (National Governors Association Center for Best Practices & Council of Chief State School Officers, 2010) specifies that the standard multidigit-division algorithm should be taught in 6th grade, so the PST was right that teachers are required to teach the U.S. algorithm whereas they are not required to teach the alternative algorithm. So, from the PST's perspective, why does she think teachers needs to learn alternative algorithms? Below are some possibilities.

Stance #1. Teachers need to learn alternative algorithms so that they might explicitly teach them to their students.

Stance #2. Teachers need to learn alternative algorithms so that they might look for these algorithms and appreciate them when they see them in their classes, and they may also seek them out among the parents and family members of their students.

Stance #3. Teachers need to learn alternative algorithms because in so doing, they can learn to see and thereby help their students see relationships among the underlying mathematical ideas of place value in various algorithms.

Stance #4. Teachers need to learn alternative algorithms so that they might help students think about the differences between mathematical principles and mathematical conventions.

My conjecture is that if the PST holds any of these stances, it is likely Stance #1, possibly Stance #2, but definitely not Stances #3 and #4. I would want all of these issues raised in the class discussion. Furthermore, associated with discussing Stance #1 about explicitly teaching an algorithm to students is the tendency I find in many PSTs to believe that, pedagogically, we should take the ends as the means. Because students are expected to learn the standard multidigit-division algorithm in Grade 6, PSTs assume that teachers ought to explicitly teach the algorithm to children, independently of whether the children have first developed the conceptual foundations for understanding it. But teachers are not required to teach the content within a grade level in a specified sequence. In fact, according to the CCSSM, students must have opportunities to develop conceptual foundations before procedures are introduced. As such, I would want the PSTs to think about how they might support a meaning-making approach for their students to develop understandings for the underlying place-value ideas and the meaning of the operation of division long before they are expected to learn the standard algorithm, and grappling with alternative algorithms is a means to support this goal.

RESPONDING TO THE DILEMMA

I end by considering how I might begin to respond to this case in my own teaching to create an opportunity to focus on the PSTs' thinking about the key mathematical and pedagogical issues. One approach I use to initiate a discussion so that all PSTs may reflect on a pertinent issue is to write on the white board, verbatim, a student's comment I consider profound, provocative, inspiring, or in need of further consideration. In this case, I might write, "This is nice, but they need to learn to do things the U.S. way," and I might ask the student to talk more about what she was thinking. I might follow up by asking, "Does your inclusion of *but* indicate a feeling of tension here?" If so, I would ask for elaboration, and other students' comments. I would ask all the students to individually write for 5 minutes their answers to this question: *On a scale of 1–5, how important is addressing algorithms other than the standard algorithm in school, and why?* I would then use the PSTs' responses as a starting point from which the issues described above might be addressed, ensuring that the principle/convention, means/end, and valorization issues are raised.

REFERENCES

Hill, H. C., Sleep, L., Lewis, J. M., & Ball, D. L. (2007). Assessing teachers' mathematical knowledge: What knowledge matters and what evidence counts? In F. Lester (Ed.), *Second handbook for research on mathematics education* (pp. 111–155). Charlotte, NC: Information Age.

National Governors Association Center for Best Practices, Council of Chief State School Officers. (2010). *Common core state standards (mathematics).* Washington, DC: Author.

Philipp, R. A., Siegfried, J. M., Jacobs, V. R., Lamb, L. C., Bishop, J. B., & Schappelle, B. P. (2015). *An analysis of knowledge for teaching mathematics.* San Diego, CA: Center for Research in Mathematics and Science Education, San Diego State University.

CHAPTER 11

USING MATHEMATICS TO INVESTIGATE SOCIAL AND POLITICAL ISSUES

The Case of "Illegal Immigration"

Mathew D. Felton-Koestler
Ohio University

INTRODUCING THE CASE

I taught two mathematics courses for prospective K–8 teachers in the mathematics department at The University of Arizona. The prospective teachers in my courses were primarily sophomores and juniors who had not yet started the education program. My courses focused on two major themes, which were summarized in the syllabus (see Figure 11.1).

For this case I focus on the second theme of the course, the social and political dimensions of mathematics. I integrated this theme into the course using a variety of methods: I had the prospective teachers respond to readings that deal with the intersection between mathematics education and a broad range of issues such as culture (D'Ambrosio, 2001; Zaslavsky, 2001), race (Tate, 1994), gender (Harris, 1997), and social justice (Frankenstein,

Cases for Mathematics Teacher Educators, pages 243–250
Copyright © 2016 by Information Age Publishing
243

Learning Mathematics with Understanding
We will take an approach to mathematics that is based on the National Council of Teachers of Mathematics' (NCTM's) Process Standards, the Mathematical Practices in the Common Core State Standards for Mathematics (CCSSM), and what we know from research about how children learn mathematics with understanding. In short we will approach mathematics with the idea that everything in mathematics makes sense, and that you (and your future students) are capable of understanding all of K–8 mathematics....

The What, How, and Who (WHW) of Mathematics
We will be asking the following questions throughout the course:
- *What* messages do we send about mathematics and its role in our world?
- *How* are mathematical **concepts** and real world **contexts** related in math problems?
- What is the relationship between people (the *Who*) and the real world contexts we use for mathematics problems?

We will use these three aspects of mathematics to consider the **social and political dimensions** of mathematics content and the teaching and learning of mathematics.

Figure 11.1 Major course themes from the syllabus.

1998; Gutstein & Peterson, 2005); I incorporated lessons that involved political topics, such as income inequality or standardized test scores (see Felton, 2012; Felton, Simic-Muller, & Menéndez, 2012); and I had the prospective teachers complete projects in which they chose a sociopolitical issue to explore mathematically, which is the focus of this case.

The goals for the projects, as explained on a course handout, are summarized in Figure 11.2. The prospective teachers were not asked to create a social-justice lesson to be conducted with students (although many of the prospective teachers ended up framing their work in that way). Instead, the purpose was for the prospective teachers to explore a social or political

---

For this project you will choose a topic to investigate using the mathematics from our course. You will research the topic, provide background information, and then demonstrate how you used mathematics to further analyze or understand that topic.

The purpose of this project is twofold.
- First, it is an opportunity for more authentic and sustained experience with the second theme of the course: the *What, How,* and *Who* (WHW) of mathematics.
- Second, as a teacher you will often need to develop your own mathematical problems/lessons/projects. While this course does not focus on the specifics of good teaching per se (e.g., what kinds of problems to give to your students and when and why to do so), this is a chance to analyze a topic mathematically and think about how you could make connections between that topic and the mathematical concepts you are expected to teach.

Figure 11.2 Project goals.

issue of their choosing using mathematics. I told the prospective teachers that although they could investigate a variety of perspectives, they would have to consider the point of view of those with the least power in society relative to the issue they explored. Ultimately, I hoped the projects would provide the prospective teachers with more time to devote to connecting mathematics to a real-world topic of their choosing than was normally available during class time.

The prospective teachers had several assignments related to their projects: (1) list one to three possible topics, include a source for each topic, and briefly describe how they planned to investigate each topic using mathematics; (2) select a topic, detail some mathematics they had done thus far with their topic, and explain what further mathematics they planned to do; (3) create a detailed outline of the major themes of their project and what mathematics they are using to understand them; and (4) hand in a complete project. I provided feedback on each assignment.

In this case I explore the work of one of the prospective teachers in my course. For her project she chose the topic of "illegal immigration." As mentioned above, this work was done at The University of Arizona, which is in Tucson, AZ, a little more than an hour from the Mexican border. The local schools have a large number of immigrant students and English Language Learners. In addition, there is a history in Arizona of legislation that is anti-immigrant and that constrains instruction for English Language Leaners in ways that contradicts research findings (Goldenberg, 2008). However, Tucson is often a site of resistance to these policies. Approximately two years before this case occurred, SB 1070, a controversial anti-immigration bill, was signed into law but was challenged by the federal government and was under an injunction as it made its way through the courts. In addition Arizona has a ban on any courses that, among other things, "are designed primarily for pupils of a particular ethnic group" or "advocate ethnic solidarity instead of the treatment of pupils as individuals" (Arizona Revised Statutes, 2015). The State Superintendent at the time, John Huppenthal, used this law to target a Mexican American Studies program in the Tucson Unified School District (TUSD), and despite a state-ordered independent audit that found no evidence that any of the courses violated the law (Cambium Learning, 2011), Huppenthal ruled the district was out of compliance and threatened to withhold 10% of the district's annual funding. In response the TUSD school board voted to end the program and removed books used from classrooms. We briefly discussed the controversy surrounding the Mexican American Studies program in our course. Although it is important to understand the context in which this work was taking place, it is also important to note that throughout the entire project, the prospective teacher never mentioned the specific controversies in Tucson or Arizona.

THE DILEMMA

Figure 11.3 shows the entirety of one of the prospective teacher's work on the first two assignments for their project (all of the quotation marks are from the original work).

As a mathematics teacher educator, this work posed a dilemma for me: *How could I respond in a way that challenged the prospective teacher's portrayal of undocumented immigrants while still supporting her in feeling a sense of ownership over her project?* A secondary concern for me was the lack of mathematics the prospective teacher had done. Thus far she had mainly stated facts that involved numbers; for these projects I wanted the prospective teachers to go beyond this by using mathematics to analyze and unpack their topics. However, I felt that the overriding issue was to address how she was framing undocumented immigration. Once I had challenged her to consider new ways of looking at the topic, she could then use mathematics to provide insight into both her original ways of thinking about the topic and any new perspectives I raised.

Portrayals of Undocumented Immigrants

The prospective teacher's work seemed to reinforce a narrative about undocumented immigrants as individuals who take resources away from Americans. This is problematic in at least three ways. First, it is, at the very least,

---

First Assignment
The controversial topic of illegal immigration has been one of the most talked about topics. The number of illegal immigrants in the U.S. today is mind blowing and the amount of money we spend on illegal immigration; whether for health care services or just to keep people from crossing, is crazy.
http://immigration.procon.org/

Second Assignment
http://immigration.procon.org/
http://en.wikipedia.org/wiki/Economic_impact_of_illegal_immigrants_in_the_United_States

- "an estimated $21.8 billion in annual expenditures on illegal aliens"
  Using the tables full and football fields full of money to show how much this is.
- "the cost of providing education to these children is about $11.2 billion"
- The money spent on their education could be spent on bettering our schools or more supplies for classrooms.
- "30 million illegal aliens" How many cross per day? Per year?
- "payment for the treatment of the uninsured may run as high as $2.2 billion."

Figure 11.3 A prospective teacher's work on the first two assignments (all quotation marks are from the original).

an incomplete view of the role of undocumented immigrants in the U.S. economy; there are a number of ways in which everyday citizens and wealthy individuals and corporations benefit financially from the presence of undocumented immigrants. Second, it looks at immigration in largely individual terms without considering the broader social, historical, and political landscape in which immigration is situated. This perspective, for example, fails to consider the role of U.S. policies in depressing economic conditions in other countries, the extent to which governmental costs are related to how the U.S. has chosen to respond (e.g., increasing border patrols versus creating an accessible path to citizenship), and what gives some people the right to a higher standard of living based on where they were born. Finally, the prospective teacher fails to consider other voices, notably those of undocumented immigrants and what their experiences with immigration are.

I view this as a specific instance of a broader dilemma that other social justice educators face. Most social justice educators have a goal of supporting learners in coming to recognize injustice and finding ways to fight against it. However, there is no guarantee that learners will interpret complex social and political issues in the same way as the instructor. Brantlinger (2005, also see Bieda, this volume) has written about how when teaching a lesson on riots in South Central Los Angeles surrounding the Rodney King Riots, he "had not expected this lesson to potentially reinforce the dominant worldview that the problem with South Central is the people who live there" (p. 100). In particular, some students (the majority of whom were non-White and low-income) seemed unsure of what Brantlinger's (who is White) stance on the issue was. Elsewhere, I have written about how exploring gaps in standardized test scores could potentially reinforce perceptions that some groups of students are simply more capable than others (Felton, 2012). Most controversial topics are complex and deeply connected to prospective teachers' broader belief systems, which makes exploring them challenging terrain to navigate.

Ownership of the Project

Gutiérrez (2007) has argued that "students need to have opportunities to see themselves in the curriculum (mirror) as well as have a view onto a broader world (window)" (p. 3). Because I chose the vast majority of the content and real-world contexts in my courses, I created the projects as an opportunity for the prospective teachers to select a topic that reflects their interests and allows them significant input into the direction of investigation. However, in my feedback to the prospective teachers I am often concerned that I end up taking over the projects in two ways. First, as in this incident, the prospective teachers may approach topics in ways that reinforce status-quo power relationships in society. I generally tell them that they are free to include this perspective, but they also have to include the alternative perspectives that I suggest. Although I hope that allowing them to include multiple points of view allows at least some of the project to serve as a mirror, I do remain concerned that pushing too far will make the project mine and not theirs.

Second, it remains quite challenging for the prospective teachers to make substantive connections between mathematics and complex real world issues. In some cases the mathematical connections are largely superficial, and in other cases the prospective teachers do not know how to start. In these cases I often worry that I end up crafting much of the mathematics, again limiting the prospective teachers' ownership over the project. Civil (2002, 2007) faced similar struggles in her work connecting everyday activities to classroom mathematics, raising the concern that "once we start mathematizing everyday situations, we may be losing what made them appealing in the first place" (Civil, 2002, p. 44).

MY RESPONSE TO THE DILEMMA

In response to the prospective teacher's work on the first assignment, I encouraged her to "look at it [immigration] in a somewhat more complex way." Among other things, I suggested that this might include looking at how costs are a function of policy—such as choosing to spend money on border enforcement and making legal immigration difficult—and how there are, in fact, a number of benefits (financial and otherwise) that U.S. citizens and companies experience from undocumented labor. In response to the prospective teacher's work on the second assignment I pushed back more explicitly. Part of my response can be seen in Figure 11.4. I then encouraged her to pursue her own ideas for investigating the second narrative but said that if she did not have any ideas, then she should definitely investigate some of the economic benefits of undocumented immigration (I had given a number of suggestions and resources related to this on the first assignment).

LOOKING BACK

In the end, the prospective teacher integrated sources and some mathematics around both "the pros and cons of immigration." Although on the surface this met my goal of pushing back against the negative portrayal of undocumented immigrants, I do not know if this prospective teacher genuinely reflected on and explored these issues or if she approached it as a hoop to jump through to meet my requirements. I also do not know how

Narrative 1
Oh poor U.S. citizens! There are all these illegal immigrants who come and steal our jobs and it costs us a ton of money on border patrol and social services, etc. to pay for them.

Narrative 2
Oh poor undocumented workers ("illegal immigrants")! They want a better life and are generally willing to pay taxes, etc. in our country but it's basically impossible to do so, so instead they are forced to do very difficult work for very low pay (sometimes less than minimum wage) and with little or no rights and protections. All the while, the U.S. citizens are benefiting from a stronger economy.

To be clear, **it does not matter what you believe**. I don't care if you believe Narrative 1, 2, neither, something else entirely, or some more complicated combination of them. My concern is that if you just focus on the examples you have given so far then your work will connect much more closely to *Narrative 1*. For the purposes of this project you also have to do some work that connects closely with *Narrative 2*.

Figure 11.4 Partial response to prospective teacher's work.

this affected her sense of ownership over the project. I continue to struggle with how to best balance providing direction—both in terms of challenging problematic narratives and in providing substantive mathematical connections—without taking over the project.

As discussed above, in assigning these projects I tell the prospective teachers that they will have to consider the point of view of those with the least power in society relative to the issue they explore. However, many of my prospective teachers struggle with taking on this perspective, and I continue to look for better ways to support them in understanding what this means so they can take up this stance in their own projects. In an ideal world I would find time to unpack what it means to take a social-justice oriented stance, in general, and in mathematics in particular, to better support my prospective teachers in asking questions such as

- Who benefits from and who is hurt by the status quo?
- What is the perspective of those who are hurt by the status quo?
- Why are things like this?
- How could things be different?

REFERENCES

Arizona Revised Statutes, § 15-112 (2015). Retrieved from http://www.azleg.gov/FormatDocument.asp?inDoc=/ars/15/00112.htm&Title=15&DocType=ARS

Brantlinger, A. (2005). The geometry of inequality. In E. Gutstein & B. Peterson (Eds.), *Rethinking mathematics: Teaching social justice by the numbers* (pp. 97–100). Milwaukee, WI: Rethinking Schools.

Cambium Learning (with National Academic Educational Partnership). (2011). *Curriculum audit of the Mexican American Studies Department Tucson Unified School District.* Miami Lake, FL. Retrieved from http://www.tucsonweekly.com/images/blogimages/2011/06/16/1308282079-az_masd_audit_final_1_.pdf

Civil, M. (2002). Everyday mathematics, mathematicians' mathematics, and school mathematics: Can we bring them together? *Journal for Research in Mathematics Education. Monograph, 11,* 40–62.

Civil, M. (2007). Building on community knowledge: An avenue to equity in mathematics education. In N. S. Nasir & P. Cobb (Eds.), *Improving access to mathematics: Diversity and equity in the classroom* (pp. 105–117). New York, NY: Teachers College Press.

D'Ambrosio, U. (2001). What is ethnomathematics, and how can it help children in school? *Teaching Children Mathematics, 7,* 308–310.

Felton, M. D. (2012). Test scores in the U.S.: Introducing the data to pre-service teachers. *Teaching for Excellence and Equity in Mathematics, 4,* 7–14.

Felton, M. D., Simic-Muller, K., & Menéndez, J. M. (2012). "Math isn't just numbers or algorithms": Mathematics for social justice in preservice K–8 content courses. In L. J. Jacobsen, J. Mistele, & B. Sriraman (Eds.), *Mathematics teacher education in the public interest: Equity and social justice* (pp. 231–252). Charlotte, NC: Information Age.

Frankenstein, M. (1998). Reading the world with math: Goals for a critical mathematical literacy curriculum. In E. Lee, D. Menkart, & M. Okazawa-Rey (Eds.), *Beyond heroes and holidays: A practical guide to K–12 anti-racist, multicultural education and staff development* (pp. 306–313). Washington, DC: Network of Educators on the Americas.

Goldenberg, C. (2008). Teaching English language learners: What the research does—and does not—say. *American Educator, 32*(2), 8–23, 42–44.

Gutiérrez, R. (2007). Context matters: Equity, success, and the future of mathematics education. In T. Lamberg & L. R. Wiest (Eds.), *Proceedings of the annual meeting of the North American Chapter of the International Group for the Psychology of Mathematics Education* (pp. 1–18). Reno, NV: University of Nevada.

Gutstein, E., & Peterson, B. (2005). Introduction. In E. Gutstein & B. Peterson (Eds.), *Rethinking mathematics: Teaching social justice by the numbers* (pp. 1–6). Milwaukee, WI: Rethinking Schools.

Harris, M. (1997). An example of traditional women's work as a mathematics resource. In A. B. Powell & M. Frankenstein (Eds.), *Ethnomathematics: Challenging Eurocentrism in mathematics education* (pp. 251–222). Albany: SUNY. (Reprinted from *For the Learning of Mathematics, 7*(3), 26–28, 1987)

Tate, W. F. (1994). Race, retrenchment, and the reform of school mathematics. *The Phi Delta Kappan, 75,* 477–480.

Zaslavsky, C. (2001). Developing number sense: What can other cultures tell us? *Teaching Children Mathematics, 7,* 312–319.

COMMENTARY 1

TENSIONS AND OPPORTUNITIES WHEN IMPLEMENTING SOCIAL JUSTICE MATHEMATICS TASKS

A Commentary on Felton-Koestler's Case

Kyndall Brown
University of California, Los Angeles

MY POSITIONALITY

I am a person of African descent, born in the midst of the civil-rights movement, who grew up in Los Angeles, California. I have been a mathematics educator for 30 years, serving as a secondary mathematics teacher, teacher leader, coach, professional-development provider, and faculty supervisor. I believe in a constructivist, sociocultural approach to teaching and learning. I support the use of culturally relevant and responsive teaching methods to eliminate performance gaps between different groups of students.

Cases for Mathematics Teacher Educators, pages 251–254
Copyright © 2016 by Information Age Publishing
All rights of reproduction in any form reserved.

251

Since the publication of *Rethinking Mathematics: Social Justice by the Numbers* (Gutstein, 2005), I have been incorporating social-justice lessons in the methods and seminar classes I teach to secondary mathematics preservice and inservice teachers, as well as the professional-development institutes and workshops I facilitate. I have taken lessons from the book, pulled together all of the resources necessary for the tasks, and engaged teachers.

INTERPRETING THE DILEMMA

After reading the "Illegal Immigrant" case study, two main ideas come to mind related to my own practice. The first idea has to do with how teachers or professional-development facilitators deal with comments or statements that might be considered insensitive or offensive. The second idea has to do with how to keep the focus of a lesson on social-justice issues.

For the last 30 years, mathematics educators have been exploring ways to address the test-performance gap between various groups of students based upon their race, ethnicity, gender, and language ability (Boykin & Noguera, 2011). One strategy that has been utilized by mathematics educators is to examine who has access to high-quality mathematics instruction and who does not, as well as whether academic resources are allocated in ways that the students who have the most need receive the most support. Over the years, these issues have been classified under the title "Access and Equity."

In order to address access and equity, educators need to discuss sensitive topics related to racism, classism, and sexism. Weissglass (1999) devised a number of communication tools intended to facilitate these sensitive discussions. These structures, along with strategies from Cognitive Coaching (Costa & Garmston, 2002) like paraphrasing, asking clarifying questions, and providing wait time, have assisted me in dealing with teachers or students who make statements or assertions similar to the teacher in this case study.

As is the case with all learners, teachers bring their own histories, cultures, and prejudices to the learning environment that influence the ways they interpret social phenomena. As someone who takes a strong social justice stance on most issues, it has been surprising for me to find teachers who take a position that would seem in opposition to my progressive leanings. By engaging in structured, equity-based dialogue, the intent is that the teachers will be able to recognize how their statements might be considered insensitive to others and in opposition to access and equity.

RESPONDING TO THE DILEMMA

One difference between my work and Felton-Koestler's case study is that I have not required teachers to create their own tasks. I have found that teachers are not necessarily that skilled at creating rich mathematical tasks. The *lesson study* process used by Japanese teachers to create engaging mathematical tasks can take years (Stigler & Hiebert, 1999). Once Japanese teachers create a lesson, they observe each other teaching the lesson, meet to debrief the lesson, then revise and reteach the lesson. This cycle continues until the teachers feel that they have a well-developed lesson. Given the sensitive nature of social justice lessons, I believe teachers need to undergo a process similar to lesson study if they are going to create lessons that are mathematically sound and serve the purpose the teacher intended.

I think that we need to be very explicit when we talk about what it means to teach mathematics for social justice. I believe the student in this case study mistook a social *issue* with social *justice*. Gutstein (2005) believed that the goals of a social justice curriculum should provide students with an opportunity to

- develop an understanding of the sociopolitical, cultural, economic, and historical dynamics of racism, along with their interconnections;
- appreciate the complexity of different forms of racism (structural, institutional, individual);
- develop and use analytical tools (such as data analysis, graphing, and mathematical modeling of social phenomena) to understand and dissect racism;
- support views, develop coherent arguments, and engage in group discussions to develop individual and collective analyses; and
- develop an appreciation for multiple, alternative, and compelling perspectives while constructing their own independent knowledge.

Using these goals as a framework, an instructor would be able to provide feedback to the teacher who posed the Illegal Immigrant project. I would ask the teacher how the project helped to develop an understanding of the dynamics of racism. I would ask the teacher how the project helps students to understand the complexity of racism. I would ask the teacher how the project developed and used analytical tools to understand racism. I would ask the teacher how the project helped students engage in group discussions to develop individual and collective analyses. I would ask the teacher how the project helped students to develop multiple and alternative perspectives. Hopefully, the conversations created by answering these questions would help the teacher reevaluate the context of their task.

I believe the author was correct in referring the prospective teacher to alternative sources of information regarding the benefits of immigrant labor to the economy of the United States. In order to avoid feeling as if he has taken over the project, a requirement could be asking a question that would require exploring both sides of an issue. In this case, the question could be "Do undocumented immigrants contribute more to the U.S. economy than they cost U.S. taxpayers?" A question like this opens up the opportunity for prospective teachers to conduct research from multiple perspectives and come to conclusions by themselves on the basis of the mathematics.

FINAL THOUGHTS AND TAKEAWAYS

Most of the mathematics in the social justice lessons I have seen are at the middle-school level. Many high-school teachers struggle to find social justice lessons that are aligned to courses above Algebra 1. The Common Core standards now require Statistics and Probability to be integrated into the secondary mathematics curriculum. Statistics and Probability offer an opportunity to explore a variety of social justice topics in the secondary curriculum. Probability can be used to study racial profiling. Scatterplots can be used to see if there are correlations between immigration and the Gross Domestic Product. It is important that the facilitator be well versed in the statistics and probability standards in order to offer suggestions to students and participants.

The Common Core Standards also require students at all levels to model real-world phenomena using mathematics. At the high-school level, students are supposed to solve problems arising from real-life contexts that have no standard solution strategies. The modeling process includes identifying variables, formulating a mathematical model to describe the relationships between the variables, performing operations on the relationships, validating conclusions and improving the model, or reporting on the conclusions (California Department of Education, 2010). The modeling process also opens up opportunities to engage students in social justice tasks, because they are based upon real-world situations.

REFERENCES

Boykin, A. W., & Noguera, P. (2011). *Creating the opportunity to learn: Moving from research to practice to close the achievement gap*. Alexandria, VA: ACSD.

California Department of Education. (2010). *California common core state standards: Mathematics*. Sacramento, CA: Author.

Costa, A. L., & Garmston, R. J. (2002). *Cognitive coaching: A foundation for renaissance schools*. Norwood, MA: Christopher-Gordon.

Gutstein, E. (2005). *Rethinking mathematics: Teaching social justice by the numbers*. Milwaukee, WI: Rethinking Schools.

Stigler, J. W., & Hiebert, J. (1999). *The teaching gap: Best ideas from the world's teachers for improving education in the classroom*. New York, NY: The Free Press.

Weissglass, J. (1999). *No compromise on equity in mathematics education: Developing an infrastructure*. Santa Barbara, CA: Center for Educational Change in Mathematics and Science.

COMMENTARY 2

THE NEED TO BE INTENTIONAL IN THE INTEGRATION OF SOCIAL JUSTICE IN MATHEMATICS CONTENT COURSES

A Commentary on Felton-Koestler's Case

Sylvia Celedón-Pattichis
University of New Mexico

MY POSITIONALITY

Felton-Koestler presents a case that focuses on social justice and political issues in his mathematics for teachers course. The case on "illegal immigration" is familiar to me in a couple of ways. One way this case is familiar to me is based on my personal experiences. As a young child, I was born in Salineño, Texas and raised in Ciudad Miguel Alemán, Tamaulipas until the age of 8. These are two towns along the border of Texas and Mexico. At the age of 8, my parents obtained a visa, and we moved to California for

Cases for Mathematics Teacher Educators, pages 255–259
Copyright © 2016 by Information Age Publishing
All rights of reproduction in any form reserved.

one year. The fact that my parents obtained a visa, because my siblings and I were born in the United States, gives me privileges that others who cross the border undocumented do not have. My journey crossing to the United States was very different than students who face so many hardships, even death, crossing the border. However, I experienced people labeling me as "other" when my status as a learner became English as a second language (ESL); I was also viewed as different because I had an accent and was raised in a low-income family. This ESL label followed me until the ninth grade, when a White English teacher advocated for my placement to be changed to college preparatory coursework (see Celedón-Pattichis & Ramirez, 2012). My family settled in Rio Grande City, Texas, another border town, in the mid-1970s. I graduated from high school, prepared for secondary mathematics education at the University of Texas–Austin, and returned to the same high school to teach mathematics where many students crossed the border from Camargo, Tamaulipas to attend school.

Another way this case resonates with me involves my professional experience. As a bilingual and mathematics teacher educator, I coconstruct with preservice teachers what it means to work with culturally and linguistically diverse students in mathematics classrooms. This work involves not only students sharing their own personal and professional experiences and their successes and challenges in learning mathematics but also my own sharing of personal stories as an ESL learner of mathematics and professional experiences in mathematics classrooms. hooks (1994) referred to this teaching as *engaged pedagogy* that is empowering to both students and teacher and involves taking risks and engaging in conversations that are tensional when addressing equity and social justice issues. Through 20 years of work in academia and in schools, I have learned that how students connect to equity issues is based on what they themselves have experienced personally and what entry points we provide for them to engage in critically examining sociopolitical issues that affect their communities.

INTERPRETING THE DILEMMA

In addressing social justice issues in his mathematics-for-teachers courses, Felton-Koestler faced two dilemmas. One dilemma Felton-Koestler encountered with integrating social or political issues with mathematics was that he wanted to challenge a preservice teacher's portrayal of undocumented immigrants while also supporting her to feel ownership of her project. Another dilemma Felton-Koestler faced was that the preservice teacher had not used mathematics to address her topic. The prospective teacher had merely used numbers to state facts that focused on one point of view. Gutstein (2006a) has discussed this tension when teaching mathematics for social justice and presents a *continuum* that moves from a focus on teaching mathematics to teaching topics of social justice. Because the preservice teacher had not included in her project discussions on the sociopolitical issues affecting Arizona (the anti-immigration law, the ban on bilingual education and ethnic studies, etc.) even after Felton-Koestler had mentioned these events in class, he saw the need to challenge the preservice teacher's one-sided argument by being *intentional* in the way he responded to her. Felton-Koestler used narratives to remind the preservice teacher that she needed to use a different perspective (that of an undocumented immigrant) because this was also part of the assignment. According to Bartell (2013), the majority of a group of teachers in her study made "students aware of the issue as they themselves interpreted it rather than thinking about the various ways students might critically examine, interpret, or respond to the issue" (p. 157). I believe this is exactly the issue Felton-Koestler is trying to address; he is preparing preservice teachers to anticipate different ways their own students might respond to these issues.

RESPONDING TO THE DILEMMA

Having said that, however, Gutstein (2006b) has discussed the need to include a *pedagogy of questioning* to help coconstruct with students different issues that are important to them and that affect their own communities. Although Gutstein's studies have focused on teaching mathematics for social justice with middle- and high-school students, we can learn what has worked for these students and consider including topics that are *more relevant* to preservice teachers. For example, asking questions such as the ones Felton-Koestler lists in "Looking Back" might be a great way to begin generating the topics that can be explored with preservice teachers. In doing so, preservice teachers will have opportunities to co-construct topics that are relevant to them. This process provides an entry point for them. After preservice teachers have themselves experienced relevant topics, then more complex topics such as undocumented immigrants can be presented as options in assignments. Gutstein (2012) has reminded us that teaching mathematics for social justice takes a *long time,* and Bartell (2013) recommended that *continuous professional development* that targets long-term goals is needed if we are to make progress in preparing teachers to teach mathematics for social justice. For example, will the integration of social and political issues be covered in one lesson, one unit, or throughout the year? These would be important questions to answer in teaching mathematics content courses for teachers and other courses that integrate these topics.

With my limited knowledge about what happened exactly with the teaching of mathematics in this case, I believe that the dilemma Felton-Koestler presents is related to who owns the mathematics—the mathematics professor, the preservice teachers, or both? I am wondering how many opportunities are provided for students themselves to co-construct the mathematics that can be used to address sociopolitical issues. It would be important to provide opportunities where students collaborate to explore the mathematics that can be used when addressing different topics. A good beginning point might be having students read through examples of different topics and the mathematics that can be used to critically examine those ideas (Gutstein, 2013).

In conclusion, I applaud Felton-Koestler's effort to integrate topics of social justice into a course focused on mathematics content. I believe more integration of this type needs to occur between mathematics departments and education to avoid multicultural education courses being viewed as the ones "students have to take" or the only ones addressing issues of equity and diversity in most elementary and secondary education programs (Gutiérrez, 2003). In order to address social and political issues in the teaching of mathematics, there will be tensions we encounter with either too much focus on the teaching of mathematics or a major focus on sociopolitical issues. Reaching a balance between the two takes time. I also believe that Felton-Koestler did what he thought was appropriate at the time—to give more attention to deconstructing one-sided arguments about "illegal immigration." As Gutstein (2012) reminded us, "If one has never taught critical mathematics, it is hard to learn how. Teachers need to be patient with themselves and students. Start slowly" (p. 66). Reflecting on lessons taught, addressing tensions that emerge, and co-constructing ideas on how we move forward to teach what matters will be critically important to change the way we view our world.

REFERENCES

Bartell, T. G. (2013). Learning to teach mathematics for social justice: Negotiating social justice and mathematical goals. *Journal for Research in Mathematics Education, 44*, 129–163.

Celedón-Pattichis, S., & Ramirez, N. G. (2012). *Beyond good teaching: Advancing mathematics education for ELLs.* Reston, VA: National Council of Teachers of Mathematics.

Gutiérrez, R. (2003). Beyond essentializing: The complex role of language in teaching Latina/o students mathematics. *American Educational Research Journal, 39*, 1047–1088.

Gutstein, E. (2006a). *Reading and writing the world with mathematics: Toward a pedagogy for social justice.* New York, NY: Routledge.

Gutstein, E. (2006b). "So one question leads to another": Using mathematics to develop a pedagogy of questioning. In N. S. Nasir & P. Cobb (Eds.), *Increasing access to mathematics: Diversity and equity in the classroom* (pp. 51–68). New York, NY: Teachers College Press.

Gutstein, E. (2012). Reflections on teaching and learning mathematics for social justice in urban schools. In A. A. Wager & D. W. Stinson (Eds.), *Teaching mathematics for social justice: Conversations with mathematics educators* (pp. 63–78). Reston, VA: National Council of Teachers of Mathematics.

Gutstein, E. (2013). *Rethinking mathematics: Teaching social justice by the numbers* (2nd ed.). Milwaukee, WI: Rethinking Schools.

hooks, b. (1994). *Teaching to transgress: Education as the practice of freedom.* New York, NY: Routledge.

COMMENTARY 3

"STRATEGIC INTRUSION"

A Commentary on Felton-Koestler's Case

La Mont Terry
Occidental College

MY POSITIONALITY

I view this case through a particular set of lenses shaped by my experience. I am a heterosexual, Black male with a bicultural upbringing—I was born into a military home in Texas and raised within a variety of family contexts (including nuclear, extended, and, eventually, single-parent households) in Los Angeles, CA, and Seattle, WA. Coming into political consciousness as a teenager in the early '90s, I have worked to make sense of what it means to be Black *and* male *and* poor in America; that said, as a U.S. citizen, I have been personally shielded from the marginalization that comes with having darker-than-white skin while being an immigrant in these United States. I am also a former middle-school mathematics teacher and math coach. In the community I taught, I developed deep personal and professional concern for my students and their families, many of whom are first or 1.5 generation (those who immigrate before their teens) immigrants from Mexico, Central America, Taiwan, and Vietnam. Currently, I am a tenured professor

Cases for Mathematics Teacher Educators, pages 261–265
Copyright © 2016 by Information Age Publishing
All rights of reproduction in any form reserved.

in an Education department at a historically White college in Los Angeles. In my current work (mostly with undergraduates), as well as in the context of inservice professional development with math teachers in urban school districts, I am regularly engaged by worldviews that position people on the margins as "problems". As such, the interaction between Felton-Koestler and the prospective math teacher in this case resonates with my experience.

INTERPRETING THE DILEMMA

The dilemma, as presented by Felton-Koestler, is how to artfully and compassionately navigate the prodding required to press prospective teachers towards a more critical sociopolitical consciousness (here, in the case of immigrants and immigration), while also not disrupting the development of students' teacher and mathematics identities. A pivotal concept in Felton-Koestler's notion of identity development is Gutiérrez's (2007) *mirrors and windows* metaphor—the idea that, from a teacher educator's perspective, supporting students' identity development with an eye towards equity requires careful attention to the balance between self and others. In this case, Felton-Koestler is concerned that his *political goal* of preparing critically aware math teachers may abrogate his students' investments of their political selves into these projects. Likewise, there is an analogous concern that his *pedagogical goal* of preparing mathematically proficient teachers may also disrupt their ownership over the mathematical dimensions of developing these projects. I do not disagree with Felton-Koestler's views about the necessity of this kind of careful attention to how students experience curricula. In fact, my sense of what is at stake here requires that we press into this framing of the situation a bit more.

It strikes me as matter-of-fact that we want to produce math teachers who are "mathematically fit." For prospective math teachers, if or when we find that they are not fit, we put them to work developing and exercising the kind of mathematical habits we believe are necessary. Here, the way in which Felton-Koestler carefully measures his influence over the substance and direction of the actual mathematics in these projects reflects both our views about math content as well as his unique pedagogy. On his view, inserting himself too much in order to conserve math content intrudes on his students' *ownership*, thereby disrupting their identity development. I understand this balancing, but I caution that there is a very thin line here between not inserting one's self too much and not providing enough exposure to powerful examples of mathematizing the sociopolitical.

At the core of my concern is that, in understanding how students develop ownership (and, thus, identity), we should be careful not to overestimate the importance of the originality of the math. Although both seem important, I would argue that developing personal competence with the topics explored is much more integral to a sense of ownership than one's unique role in producing original math connections. I'm not advocating that Felton-Koestler "do the math" for his prospective teachers; rather, I'm suggesting that erring on the side of intrusion, as it were, may at least secure a sense of *competence* with the process of mathematizing phenomena, even if to the detriment of *originality*. Among those prospective teachers who struggle to draw more substantive mathematical connections in the process of mathematizing our social worlds, originality may be less integral to their sense of ownership (and, therefore, their math identities) than we think.

In that spirit, I also see an equivalent need to advocate for a more strategic intrusion into the habits of mind that position prospective teachers to recognize the politics of our social worlds. Specifically, I am referring here to a student's ability to critically reflect upon power—their own, and that of others (Apple, 2004; Diversity in Mathematics Education Center for Teaching and Learning, 2007). Much in the same way that students come to our classrooms with varied mathematical experiences, so too they come with varied experiences in critical reflection. The political goals we have as social justice math educators are contingent upon structuring experiences that facilitate and encourage this habit of reflecting on the differential power embedded in our relationships. However, if we are seeking to create change, then we should expect teacher preparation to *always* be intrusive—or, at least, to always be perceived so by those in privilege and power. That said, this approach to teacher preparation will always be problematic if our fundamental view of supporting identity requires that we refrain from intruding.

RESPONDING TO THE DILEMMA

In my read of Gutiérrez's (2007) piece, the mirrors/windows metaphor is useful precisely because it drives a notion of equity and access designed to change the experiences *of the marginalized* in the classroom. On this view, it is more important for those with marginal power to see themselves mirrored in the curriculum than for others—and more important for those from dominant groups and normative experiences—to perceive the Other through windows in the curriculum. These two imperatives, however, have significant implications for the experience of those in privilege and power in our classrooms. *How can we expect to change the experiences of the marginalized without also expecting to change the experiences of those who are marginalizing?* Math teacher educators whose view of identity development does not embrace critical engagement of the Self will be hard-pressed to accomplish the political and pedagogical objectives in authentic social justice.

Intruding is difficult. Because critical race theory invites us to think about race as a central hub for exploring multiple identities and politics, my professional work often involves intruding where race and racial identity concerns teaching math (Crenshaw, 1995). For me, the conversation often begins with tackling Whiteness. Although many I encounter consider themselves allied to progressive struggles, many more react to my political and pedagogical goals with a certain *fragility* (Adler-Bell, 2015; DiAngelo, 2011). If I weren't absolutely convinced of the possibilities inherent in challenging White students to critically examine Whiteness, for example, then I may be dissuaded by their reactions to my intrusion. *But I am convinced.* Although we only get a glimpse into Felton-Koestler's course here, the wonderful thing about this assignment is that he intentionally embedded an expectation that his prospective math teachers thoughtfully consider the views of "those with the least power." Not only should this aspect be considerably weighted in the determination of the grade for this project (students attend to what is valued) but, more importantly, we must recognize that unless critical self-engagement and reflection is a part of the fabric of the entire course (more appropriately, an entire program), we set these students up for either failure or frivolous engagement with social justice. Math educators can learn much, for example, from those committed to *intergroup dialogue* as a practice (Hurtado, 2001; Schoem & Hurtado, 2001; Zúñiga, Nagda, Chesler, & Cytron-Walker, 2011). Democracy, at the end of the day, is not simply a product—it is process.

REFERENCES

Adler-Bell, S. (2015). Why white people freak out when they're called out about race. *Alternet.* Retrieved from http://www.alternet.org/culture/why-white-people-freak-out-when-theyre-called-out-about-race

Apple, M. (2004). *Ideology and curriculum.* New York, NY: Routledge.

Crenshaw, K. (Ed.). (1995). *Critical race theory: The key writings that formed the movement.* New York, NY: The New Press.

DiAngelo, R. (2011). White fragility. *International Journal of Critical Pedagogy, 3*(3), 54–70.

Diversity in Mathematics Education Center for Teaching and Learning. (2007). Culture, race, power, and mathematics education. In F. Lester (Ed.), *Second handbook of research on mathematics teaching and learning* (pp. 405–433). Charlotte, NC: Information Age.

Gutiérrez, R. (2007). Context matters: Equity, success and the failure of mathematics education. In T. Lamberg & L. R. Wiest (Eds.), *Proceedings of the annual meeting of the North American Chapter of the International Group for the Psychology of Mathematics Education* (pp. 1–18). Reno, NV: University of Nevada.

Hurtado, S. (2001). Research and evaluation on intergroup dialogue. In D. L. Schoem & S. Hurtado (Eds.), *Intergroup dialogue: Deliberative democracy in school, college, community, and workplace* (pp. 22–36). Ann Arbor, MI: University of Michigan Press.

Schoem, D. L., & Hurtado, S. (Eds.). (2001). *Intergroup dialogue: Deliberative democracy in school, college, community, and workplace.* Ann Arbor, MI: University of Michigan Press.

Zúñiga, X., Nagda, B. R. A., Chesler, M., & Cytron-Walker, A. (2011). *Intergroup dialogue in higher education: Meaningful learning about social justice: ASHE higher education report, Volume 32, Number 4* (Vol. 115). San Francisco, CA: Jossey-Bass.

CHAPTER 12

SEARCHING FOR COHESION IN A MATHEMATICS COURSE FOR SOCIAL ANALYSIS

Jean M. Mistele and Laura J. Jacobsen
Radford University

INTRODUCING THE COURSE

Our mathematics content course, *Math for Social Analysis*, is a junior-level course for elementary- and middle-school preservice teachers (PSTs) taught in the Department of Mathematics and Statistics at Radford University. Math for Social Analysis is the last of three mathematics courses that PSTs take before they apply for entrance to the education program. The course integrates mathematics content knowledge with the pedagogical knowledge needed for teaching mathematics while providing the PSTs experiences using mathematics to explore timely social issues. The course has a semester project in which one option is service learning. This creates a connection between mathematics in the classroom and in an afterschool program in our local community. The second option is research based and provides an alternative for PSTs whose schedules do not align well with the needs of the afterschool program.

Cases for Mathematics Teacher Educators, pages 267–274
Copyright © 2016 by Information Age Publishing

Math for Social Analysis was part of the *Mathematics Education in the Public Interest* (MEPI) project that was supported by a 2-year National Science Foundation grant in 2008. The goals for the project included: (1) supporting equity and social justice in mathematics education, (2) diversifying student interest and participation in mathematics, and (3) broadening and enriching the ways mathematics is viewed as a discipline. These goals were informed by Gutstein (2006) and Spielman[1] (2008) and were used to guide the development of the course and its curriculum. In the course, PSTs examine social issues using mathematics, with learning outcomes focused on deepening their mathematical knowledge for teaching. They also create mathematics activities, lessons, or units designed to help children understand mathematics in the context of exploring important social issues.

Mathematically, we have structured the course around the five National Council of Teachers of Mathematics (NCTM) standards (2000) content strands for Grades K–8: (1) number and operation, (2) algebra, (3) geometry, (4) measurement, and (5) data analysis and probability. We drew on and expanded the mathematics that the PSTs learned in their two prerequisite courses that used Sybilla Beckmann's (2011) *Mathematics for Elementary Teachers with Activities* textbook. Part of the reason for the continued, broad mathematical emphasis in Math for Social Analysis across all five strands is because the community colleges in our state do not offer parallel mathematics courses to either of the two prerequisite courses designed specifically for prospective teachers. Transfer students generally bring liberal-arts mathematics coursework in lieu of our two mathematics prerequisites, and we still want them to gain at least some experience with mathematics content for teaching across the strands.

The Math for Social Analysis course was designed to present many different social issues using mathematics content at all grade levels (see Table 12.1). We aimed to limit duplication of mathematical ideas across social issues in order to have adequate time to meet the wide-ranging mathematical learning outcomes for the course. We provided multiple examples of serious social issues with meaningful mathematics at various grade levels as examples for all of our PSTs. The large majority of our students are elementary PSTs, and the remainder are middle-school PSTs and K–12 special-education PSTs.

Over the years, we found that the content strands, such as number and operations, statistics, and algebra naturally align with most social issues. However, given the course's comprehensive set of mathematics content learning outcomes, the fit between various social issues and the mathematics at times became contrived, making either the mathematics or the social issue feel forced. Other times due to the serious nature of the social issues, we found our PSTs became emotionally drained from exploring the many injustices on a local, national, and global level. They sometimes expressed just wanting to do "math," without these applications.

TABLE 12.1 Content Strands and Social Issues	
Mathematics Content Strands	**Social Issue**
Number and Operations	Environmental Devastation (Conservation, Endangered Species, Rainforest–Trees, Arable Land, Mountain Top Removal, Alaskan Pipeline)
Algebra	Distribution of Wealth (Poverty—Global and National, Population Density)
Measurement	Racism, Sexism, Bullying (Police, Achievement Gap, Name Calling)
Geometry	Child Labor (Sweat Shops, Mining)
Data Analysis and Probability	Global Conflict/War (Iraq, Government Spending)
Multiple Content Areas	Media Influence (Self-Image)
Multiple Content Areas	Health (Childhood Obesity, Nutrition)
Multiple Content Areas	Pollution (Water, Air, Land)

Through several iterations of offering the course, the course was structured in part around the book edited by Rico Gutstein and Robert Peterson (2013), *Rethinking Mathematics: Teaching Social Justice by the Numbers,* now in its second edition. We want our PSTs to understand that the changing demographics of the K–8 classroom require that they create mathematics activities that link mathematics to the real world as experienced by the diverse population of students. We also want our PSTs to view schooling through a wider lens, a lens that promotes engaged citizenship.

THE DILEMMA—HOW SHOULD WE ORGANIZE THE COURSE?

Our goal is to organize our mathematics lessons by joining two ingredients: (1) mathematics content knowledge for teaching and (2) social issues. We model standards-based mathematics so that the PSTs gain a deeper and more flexible understanding of the mathematics. Simultaneously, we want to use the mathematics to critique important social issues to elevate the PSTs' awareness about these issues. We aim to also increase our PSTs' interest in mathematics using these applied lessons.

In practice, however, we found that the mathematics and the social issues compete for attention. We also found our PSTs were impacted by the course in two different ways: (1) they were concerned the social issues were inappropriate for elementary students, and (2) we noticed they were emotionally impacted by the serous social injustices we exposed to them in each class. We noted that at times the social issue took center stage, and the mathematics became mechanical as it supported the social issue. Other times the mathematics took center stage, and the social issue became contrived to support the mathematics. In sum, tension exists between organizing a mathematics course in a mathematics department that must teach a specific content base of rich mathematics and simultaneously demonstrate the power of mathematics to unpack and critique real-world issues. In the following paragraphs we describe these two outcomes in the course and our perspective on this dichotomy.

Mathematics Centered

When we organized the course around the five mathematics content strands the social issue linked to some of the content became contrived. The PSTs perceived the mathematics as merely complex word problems with multiple parts—as another worksheet. Many times, the attention placed on learning mathematics sidelined the social issue. As a result, the PSTs' understanding about the social issue developed only superficially because the social issue came across as "window dressing" (Frankenstein, 2009) for the mathematics. Other times, the mathematics was insufficient to dig deeper into the social issue. This had two manifestations: (1) the PSTs enjoyed the math activities but did not realize they did not use the mathematics to understand the social issue on a deeper level, or (2) the PSTs believed social issues have no place in a mathematics course, because the social issue became a distraction from doing and learning mathematics. In the following paragraphs, we show an example of a mathematically rich activity that fails to advance the serious issues of habitat destruction that leads to species extinction and our reason for including the activity.

Example
We created a unit lesson that investigated the destruction of natural habitats and their local communities. The unit includes mountain-top removal in the Appalachian Mountains, which has significant impacts on the environment and surrounding community, and rainforest depletion. We will address the Rainforest unit in which the PSTs learned about the negative impact of deforestation and the positive impact via financial gains to the governments and the local people. In particular, one lesson from the unit targeted an endangered species that lives in the rainforest, the poison dart frog. This lesson is an example of a rich mathematics activity that failed to advance the PSTs' understanding of the plight of the endangered poison dart frog. This lesson came later in the semester when we noted the emotional impact on our PSTs. We were receiving comments directly from the PSTs and from their reflections in which they stated that they would not use social issues when they teach. Such comments, summarized, included

"I don't believe social issues should be taught to younger students. It is for high school or college age students" and "Children should not have to learn about this, learning should be fun." As a result, we included the following activity that did not strongly connect the mathematics to the social issue, but highlighted an interdisciplinary design that might be more "fun."

The lesson begins by introducing the PSTs to the poison dart frog by watching a short YouTube video (Discovery, 2009). In the video they observe and learn about the frog's life cycle and habitat from a herpetologist from the National Zoo in Washington, DC. The National Zoo has a constructed tropical rain forest that is home to some of the poison dart frog species. In the video, the herpetologist explains the frog's toxicity and describes how the indigenous people collect and use the toxins. We have a class discussion about the poison dart frog and locate the rainforests on a world map.

The PSTs make an origami poison dart frog from a 3X5 index card. The paper frog is the data-generating tool. The frog is designed to jump when its back end is gently pressed down. A tape measure is laid out on the desk, and the PSTs record 20 jumps. They first measure the jumps using nonstandard units followed by standard measurement units. They use the nonstandard unit of measurement (e.g., toothpicks, paper clips) and the standard measurements (inches or centimeters) to calculate the mean and the mode number of jumps. As a class, we discuss the standard units of measurement and the nonstandard units of measurement they used in their calculations, the variation of jump distances, and the longest jump. Although the PSTs enjoy the activity, the mathematics activity itself does not address the social issue–the poison dart frog's predicament. When we asked the PSTs to critique the lesson, many like Jessica said, "I liked this a lot. It was fun. I will use this in my class," implying she would use it as a mathematics lesson. We agree it is a rich mathematics activity, but the relevance to the social issue is diminished, aside from the discussions following from the video display.

Social-Issue Centered

In contrast to the mathematics-centered model, when course lessons or units were instead structured around one or more social issues, this oftentimes led us to incorporate a wide range of different mathematical concepts from multiple mathematics strands, often within the same lesson. Some mathematics topics would be addressed only in superficial ways, given the focus on asking meaningful questions required to analyze the social issues. That is, the mathematics rigor was weakened by the analysis of the social issue simply by not requiring every mathematics question to be challenging. For the same reason, centering lessons on social issues did not always lead to our asking mathematics questions in a well-designed order to scaffold deeper conceptual understandings or to build on the mathematics complexity. As a result, the PSTs sometimes lost sight of the mathematical purposes of the lesson. In the following paragraphs we show an example of a socially rich activity that fails to meaningfully advance our PSTs' mathematical understandings.

Example

When exploring the distribution of wealth in the world, typically the PSTs enter the classroom and each selects, at random, a slip of folded paper from a jar labeled "I was born in..." (Gutstein & Peterson, 1998). The slip of paper identifies their birthplace from one of six regions: Africa, Asia, Europe, Latin America, North America, and Oceania (Gutstein & Peterson, 1998). The number of slips of paper for each region is based on the population proportion for each region that is calculated by the instructor before class. The PSTs sit at a table labeled with their birthplace. A large map is located at the front of the room, and each group locates their birthplace. They use their laptops to learn more about their regions and share that information with the rest of the class.

Next, each PST is given a chart that shows two sets of information: population and wealth by region. The chart shows the population in numbers for each region and the wealth in dollars for each region, including the total for the world (Gutstein & Peterson, 1998). PSTs are asked to find the percentage of the population and the percentage of the wealth for their location. On the front board, there is a chart that identifies each region. One member from each region fills in the population percentage and the wealth percentage for their region. The PSTs are asked to compare and critique these percentages. The PSTs discuss "fairness." Some PSTs believe that wealth distribution should be consistent with the population percentage. Some PSTs believe otherwise. In other course sections, sometimes we have distributed cookies proportionally to wealth distribution to drive the point home. We then turn our attention to the mathematics, walking around the classroom and observing the PSTs calculating the percentages. We notice some PSTs struggle to identify the appropriate numbers to use. Most of the PSTs use a procedure they remember from elementary school for converting fractions into percents via decimals. We ask probing questions from these PSTs about their conceptual understanding of the mathematics.

Overall, the lesson is perceived as a social issue lesson addressing world's wealth distribution, more so than a mathematics lesson. The PSTs generally enjoy the activities as well as learning about this issue. However, the lesson places less attention on learning mathematics for deep understanding than we would prefer, partly due to limited time. Although it would have been possible to do much richer lessons on percentages from a pedagogical standpoint, such an instructional approach would have likely come across to our PSTs as a detour from the focus on the social issue being addressed. Meaningfully addressing *both* the content and the application simultaneously in our lessons proved difficult for many lessons in the course.

OUR REFLECTION

When we teach mathematics within the context of social issues, there is tension between the mathematics content and the issues. Oftentimes, the PSTs walk away scratching their heads, wondering, "Was this a math class?" Although we have had many successes in the course, we still struggle to find or to create rich mathematics activities that address the social issues head on for all Grades K–8 content strands. It remains a challenge to design curricula that meet all of our mathematical learning objectives across the content strands without trivializing the social issues, scattering the course's focus, or undermining the mathematics experience.

Consequently, we have found that some social issues align more easily with certain content strands rather than others. As a result, maintaining smooth course transitions from content strand to content strand, or from social issue to social issue, poses challenges. In the end, some PSTs lose sight of the purpose of the course because of the tensions that exist between the mathematics and the social issues.

Please note that we did include many mathematics activities in the class that more closely and more effectively tied the mathematics content to the social issues than the illustrations we provided above. Interestingly, in those cases, the PSTs generally perceived the goal to be to teach the social issue. They frequently questioned the purpose and appropriateness of teaching this way in the mathematics classroom. Specifically, they anticipated mathematics to be "neutral" (Gutstein & Peterson, 2013, p. 6), and in using the mathematics to analyze current issues and events, this colored their view of the classroom goal to be more politicized. Our success at making lessons interdisciplinary in nature, such as by looking at maps while also doing mathematics, was relatively strong. However, our success at weaving rich mathematics with meaningful social issues was less consistent. We continue to revise the course to build cohesion.

NOTE

This paper's coauthor, Laura Jacobsen, was formerly Laura Spielman.

REFERENCES

Beckmann, S. (2011). *Mathematics for elementary teachers, with activities* (4th ed.). Boston, MA: Addison Wesley.

Discovery. [Jorge Ribas]. (2009, May 4). *Poison Dart Frog* [video file]. Retrieved from https://www.youtube.com/watch?v=bqBc80FiJB4

Frankenstein, M. (2009). Developing a critical mathematical numeracy through real real-life word problems. In L. Verschaffel, B. Greer, W. Van Dooren, & S. Mukhopadhyay (Eds.), *Words and worlds: Modelling verbal descriptions of situations* (pp. 111–130). Boston, MA: Sense.

Gutstein, E. (2006). *Reading and writing the world with mathematics: Toward a pedagogy for social justice.* New York, NY: Routledge.

Gutstein, E., & Peterson, B. (Eds.). (1998). *Rethinking mathematics: Teaching social justice by the numbers.* Milwaukee, WI: Rethinking Schools.

Gutstein, E., & Peterson, B. (2013). *Rethinking mathematics: Teaching social justice by the numbers* (2nd ed.). Milwaukee, WI: Rethinking Schools.

National Council of Teachers of Mathematics. (2000). *Principles and standards for school mathematics.* Reston, VA: Author.

Spielman, L. J. (2008). Equity in mathematics education: Unions and intersections of feminist and social justice literature. *ZDM—The International Journal on Mathematics Education, 40,* 647–657.

COMMENTARY 1

EMBRACING TENSIONS

A Commentary
to Mistele and Jacobsen's Case

Jessica Pierson Bishop
Texas State University

In the case that Mistele and Jacobsen present, they discuss the dilemmas they encountered when designing and implementing a new mathematics content course, *Math for Social Analysis*. This content course was a capstone course developed for prospective elementary- and middle-school teachers with a primary goal of using mathematics to explore serious social issues. Although my personal context is different than that of the authors, the issue of balancing competing demands when designing formal courses and other kinds of learning opportunities is a challenge I also face in my work. My perspective on this case is informed by my background as a high school mathematics teacher, as a mathematics educator in a college of education, and my personal work with teachers and students in elementary- and middle-grades classrooms.

Cases for Mathematics Teacher Educators, pages 275–279
Copyright © 2016 by Information Age Publishing
All rights of reproduction in any form reserved.

INTERPRETING THE DILEMMA

Broadly speaking, the issue the authors raised is one that all teachers face. Deciding which topics and big ideas to foreground and background and at what times is an issue encountered not only in mathematics content courses but also in methods courses, professional-development contexts, and K–12 classrooms. In their case, Mistele and Jacobsen consider the complementary and, at times, competing goals of addressing mathematics content through the use of realistic situations that problematize and increase awareness of equities, inequities, and broader societal questions. They give us their personal perspective on course design and the inherent trade-offs involved by sharing some of the difficult decisions they made and the challenges that emerged from those decisions. For them, the dilemma lay in minimizing either the mathematics or the social issues.

In order to understand the dilemma Mistele and Jacobsen present, we first need to consider the context and conditions that gave rise to the dilemma. My perspective is that the challenges they faced were directly related to several of the design features of their course; however, these design features were deliberately chosen based on the goals they had. First and foremost, the course was meant to investigate and critique timely social issues by using meaningful mathematics as the tool for those investigations. In addition to the focus on social issues, another goal of the course was to deepen prospective teachers' mathematical knowledge for teaching. And, finally, the course was designed to help prospective teachers learn how to develop activities that "link mathematics to the real world as experienced by the diverse population of students" (p. 269) so they would come to value and, hopefully, implement these kinds of lessons in their future classrooms. Based on these goals, one of their design features was that all mathematics learning outcomes would be embedded in a social issue. In other words, mathematics was not done for the sake of mathematics. Second, the course curriculum was purposefully selected to ensure that each of the National Council of Teachers of Mathematics's (NCTM's) five content strands was addressed. Comprehensive content coverage was a critical design feature driven by the fact that this course would be the only exposure for many of their students to mathematics content for teaching from a mathematics-education standpoint.

I contend that these two design decisions—embedding all mathematics in social issues and comprehensive content coverage—had a direct bearing on the challenges that emerged for the authors. Their primary challenge was designing a curriculum and corresponding mathematics activities that addressed substantive mathematics in service of investigating social issues in meaningful and developmentally appropriate ways. In particular, the design of the course around a comprehensive set of mathematics learning outcomes presented challenges in terms of identifying appropriate social issues that could feasibly be addressed with the given mathematics. Because of the desire for coverage of all five content strands, the social issues the mathematics was situated within were, at times, contrived and superficial. At other times, when the social issues were foregrounded, the mathematics content addressed was either minimal or insufficient to adequately investigate the social issue. Although the authors allude to successes they had in integrating these two goals, they focus their case on problematic instances that arose when they did not successfully blend the mathematics and social issues.

RESPONDING TO THE DILEMMA

In considering how I might have responded to this challenge, I reflected on the importance of the course goals that drove the design as well as the context in which it was taught. Although I understand and appreciate Mistele and Jacobsen's design principles of comprehensive content coverage and situating mathematics within social issues, I believe these course features acted, at times, as unnecessary constraints on the instructors' decision making and course activities. Consider the poison-dart-frog lesson, which they share as an illustration of a rich mathematical activity that does not provide adequate mathematical tools to engage deeply with the larger social issue of deforestation and habitat destruction. Although this activity served as a realistic context for investigating mathematical topics related to measurement, the mathematical skills and knowledge developed did not help students to grapple with the underlying ethical, economic, and environmental arguments surrounding logging practices, profit margins, protecting endangered species, and habitat conservation. Mistele and Jacobsen characterized this example as problematic. But I did not necessarily see it as such. I wondered if all of the mathematics in this course needed to be situated in a social issue in order to still meet their broader course goal of using mathematics as a tool to investigate and critique timely social issues. Relaxing this design feature so that *most or much* (but not *all*) of the mathematics is situated in social issues would, in my opinion, still meet the goals of the course but alleviate the problem of trivializing some of the social issues. This design feature also raised questions for me regarding the nature of the mathematics presented to the students. I realize that all too often students learn mathematics procedurally without seeing the benefits or usefulness of the subject, and part of the course design was meant to counteract these effects. But isn't it also valuable for students to be exposed to mathematics that is decontextualized and more abstract, especially if it is done in a thoughtful way so that students can experience some of the internal consistency and structures that are part and parcel of mathematics? After all, much mathematics is done without concern for whether it has a real-life or social application.

Alternatively, I would also consider relaxing the constraint of comprehensive content coverage. Because one of the major difficulties was incorporating all mathematics content areas, I might only address content that lends itself more easily to social issues. (The authors note that the content strands of algebra, number and operations, and data analysis are better suited to address social questions.) The broad goal of developing mathematical knowledge for teaching would still be a focal point, but it would not be addressed within all content domains. From my perspective, this change would modify one of the goals of the course, but not substantially; the trade-off is that it would allow the authors to maintain a focus on *substantive* social issues.

Mistele and Jacobsen openly discussed the difficulties, successes, and dilemmas they faced as they navigated the design process for a new course meant to balance mathematics content with social issues. As mentioned earlier, this case raised a broader issue for me—that is, the unavoidable dilemmas, contradictions, and trade-offs involved in teaching. In general, dilemmas arise when equally important but competing instructional goals are in contradiction. In her personal reflection on teaching, Lampert has described this tension: "I felt I could not choose a solution without compromising other goals I wanted to accomplish" (1985, p. 181). Her statement echoes other research that explores the tensions and compromises teachers confront during instruction (Adler, 1999; Sherin, 2002). For example, in Adler's (1999) study, the challenge was not in balancing mathematics and social issues but in balancing mathematics content and explicit instruction on how to use the language of mathematics; she described this as the "dilemma of transparency," which is particularly salient for teachers in multilingual classroom contexts. Similarly, Sherin (2002) highlighted the "process-content tension" a secondary teacher faced as he worked to develop a mathematical discourse community. The dilemma for him was how to support student-centered discourse processes where student ideas were the basis for discussions while, at the same time, ensuring discussions of significant math content. In closing, Mistele and Jacobsen have reminded us of the importance of thoughtfully considering the paradoxes that arise in the work of teaching; there may be no "right" response to these dilemmas. Sometimes, as Lampert (1985) advised, the best decision is to embrace the tensions and contradictions without trying to resolve them.

REFERENCES

Adler, J. (1999). The dilemma of transparency: Seeing and seeing through talk in the mathematics classroom. *Journal for Research in Mathematics Education, 30*, 47–64.

Lampert, M. (1985). How do teachers manage to teach? Perspectives on problems in practice. *Harvard Educational Review, 55,* 178–194.

Sherin, M. G. (2002). A balancing act: Developing a discourse community in a mathematics classroom. *Journal of Mathematics Teacher Education, 5,* 205–233.

COMMENTARY 2

LESS IS MORE

A Commentary
on Mistele and Jacobsen's Case

Anthony Fernandes
University of North Carolina at Charlotte

I was born and raised in India and moved to the United States to attend graduate school. My beliefs about mathematics were shaped through my experiences as a student in India. The teacher would introduce a topic, work out a few examples, and assign challenging homework problems. Though the homework problems pushed us beyond the examples that we had done in class, they were not completely beyond reach. I enjoyed the challenge and engaged in many interesting discussions around elegant solutions with my peers. My graduate experience at the University of Arizona was also similar to my school experiences. I thought that I had a clear understanding of the subject and the ways one could be successful at it.

My first introduction to mathematics for social justice was at a summer school course organized by the National Science Foundation-funded *Center for the Mathematics Education of Latinos/as* (CEMELA). Rico Gutstein, the instructor for the course, introduced us to the ideas of social justice based

Cases for Mathematics Teacher Educators, pages 281–284

on the readings of Paulo Freire (1970/1998). Rico discussed how he integrated Freire's ideas in designing and teaching mathematics classes in a Chicago school that served mostly minority students. The mathematics was developed around social issues that were raised by his students and impacted their lives (Gutstein, 2006). After the class I was not convinced that these social issues, which were controversial in nature, should be discussed in the mathematics class because it took away from the "real mathematics." I remember engaging in heated discussions with other CEMELA fellows as they tried to convince me of the merits of this approach that prepared students mathematically and at the same time developed their critical consciousness. Over time I thought more about the ideas and gradually came to understand the importance of teaching mathematics for social justice and the way this approach empowered and engaged students who felt that they were marginalized by the system. It took me a while to understand and acknowledge my own biases about the nature and purpose of mathematics and to appreciate other points of views. Later I invited Rico to share his ideas with my colleagues and the broader community at my institution, the University of North Carolina, Charlotte. Much like my own experience, Rico's visit generated a debate about the appropriateness of incorporating mathematics for social justice in the classroom.

INTERPRETING THE DILEMMA

The authors of this case bring up a dilemma they had about maintaining a focus on both the mathematics content and social issues in their course *Math for Social Analysis*. The goals of the course were for the preservice teachers to (1) "gain a deeper and more flexible understanding of the mathematics" and also (2) "use mathematics to critique important social issues." The authors found it challenging to implement these goals; this is similar to the experience that other mathematics education colleagues have shared with me. Given that this was a content course, the authors felt that it was necessary to cover certain topics. The authors attempted to develop the preservice teachers' social awareness throughout the course by aligning every content topic with a social issue (see Table 12.1 in Mistele & Jacobsen, this volume), even if these connections appeared to be contrived in some topics. In this process they found that in some lessons the preservice teachers focused on the mathematics content with little or no awareness of the social issue that was mentioned in the problem. In other lessons the preservice teachers focused on the social issue at hand with only minimal focus on the associated mathematical concepts. The dilemma discussed in this case brings to mind some themes that would guide my thinking if I were teaching a similar course to my preservice teachers. I discuss these themes below.

RESPONDING TO THE DILEMMA

Less Is More

Balancing both the content and the social issues in every lesson in a single course can be challenging. From my point of view, the eventual goal of teacher preparation is to develop teachers who can think critically about social issues through a mathematical lens. With this is mind, I would keep the major focus on the mathematics and blend the social issues in some lessons where they are well integrated with the content. By promoting a stronger mathematical foundation, the preservice teachers will be better positioned to realize connections between mathematical concepts and the social issues when they are discussed. Further, developing automaticity with the mathematics will allow the preservice teachers to focus on the social issues and use the mathematics they have learned to analyze the issues in depth. Along with careful integration of social issues within a course, I would spread the integration of these issues over the duration of the teacher preparation. This brings me to my next point of time.

Time

The reaction of the preservice teachers in the course was very similar to the reaction I had after being introduced to the ideas of mathematics for social justice—"They frequently questioned the purpose and appropriateness of teaching this way in the mathematics classroom." Instead of concentrating all the social issues into one course, I would seek to infuse well-integrated social issues in all the mathematics content courses in the program. Thus the preservice teachers would get regular opportunities to revisit these connections and gradually internalize them over time.

Managing Reactions in the Classroom

Discussing social issues in the class is bound to generate strong reactions from preservice teachers. For example, in this case, the preservice teachers thought that social issues were not appropriate to discuss with younger children. As these issues arise in the class, I would have a list of examples that could convince the preservice teachers that even younger children have the capability of understanding complex ideas when these ideas are tied to their own experiences. For example, a parallel context to the wealth-distribution activity in this case could be designed through the unequal sharing of cookies with young children (Gutstein & Peterson, 2013). As the instructor, I would also pay careful attention to the mood of the class. Sometimes preservice teachers could perceive that the instructor is attempting to push her or his political views on them. Managing these tensions carefully will determine if the classroom discussions remain productive.

Using mathematics to examine social issues is important for preservice teachers as they prepare to work with diverse students in their future classes. In addition to developing a critical consciousness, it is important that the school students also develop a deeper understanding of the mathematics in the process. Though challenging, continuous effort towards balancing the focus on the mathematics content and social issues throughout teacher preparation can help preservice teachers acquire the knowledge and beliefs needed to apply these ideas in their future teaching.

REFERENCES

Freire, P. (1998). *Pedagogy of the oppressed* (M. B. Ramos, Trans.). New York, NY: Continuum. (Original work published 1970)

Gutstein, E. (2006). *Reading and writing the world with mathematics: Toward a pedagogy for social justice.* New York, NY: Taylor Francis.

Gutstein, E., & Peterson, B. (2013). *Rethinking mathematics: Teaching social justice by the numbers.* Milwaukee, WI: Rethinking Schools Ltd.

COMMENTARY 3

RESPONDING TO STUDENTS' NEEDS

A Commentary on Mistele and Jacobsen's Case

William Zahner
San Diego State University

I am a mathematics educator at San Diego State University (SDSU). My primary responsibilities include conducting research and teaching. In my research I study how *all* students, and particularly English Learners, can learn important algebraic concepts through classroom discussions. This research focus reflects a personal commitment to promoting equity in and through mathematics. In fact, I was inspired to enter the field of mathematics education after reading *Radical Equations* (Moses & Cobb, 2001), a book that powerfully makes the case that broadening access to mathematics is a social justice and civil-rights issue. My main teaching responsibilities include teaching courses for preservice secondary teachers. One of my courses focuses on the development of algebra and algebraic reasoning in secondary (Grades 7–12) mathematics.

Cases for Mathematics Teacher Educators, pages 285–288
Copyright © 2016 by Information Age Publishing
285

My students at SDSU are relatively diverse: There is no single racial or ethnic majority among SDSU students, and the students come from a wide spectrum of economic backgrounds. Some of my students are "traditional" college students who attend college full time and who live in dormitories. Others, including some students taking a full-time academic load, work to support themselves and their families while taking classes. On the basis of our common experience of teaching prospective teachers, I could identify with the tensions that Mistele and Jacobson describe in their reflection on teaching Mathematics for Social Analysis.

INTERPRETING THE DILEMMA

Mistele and Jacobsen describe two major dilemmas arising from teaching a mathematics course for prospective elementary teachers focused on applying school mathematical concepts for social analysis. First, some of Mistele and Jacobsen's prospective students have resisted or even rejected the notion that mathematics teachers at the elementary-school level should use mathematics to engage in social analysis. This objection seems to follow from the common idea that mathematics is about abstract things (e.g., numbers and shapes). Although mathematics can be applied to analyze real-world situations, learning through applications may not match the prospective elementary teachers' expectations for a mathematics course. (Compounding this issue is the fact that some of Mistele and Jacobsen's prospective teachers take the prerequisite courses at community colleges, so there is likely significant diversity in their prior mathematical experiences at the college level.) Following this line of reasoning, the prospective teachers in Mistele and Jacobsen's class also voiced that they did not think they would use social analysis in their future teaching of mathematics, and they resisted learning mathematics in a way that they would not teach.

A second dilemma that Mistele and Jacobsen identify is the fact that teaching mathematics through social analysis presents a design challenge: developing tasks that build understanding of *both* important mathematical ideas as well as social issues. Meeting this design challenge is difficult, to say the least. Mistele and Jacobson write that they were left with the feeling that either the mathematics or the social analysis was given short shrift in many of their activities. They illustrate this design challenge using the example of a measurement lesson. The social issue was about conservation and deforestation. The lesson started with a discussion of deforestation and the destruction of a threatened frog's habitat. The discussion of the frog then led to a hands-on activity in which the prospective teachers constructed a toy frog from an index card and made the frog "jump." The measurement portion of the activity involved measuring how far the frog jumped using nonstandard and standard units. Despite high levels of engagement and satisfaction from the prospective teachers, Mistele and Jacobsen describe their dissatisfaction with this lesson because the students did not seem to engage with the social context of the problem in an in-depth way and instead focused on how fun the jumping-frog activity was.

RESPONDING TO THE DILEMMA

This case study of tensions in teaching for social justice provides an opportunity for teacher educators to pause and reflect on the assumptions we bring to our work. Although it may be tempting to dismiss the students' objections to the social-analysis tasks, I believe this would be a mistake. First, the teachers are expressing a commonly held idea that mathematics is abstract. Behind this sentiment might be the belief that students need to learn the basics before trying to apply their knowledge. Although I (and I presume Mistele and Jacobsen) disagree with this common belief, we ignore it at our own peril. So how do we address this objection in a way that respects our students?

First, I think we can use state standards to bolster the argument that school mathematics *ought* to include applied problems. Most mathematics teachers in the U.S. are currently aligning their instruction to the Common Core State Standards for Mathematics ([CCSSM] National Governors Association Center for Best Practices & Council of Chief State School Officers, 2010). In non-CCSSM states, mathematics teachers must be responsive to their state standards. Both the practice standards in the CCSSM and (for Mistele and Jacobson in a non-CCSSM state) the five core "goals for students" in the Virginia Standards of Learning (Board of Education, Commonwealth of Virginia, 2009) highlight modeling and application as central to learning mathematics. Thus, one way to engage prospective teachers in a dialogue around the appropriateness of learning mathematics through applications is to highlight how this is precisely what the authors of standards documents expect teachers to do. Although it is unlikely that citing the standards once will convince teachers to change a deeply held belief, the standards can be used to frame the instructors' decisions as responsive to the demands of the broader field and not simply an individual choice.

The second tension—balancing the depth of social analysis and mathematics learning for each task—presents a formidable challenge. Designing good and engaging instructional tasks is difficult. It is even more difficult to do this with applied modeling tasks, where either the problem can be trivialized (as often happens in traditional word problems) or the complexity can spin out of control. Given that there is a finite amount of time in a semester, one way I might approach this challenge is through focusing on fewer social issues throughout the unit and returning to different aspects of the same social issue in different units. For example, social issues related to race and racism could be examined through multiple mathematical lenses. Racial profiling in police stops might be analyzed using data analysis and probability, with connection to number and operations. The topic could then be revisited using concepts from geometry (e.g., analyzing patterns of residential segregation using maps and census data). Such thematic development would decrease the overhead of introducing a new social issue every 2 weeks.

In reading the case and imagining my own students' reactions, I had the sense that implicit in the teachers' critique of using mathematics for social analysis was the sense that the social issues in the curriculum do not necessarily apply to them. Therefore, one way to resolve the design challenge of relevance might be to pick topics of high interest and relevance to the prospective teachers who are students in Math for Social Analysis. Perhaps the best way to do this would be to learn what issues are most relevant to undergraduates who are preparing to be elementary teachers. In my local context, I know that finances and loans are a major issue for my students. Therefore, if I were designing a unit in this class, I might focus on the mathematics of loans and the social analysis around the student-loan crisis and the affordability of higher education. I suspect this focus would lead to engagement among a wide swath of current college students!

In sum, one of the main insights I take from reflecting on Mistele and Jacobsen's case is the importance of responding to the needs of students. The goals of Math for Social Analysis are important, and the goals are clearly aligned with the relevant standards. However, motivating the need to learn mathematics this way is difficult and can degenerate into activities where the mathematics is trivial. Thus, following the deeply held interests and needs of students in an authentic way may be a more effective way for motivating the deep connection between mathematics and social analysis. In this thought I am inspired by Freire's (1970) notion of building curriculum around the questions that students bring and responding to students' reality in the development of curriculum. This may be more work, but ultimately, it will likely result in a much richer experience of both mathematics and social analysis.

REFERENCES

Board of Education, Commonwealth of Virginia. (2009). *Mathematics standards of learning for Virginia public schools.* Retrieved from http://www.doe.virginia. gov/testing/sol/standards_docs/mathematics/2009/stds_math.pdf

Freire, P. (1970). *Pedagogy of the oppressed.* New York, NY: Continuum.

Moses, R., & Cobb, C. (2001). *Radical equations.* Boston, MA: Beacon Press.

National Governors Association Center for Best Practices & Council of Chief State School Officers. (2010). *Common core state standards for mathematics.* Washington, DC: Authors.

CHAPTER 13

NOT CALLED TO ACTION (OR CALLED UPON TO ACT)

Can Social Justice Contexts Have a Lasting Impact on Preservice Teachers?

Ksenija Simic-Muller
Pacific Lutheran University

INTRODUCING THE CASE

Mathematics educators inspired by the work of Paolo Freire (1970) have argued that mathematics is a powerful tool for reading and writing the world (Frankenstein, 2009; Gutstein, 2006) and see it as a humanistic endeavor that holds great potential not only for helping understanding the world, but also for changing it. According to Frankenstein, "The main goal of criticalmathematical [sic] literacy is not to understand mathematical concepts better, although that is needed to achieve the goal. Rather it is to understand how to use mathematical ideas in struggles to make the world better" (p. 112). Similarly, Gutstein and Peterson (2013) called on teachers to "position teaching and learning mathematics in the service of humanity and nature" (p. xii). Frequently called *teaching mathematics for social justice*

Cases for Mathematics Teacher Educators, pages 289–296
Copyright © 2016 by Information Age Publishing
All rights of reproduction in any form reserved.

, 2006), this approach to teaching has been used in mathematics rooms at all levels: from K–12 classrooms and afterschool programs .g., Gutstein, 2006; Turner, Varley Gutiérrez, Simic-Muller, & Díez-Palomar, 2009) and adult education and community-college courses (e.g., Frankenstein, 2009), to mathematics methods (e.g., Koestler, 2012; Rodriguez & Kitchen, 2005) and content courses for preservice teachers (Felton, Simic-Muller, & Menéndez, 2012; Mistele & Spielman, 2009). In particular, topics covered in mathematics content courses for preservice K–8 teachers, such as proportional reasoning and statistics, lend themselves to exploring a variety of social justice topics.

Gutstein (2006) described three types of mathematical knowledge: classical, critical, and community. Classical mathematical knowledge is what is typically meant by "knowing mathematics" and is often valued above others in mathematics classrooms; community knowledge is the knowledge that students bring to the classroom from their homes, communities, and cultures; critical mathematical knowledge is what Gutstein referred to as "reading the world with mathematics," and is the ability to use mathematics to understand the sociopolitical circumstances that affect one's life and the world at large. In social justice pedagogy, the three knowledge bases are interconnected and valued equally (Gutstein, 2006). In contrast, mathematics content courses for preservice teachers typically focus on developing classical knowledge while neglecting the other two knowledge bases. Many, if not most future teachers will teach in diverse communities, urban neighborhoods, and underfunded schools to English Language Learners and students whose low test scores are too often attributed to faults in their personalities, families, or cultures. For this reason, preservice teachers need community and critical awareness to understand the institutional and systemic forces that shape the experiences of their future students. They need to read the world with mathematics, but in order to be advocates for their students, preservice teachers need to understand that they have agency: They need to write the world with mathematics as well.

CONTEXT

I teach in a mathematics department of a medium-sized, private, liberal-arts university with a strong commitment to social justice. The university offers two mathematics content courses for preservice K–8 teachers, housed in the mathematics department. The first course focuses on number and algebraic sense, and the second on geometry, measurement, probability, and statistics. One section of each course is offered every semester, and I typically teach both. There are between 15 and 25 students enrolled in each section, primarily White females in their late teens or early twenties interested in teaching lower elementary grades, who come from the region but typically from wealthier backgrounds rather than the low-income suburb that surrounds the campus.

As the primary instructor for the courses, I have much freedom in choosing the curriculum. The main goal of the courses is to deepen the preservice teachers' understanding of the K–8 mathematics curriculum—to develop their classical mathematical knowledge—and this goal determines the largest part of the content. The courses are built around problem solving and discovery, with an emphasis on working with manipulatives and in groups. In addition to fostering the development of classical mathematical knowledge, I am committed to exposing preservice teachers to ideas of social justice and strengthening primarily their critical but also community knowledge bases. Over the years that I have taught at my current institution, I have developed problems, lessons, and assignments about a variety of issues: I have posed short problems, for example about the number of plastic bottles that end up in landfills; created short lessons, for example about the racial and ethnic makeup of the U.S. Congress; and created longer assignments about topics such as sweatshop labor and homelessness. The contexts are relevant to me as a human and citizen, not just educator, but I also continuously make an effort to allow the preservice teachers to draw conclusions about the issues rather than telling them what to think. At the same time, I agree with Gutstein and Peterson (2013) in the assessment that no teaching is neutral and am aware that my choice of topics, problems, and discussion questions positions me ideologically.

My goals in incorporating social justice contexts into the curriculum are the following:

1. Help preservice teachers develop an appreciation for real-world mathematics problems. My students enter mathematics courses with apprehension, dislike "story problems," and are often unable to give examples of real-world contexts relevant to K–8 mathematics. At the same time, the Common Core State Standards in Mathematics (Common Core State Standards Initiative, 2010) place an increased emphasis on real-world problem solving, which is especially visible in the modeling standard for mathematical practice. Some form of modeling should be seen at every grade level, and future teachers need to experience it in their own learning.

2. Help preservice teachers read the world with mathematics. The particular type of contexts that I propose, social justice contexts, are relevant to preservice teachers because, as primarily White, middle-class females who grew up in monocultural suburban areas, many of my students have had little exposure to issues that their future students likely face. By engaging with mathematics that helps

them read the world, preservice teachers gain knowledge of the experiences of others.

3. Offer teaching mathematics for social justice as a model for the preservice teachers to use in their future teaching. For teachers to be able to use social justice contexts in their teaching, they have to experience them as learners first.

4. Help preservice teachers see themselves as global citizens who have a responsibility to society. Learning about social justice issues helps preservice teachers see their place in the world, encouraging them to act responsibly and ethically.

For the dilemma I will describe here, the fourth goal is the most relevant.

THE DILEMMA

On the basis of feedback from their reflections, preservice teachers usually react positively to the social justice contexts in my courses, as the following relatively typical quote shows:

> I believe that it is important for social justice topics to be incorporated into any college level class, especially in college level math classes. It is important for students to be informed about real world issues and ways to incorporate understanding of those issues into our own classrooms. Mathematics is one way to use real world problems in the classroom, and I believe that it is appropriate to discuss and incorporate these issues into a college math class.

Though informative and encouraging, this feedback comes from graded assignments and therefore may not be completely honest. In addition to reflections, course evaluations have been helpful in gauging the preservice teachers' attitudes towards the social justice emphasis. In my first few years of teaching, evaluations occasionally included accusations that I was pushing my political agenda; though rare, these comments were fierce and showed obvious disagreement with this approach to teaching. In recent years I have made an explicit connection between the mission of the university and social justice work in my classroom, and strongly negative reactions have mostly disappeared. Each semester, two to three preservice teachers comment on the presence of social justice contexts in the course, stating that my commitment to social justice is inspiring and that it helps them learn mathematics. Most neither criticize nor praise, and I wonder if to them the social justice curriculum is just another requirement for passing the class.

With decreased resistance to the social justice contexts, and high quality of the mathematical work that preservice teachers produce, it may seem that my approach to teaching mathematics for social justice is successful.

However, if one of the goals for teaching mathematics for social justice is to create responsible global citizens or, in the words of Frankenstein (2009), to learn "how to use mathematical ideas in struggles to make the world better" (p. 112), then completing the assignments is only the first step: The learning that takes place in the assignments should be transformative and should inspire some kind of action or, what Gutstein (2006) called writing the world with mathematics. This is relevant not just from the point of view of social justice pedagogy, but also for the preservice teachers' future teaching: If they are not moved by injustice as learners, it seems unlikely they will be compelled to address it as teachers. In my teaching, I have not seen preservice teachers called to action. They are willing to engage with mathematics but are either not affected by the social justice issue or unsure how to react to it. I have observed this in many instances but will only share one example here.

Example: Sweatshops

College students across the nation have taken bold and often successful steps to eliminate sweatshop-made apparel from their campuses. Assuming the topic is generally relevant to college students, in addition to it being personally meaningful to me, I have created an assignment about sweatshops—in particular, about soccer-ball production in Pakistan and toy production in China. This is a longer homework assignment that consists of multiple parts and contains some more sophisticated uses of proportional reasoning and percentages. It shows that sweatshop wages cannot provide a living to the laborers and that their working conditions are inhumane. The assignment ends with a reflection paragraph: Preservice teachers write an argument for or against sweatshops using mathematics. The reports give emotional and rational arguments against sweatshops, with statements such as, "it's hard to see how anyone could agree with the sweatshop working conditions," "this is completely unfair and something has to be changed," and "we need to end sweatshops now." The mathematical arguments are also strong, and preservice teachers are able to use the power of simple mathematics to argue against this inhumane practice. However, after professing shock and dismay over the state of affairs, they do not feel moved to action: When I ask if this knowledge will change their own buying practices, they shake their heads and give a negative reply. Not a single one says she will at least consider what she buys and where; a few have even stated there is nothing they can do as individuals to make a difference. Can it be that the outrage expressed in the reports is there because my students know I expect it? In their reflection paragraphs, most call for the immediate abolishment of all sweatshops but never ponder small actions they can personally take or express interest in learning more about the issue. I do not expect that they will become involved in antisweatshop activism or quit buying clothes altogether: A possible action could be as simple as investigating labor practices of their favorite brands.

It is possible that sweatshops do not move preservice teachers, but the situation is similar with other topics. When we discussed youth homelessness and a youth advocate for a nearby school district spoke to the class, despite the similarly expressed outrage, only two students approached me afterwards to discuss the issue and ask to get involved.

As one of my goals in in teaching mathematics for social justice is to encourage preservice teachers to act on issues they consider unjust, this situation raises a number of questions for me: How can I structure the class so that students embrace the focus on equity and social justice instead of just doing the work assigned to them? How do I help preservice teachers build sociopolitical consciousness so they will be called to action, not necessarily in the context of sweatshops, but any topic that speaks to them? Perhaps the most important question, and the heart of the dilemma, is this: *If the preservice teachers are not moved to action by my assignments, what, if any, is my role in encouraging them to take action?*

I do not explicitly include calls to action in my assignments. I am reluctant to do so, feeling that this is where the boundary between academics and activism is crossed. I fear alienating students and colleagues, and jeopardizing my professionalism. If I require that actions be taken, am I overstepping my rights and responsibilities as instructor by placing such demands?

REFLECTION

Thus far, I have not directly addressed the issue I have presented, and instead continue to create social-justice lessons that result in some student interest but no action. I realize that my expectation that preservice teachers will instinctively be called to action is an unrealistic one, and that learning about injustice is often overwhelming for those being exposed to it for the first time. I can take certain actions to ensure that the lesson outcomes are more aligned with my expectations, but first I have to resolve the central dilemma about my role as instructor. For example, if I take on more of an activist role, I can do one or more of the following:

1. In each assignment require that an action be taken, no matter how small.
2. Provide scaffolding in early assignments by offering a list of possible actions to take.

3. Allow for agency in the assignment by requiring preservice teachers to research ways to take action.

If I choose to maintain a purely academic role, but with the hope that the students themselves will see the need to act under the right circumstances, I can make the following adjustments:

4. Switch contexts to a topic that preservice teachers feel passionate about, such as education. Topics related to social justice and education abound and may motivate preservice teachers to take action.
5. Allow preservice teachers to choose their own social justice contexts, ones that are personally relevant to them.

These action steps also create more questions and dilemmas, such as Are preservice teachers really more likely to act if they engage with lessons related to their interests? If students choose contexts for the lessons, will the focus on equity be lost? And if I require actions to be taken, how authentic and influential will they be?

REFERENCES

Common Core State Standards Initiative. (2010). *Common core state standards for mathematics.* Washington, DC: National Governors Association Center for Best Practices & Council of Chief State School Officers.

Felton, M. D., Simic-Muller, K., & Menéndez, J. M. (2012). "Math isn't just numbers or algorithms": Mathematics for social justice in preservice K-8 content courses. In L. J. Jacobsen, J. Mistele, & B. Sriraman (Eds.), *Mathematics teacher education in the public interest: Equity and social justice* (pp. 231–252). Charlotte, NC: Information Age.

Frankenstein, M. (2009). Developing a critical mathematical numeracy through real real-life word problems. In L. Verschaffel, B. Greer, W. Van Dooren, & S. Mukhopadhyay (Eds.), *Words and worlds: Modeling verbal descriptions of situations* (pp. 111–130). Rotterdam, The Netherlands: Sense.

Freire, P. (1970). *Pedagogy of the oppressed.* New York, NY: Herder and Herder.

Gutstein, E. (2006). *Reading and writing the world with mathematics: Toward a pedagogy for social justice.* New York, NY: Routledge.

Gutstein, E., & Peterson, B. (Eds.). (2013). *Rethinking mathematics. Teaching social justice by the numbers* (2nd ed.). Milwaukee, WI: Rethinking Schools.

Koestler, C. (2012). Beyond apples, puppy dogs, and ice cream: Preparing teachers to teach mathematics for equity and social justice. In A. A. Wager & D. W. Stinson (Eds.), *Teaching mathematics for social justice: Conversations with educators* (pp. 81–97). Reston, VA: National Council of Teachers of Mathematics.

Mistele, J., & Spielman, L. J. (2009). The impact of "math for social analysis" on mathematics anxiety in elementary preservice teachers. In S. L. Swars, D. W.

Stinson, & S. Lemons-Smith (Eds.), *Proceedings of the 31st annual meeting of the North American Chapter of the International Group for the Psychology of Mathematics Education* (Vol. 5, pp. 483–487). Atlanta, GA: Georgia State University.

Rodriguez, A. J., & Kitchen, R. S. (Eds.). (2005). *Preparing mathematics and science teachers for diverse classrooms: Promising strategies for transformative pedagogy.* Mahwah, NJ: Erlbaum.

Turner, E. E., Varley Gutiérrez, M., Simic-Muller, K., & Díez-Palomar, J. (2009). "Everything is math in the whole world": Integrating critical and community knowledge in authentic mathematical investigations with elementary Latina/o students. *Mathematical Thinking and Learning, 11,* 136–157.

BECOMING POLITICAL IN MATHEMATICS EDUCATION CLASS

A Commentary on Simic-Muller's Case

Eric (Rico) Gutstein
University of Illinois at Chicago

INTERPRETING THE DILEMMA

I really appreciate the issues and dilemmas Simic-Muller raises in this thoughtful, reflective piece. The key questions I hear her pose are: How do people become political actors in the world, and what are possible contributions of (mathematics) education towards that end? As a fellow teacher educator, these issues are familiar to me. These questions are central in *teaching mathematics for social justice* (which I refer to as *reading and writing the world with mathematics*, building on Frankenstein, 1983, and Freire, 1994). I have been practicing, studying, and learning how to develop and teach mathematics for social justice for 20 years, as well as trying to support others to learn how. By no means have we—mathematics teachers and teacher educators committed to social justice—figured it all out. And as a White

Cases for Mathematics Teacher Educators, pages 297–300
Copyright © 2016 by Information Age Publishing
All rights of reproduction in any form reserved.

male professor, I consistently strive to be aware of the issues of power, privilege, and race that Simic-Muller brings up.

RESPONDING TO THE DILEMMA

I believe many mathematics teachers and mathematics teacher educators are concerned with the issues Simic-Muller discusses, which I think go beyond teaching mathematics or a university setting. But I feel that we sometimes do not understand how to approach these questions. I will try to explain how I understand these dilemmas. My understanding is informed by my experiences as a teacher and researcher, and my history as a political activist. I also view her case from the perspective of someone who came of age and became politically active in the 1960s and 1970s, started in academia late (because of that activity), and has been doing various forms of political organizing outside of university settings since that time (e.g., I was a cofounder in 1998 of *Teachers for Social Justice* in Chicago and have been much involved in the fight against education privatization in Chicago for a long time).

First, it is often difficult to assess the relationships of what happens in school (at any grade level) and students' political consciousness and activity (in or out of school). You do not often see students quickly getting involved in political actions, and one of my favorite things to tell students is that "it [change] isn't a light switch." People's lives are complicated, and they generally transform slowly, except under dramatic circumstances. Furthermore, even when they appear to change, you cannot easily fully understand why or draw causal links to schooling. But that does not mean that students are unaffected by studying sweatshop conditions (for example), even if they do not immediately swear off Wal-Mart. Freire (1970/1998) spoke to that:

> Students, as they are increasingly posed with problems relating to themselves in the world and with the world, will feel increasingly challenged and obliged to respond to that challenge.... Their response to the challenge evokes new challenges, followed by new understandings; and gradually the students come to regard themselves as committed. (p. 62)

Because experiences often gradually accumulate for (young) people, change is usually slow. I learned long ago to be "patiently impatient" (Freire, 1978) when doing social justice mathematics projects with middle-school students. Some students would not respond after a project and did not demonstrate that what we had done in class had impact on who they were. However, that did not necessarily mean that they were unaffected. Later, in their writings, or after some time (even the next year or much later), these same students often expressed thinking deeply and being moved about the issues. For example, just yesterday (May 4, 2015), I received an email from a student to whom I taught 7th- and 8th-grade (social justice) mathematics (1997–1999). She mentioned learning in class about Freire (*Pedagogy of the Oppressed*) and Howard Zinn (*A People's History of the United States,* 1997), and wrote in the email, among other things, how she still thought about and reflected on the learning from her middle-school math class. The lesson for me, repeatedly, was that we cannot easily know the influences of social justice education.

Second, for people who have been little involved in social movements, what are appropriate ways to begin? Simic-Muller writes that few of her students changed their buying habits after studying sweatshop economics. Clearly it would have been good if they had (though again, we cannot know their future). But when you lack political experience, it can be hard to start. I have found that it helps to be involved myself in social movements and invite students to participate in meaningful and appropriate ways. For example, when I taught middle school in Chicago (Gutstein, 2006), I was part of the struggle against gentrification in the school community. I attended rallies, sit-ins, public hearings, and community meetings, and I invited students (and their families) to become involved; many did. Several students and I attended a city-council hearing, but I had not planned to speak. However, one of my students pushed me to do so and did not let up. We had not come all this way, she said, to be silent. Though she herself, at age 13, did not want to speak up, she made sure that her teacher, with more experience and public confidence, did not evade responsibility. She spoke through me and became involved in a way that worked for her at her age.

More recently, I have been involved in a fight in Chicago to reopen a closing[1] public neighborhood high school in an African American community beset by school closings and education privatization. In fall 2014, I brought a dozen of my preservice students to a press conference and rally at the mayor's office in City Hall. At the event, high-school students and parents spoke movingly about the conditions of their community and school— for example, students are forced to enter the school through the back door and must take both art and gym online because of budget cuts. Then, in full view of my students, 11 people (adults) chained themselves together and to a statue of George Washington just outside the mayor's office, to be arrested in protest of the school being closed. The testimony and committed activism powerfully affected my students, who previously had read about the issue in our class, but as they said, reading about it was one thing, and seeing and hearing it for themselves was totally different. Clearly, not every teacher educator can bring her students to something as dramatic as an event where parents are willing to go to jail for education. But my point is that students need to see their own teachers acting for the things they care about, at whatever level of engagement, and this can serve as a concrete example of how students too can participate.

Finally, we should appreciate that *writing the world* (Freire & Macedo, 1987) means different things for different people at different times. There is not "one way" to do this, anymore than there is one way to solve a mathematics problem. And, like good mathematics problems, social movements have multiple entry points. Besides the fact that activism "isn't a light switch," it can mean everything from getting arrested, to not speaking up yourself but pushing someone else to do it for you, to just *beginning* the process of taking seriously that our buying habits have real impact across the world—and everything in between. In contrast, it would be ahistoric and unrealistic to assign an action for every social-justice mathematics project, although teachers should not miss opportunities when and if they emerge. And although we should push and challenge students to act, as Simic-Muller points out, students cannot be forced into political activity or to take on particular ideological commitments. They, like all of us, need to grow into becoming political actors.

The work of developing *subjects* of history (Freire, 1970/1998) that Simic-Muller takes on in her case is a critical component of fighting for a better world. It is complicated and complex, as her case clarifies, but her reflective stance and thoughtful commitment contributes to our collective understanding. Mathematics education, also, can play a role. That is the essence of the meaning of reading and writing the world with mathematics.

NOTE

1. The Chicago Board of Education voted in 2012 to "phase-out" Walter H. Dyett High School, which was slated to close in June 2015. However, due to sustained community resistance, in December 2014 the Board rescinded its decision—but the school's final fate remains uncertain as of this writing (May 2015).

REFERENCES

Frankenstein, M. (1983). Critical mathematics education: An application of Paulo Freire's epistemology. *Journal of Education, 165,* 315–339.

Freire, P. (1978). *Pedagogy in process: The letters to Guinea-Bissau.* (C. St. John Hunter, Trans.). New York, NY: Continuum.

Freire, P. (1994). *Pedagogy of hope: Reliving* Pedagogy of the Oppressed. (R. R. Barr, Trans.). New York, NY: Continuum.

Freire, P. (1998). *Pedagogy of the oppressed* (M. B. Ramos, Trans.). New York, NY: Continuum. (Original work published 1970)

Freire, P., & Macedo, D. (1987). *Literacy: Reading the word and the world.* Westport, CN: Bergin & Garvey.

Gutstein, E. (2006). *Reading and writing the world with mathematics: Toward a pedagogy for social justice.* New York, NY: Routledge.

Zinn, H. (1997). *A people's history of the United States.* New York, NY: The New Press.

COMMENTARY 2

TEACHING MATHEMATICS FOR SOCIAL JUSTICE AS ENGAGING IN JOINT ACTION WITH STUDENTS

A Commentary on Simic-Muller's Case

Arthur B. Powell
Rutgers University–Newark

I am African American, a native of Brooklyn, New York, a father, and an associate professor of mathematics and mathematics education at Rutgers University-Newark who, at present, researches how to construct mathematical tasks that promote productive mathematical discourse among teachers and among students working in computer-supported, collaborative-learning environments equipped with dynamic mathematics resources. As a social-justice issue, this work aims to increase access for urban students of color to technology and opportunities for learning cognitively demanding mathematics. In general, I believe that *non satis scire*,[1] "to know is not enough," and that social action must accompany knowing.

Cases for Mathematics Teacher Educators, pages 301–305
Copyright © 2016 by Information Age Publishing
All rights of reproduction in any form reserved.

In an instructional act similar to the one discussed in this case, I taught a course on methods of teaching mathematics for prospective teachers of urban elementary schools and engaged my students in social-justice action.

In 1992, the National Security Agency (NSA) produced a video for the Mathematical Association of America (MAA) and other organizations. Before its general distribution, among other mathematics educators, I received the video, "You're Gonna Need Those Numbers." Its audience was to be teenaged African American, Latino, and female students, and its message emphasized the importance of studying mathematics to obtain eventually meaningful employment. The video included interviews with a laboratory technician, a professional athlete, a sports statistician, and an employee of the National Security Agency; computer-generated graphics; as well as a popular-music inspired soundtrack and dancing.

I screened the video with my students, and we were appalled. Stereotyped images abounded. The only woman interviewed had a seemingly low-level job as a lab technician at an aquarium and described her job as needing only "basic" mathematics. The other view of women was as provocatively clothed dancers accompanying a Black male lead singer in a song-and-dance routine. The African Americans depicted had jobs related to the entertainment industry, including the singer-dancer and a professional athlete. In his interview, the athlete insisted that young people should "stay in school" but offered no examples of how mathematics applied to his sport.

After critically analyzing the video, my students found that it contained blatant sexist and racist stereotypes. They decided on a critical collective action to disrupt and discontinue the distribution of the video (Baldassarre, Broccoli, Jusinski, & Powell, 1993). Three of the 21 students and I wrote on behalf of the class its critique to the MAA. We urged and organized other concerned students and educators to do the same, which included participants of the Tenth Annual Socialist Scholars Conference at the Borough of Manhattan Community College in New York City. As a consequence of these efforts along with those of others who wrote to protest the content and form of the video, the MAA and the NSA canceled the distribution of "You're Gonna Need Those Numbers."

INTERPRETING THE DILEMMA

Simic-Muller's case highlights two problematic, important issues for the theory and practice of teaching mathematics for social justice. One concerns the tension between teaching methods for enacting an official mathematics curriculum and pedagogy for implementing an alternative curriculum based on ideas of mathematics for social justice. The second problem turns on the tension between the call for action within the theory of mathematics for social justice and the learned apathetic response of many young adults, including those seeking to be mathematics teachers, to the reality and the violence of social injustices. Though prospective teachers comprehend intellectually and respond indignantly to social injustices, they act as if powerless to challenge the social injustices.

These two problematic issues exist within different variants of teaching mathematics for social justice. They contain persistent and ever-changing tensions inherent in teaching mathematics for social justice. Powell and Brantlinger (2008) explored aspects of these tensions, framing them within a perspective of critical mathematics that admits varied forms—each characterized by attention both to mathematics learning and teaching and to social critique—and each contributing to movements for greater social justice. What varies among the forms is the balance between mathematics and critique—or what is highlighted and what is implicit. Nevertheless, action is common. Further, Powell and Brantlinger contended that such a perspective implies a pluralistic, nuanced yet political view of what constitutes critical mathematics. It is worth noting that critical mathematics or, as I prefer, *criticalmathematics*, as coined by Frankenstein, Volmink, and Powell (1990), is defined by an eight-point platform, one point of which includes the following understanding: "No definition is static or complete, all definitions are unfinished, since language grows and changes as the conditions of our social, economic, political, and cultural reality change" (p. 6). Like language and the definition of criticalmathematics education or teaching mathematics for social change, there is a constantly changing tension between mathematics pedagogy and social action on the basis of mathematical analyses of social injustices.

RESPONDING TO THE DILEMMA

As teaching mathematics for social justice emerges from critical pedagogy, it encompasses a Freirean notion of praxis. He defined it as "reflection and action directed at the structures to be transformed" (Freire, 1970, p. 126). Praxis is a cycle of theory and action followed by evaluation and reflection, resulting in potential contributions to theory.

The nature of the assignment and the involvement of prospective teachers with its selection may shape heavily the extent of their alienation from the assignment. If the assignment is not theirs but the teacher's, then it is likely that prospective teachers may feel manipulated, less likely to go beyond the mere requirements of the assignment, and unsure of why or whether they should act. To this point, the case author posed a number of intriguing questions:

1. How can I structure the class so that students embrace the focus on equity and social justice instead of just doing the work assigned to them?
2. How do I help preservice teachers build sociopolitical consciousness so they will be called to action, not necessarily in the context of sweatshops, but any topic that speaks to them?

Perhaps the most important question, and the heart of the dilemma, is this: *If the preservice teachers are not moved to action by my assignments, what, if any, is my role in encouraging them to take action?*
These questions exist within a certain pedagogical paradigm, one in which the teacher and the teacher's political issues are the focus of the classroom work. It ignores the generative themes of the students themselves (Freire, 1970) and, as such, alienates them. A challenge for criticalmathematics educators is to counter hegemonic narratives about who can do mathematics and to reconstruct the role of mathematics in the struggle to empower learners whose mathematical powers and social activism have been underdeveloped. What seems necessary is for mathematics-for-social-justice assignments to include the teacher as an actor who collaboratively with students (in this case prospective teachers) interrogates and analyzes an unjust social situation and decides on a joint course of action. Mathematics-for-social-justice assignments need to invert the logic of schooling. Such assignments are not for students to act alone but rather for students and teacher to act jointly against injustices.

Mathematics for social justice is constantly subject to praxis, to critical reflection and revision. I want to reflect on how teaching bestows power. Power privileges certain ways of speaking, living, being, and acting in the world—cultural capital—and teaching is not free of the contaminating effects of oppressive and institutional relations of power. Teachers exercise power when they assign students to do something. From a praxis perspective, what I might have done if I were in Simic-Muller's position is to reflect on how to mitigate the power bestowed on me as a teacher and join with my students (prospective teachers) to interrogate a situation so as to decide on collective action, action in which I, too, would participate.

NOTE

1. This is the motto of the institution at which I earned my undergraduate degree, Hampshire College.

REFERENCES

Baldassarre, D., Broccoli, M., Jusinski, M. M., & Powell, A. B. (1993). Critical thinking and critical collective action in education: Race, gender, and the mathematical establishment. In W. Oxman & M. Weinstein (Eds.), *Critical thinking as an educational ideal: Proceedings of the fifth annual conference of the Institute for Critical Thinking, October 22–24* (pp. 154–158). Montclair, NJ: Institute for Critical Thinking, Montclair State.

Frankenstein, M., Volmink, J., & Powell, A. B. (1990). *Criticalmathematics Educators Group Newsletter, 1,* 1–6.

Freire, P. (1970). *Pedagogy of the oppressed.* New York, NY: Seabury.

Powell, A. B., & Brantlinger, A. (2008). A pluralistic view of critical mathematics. In J. F. Matos, P. Valero, & K. Yasukawa (Eds.), *Proceedings of the fifth international conference on mathematics education and society (Albufeira, Portugal, 16–21 February)* (Vol. 2, pp. 424–433). Lisbon, Portugal: Centro de Investigação em Educação and Department of Education, Learning, and Philosophy.

COMMENTARY 3

MATHEMATICS AND ACTIVISM

A Commentary on Simic-Muller's Case

Judith Quander
University of Houston-Downtown

I am a mathematics educator at the University of Houston-Downtown (UHD)—a midsize commuter college in the center of downtown Houston. Like the author, I teach content courses for preservice teachers. As one of two mathematics educators in the department of Mathematics and Statistics, I am responsible for our K–8 content courses and two of our 9–12 content courses. Over the course of a year, I teach about 150 preservice teachers. Unlike the author, my student population is largely Hispanic and African American. Additionally, at least half are nontraditional students who are over the age of 25. As a doctoral student, I worked with mostly White, middle-class preservice teachers from Georgia. Moving to UHD and a more diverse preservice teacher population, I have focused less on helping my preservice teachers understand issues of equity and privilege than I did while in graduate school. My thinking, right or wrong, is that my students are well aware of inequities in education because of the problems that they themselves have encountered with respect to race, language, and socioeconomic status. When I ask the question about why they want to teach, I often get answers about making a difference to students like themselves

Cases for Mathematics Teacher Educators, pages 307–309
Copyright © 2016 by Information Age Publishing
All rights of reproduction in any form reserved.

and giving back to their communities. In some ways, I feel that my students and I can start from a common place when talking about issues of equity.

INTERPRETING THE DILEMMA

Simic-Muller provides an interesting dilemma in the work of helping teachers understand inequities in their world and present solutions using mathematics as a powerful tool. I agree with her that giving teachers agency should influence them to engage in this kind of empowering work with their own students. However, as she questions, how far are we to go in pushing our students into action? If the goal of social justice education is to impart this understanding ultimately to K–12 students, wouldn't we want preservice teachers to engage in activism?

In a *Topics in Secondary Mathematics* course, one of the two content courses that I teach for 9–12 preservice teachers, the students complete a statistics project in which they collect data and use Excel to analyze it. I provide some websites such as the NCES (National Center for Education Statistics) or the CDC (Centers for Disease Control) that I know will have interesting social data, but I do not pick the data set for them. Although not all will do so, I find that most will pick data that supports an inquiry that is social justice in nature. For example, one student looked at HIV rates for African Americans in urban areas. Another looked at data on diabetes in the Hispanic community. However, although my students might naturally chose a social-justice topic because of their own background, reading this case made me question exactly that which the author is questioning, what do I do from there? Is it enough that they used mathematics to uncover patterns of racism, or do I really want their findings to spur action? Like Simic-Muller, I am lucky that my university is committed to social justice, particularly as it affects our local community. Beyond that, faculty are encouraged to incorporate community engagement into our courses. Simic-Muller and I are fortunate that we are part of institutions that support this work, as it gives even greater credence to our expectations of our students. With such administrative support, and given the goal of teaching mathematics-for-social-justice education to teach students to use mathematics as a tool to counteract social injustice, it seems natural to require the students to go the next step and actually do something with the information that they have uncovered in such projects.

RESPONDING TO THE DILEMMA

I was struck by one comment, in particular, made by Simic-Muller regarding her students. She says that in their reflection pages, most students "call for the immediate abolishment of all sweatshops but never ponder small actions they can personally take." This is an interesting statement, and I wonder if her students see the problem as beyond their control. As teacher educators, we should be responsible for helping our students think about what small steps towards activism look like. We help them make sense of activism in present day. Even more, as part of helping them to develop their professional identity as a teacher, we should help them understand what activism can look like for a teacher. Boycotting, marches, and walkouts are all forms of activism, but in the current day, we are more likely to see activism via social media. Sites like Twitter and Facebook are a means of publicly protesting or supporting a political or social movement. Some examples include a 2011 Chilean student movement that used Facebook to organize, communicate, and educate (Cabalin, 2014) or a recent movement to pressure college campuses to act on incidences of sexual violence that relied heavily on social media to gain publicity for their movement (Sander & Lipka, 2013). Teachers might strike in protest such as teachers in Chicago did in 2012 (Kaplan, 2013), but they also might use print or electronic media such as teacher journals and other education publications or websites to voice concern. Or they might be activists in their own classrooms through the lessons that they use—similar to the way that Simic-Muller does in her case. These "smaller" forms of activism could easily be built into a course project. For example, I could ask my preservice teachers students looking at HIV rates in urban communities to write an article for a local teacher journal on how this topic could be addressed in a mathematics classroom. Or, they could conduct the project with their school-age students during field experience. If the goal is to educate preservice teachers of mathematics for social justice as a means of teaching mathematics, I think it reasonable to have them think about how to connect ideas about equity and the use of mathematics to examine social injustice to their future career.

REFERENCES

Cabalin, C. (2014). Online and mobilized students: The use of Facebook in the Chilean student protests. *Comunicar, 22*(43), 25–33.

Kaplan, D. (2013). The Chicago teachers' strike and beyond. *Monthly Review: An Independent Socialist Magazine, 65*(2), 33–46.

Sander, L., & Lipka, S. (2013). Quiet no longer, rape survivors put pressure on colleges. *Chronicle of Higher Education, 59*(45), A20–A22.

CHAPTER 14

WHO COUNTS AS A MATHEMATICIAN?

Sharon Strickland
Texas State University

INTRODUCING THE CASE

I teach content courses in the mathematics department at Texas State University that are required for preservice teachers and sometimes include students from other programs. Specifically, though, this case will address History of Mathematics, which is open to all students who have had Calculus II and Modern Geometry, and required for students seeking 4–8 or 8–12 mathematics certification. Students outside the certification program rarely take the course because although it counts as an elective credit in general, it does not count towards elective mathematics hours for the "pure" or "applied" mathematics degrees. In the semester this case took place, all students were preservice teachers. The course is also designated as "writing intensive," and therefore 70% or more of the final grade is determined by full papers, brief writing prompts, journals, proofs, and so forth. Most preservice teachers in the course are juniors or seniors and are usually also in, or have already had, their methods courses through the College of Education and their semester-long observation (not student teaching) placements in middle- and high-school classrooms.

Cases for Mathematics Teacher Educators, pages 311–318
Copyright © 2016 by Information Age Publishing
All rights of reproduction in any form reserved.

The course runs each semester and during one summer semester as an online course. My first experience teaching it was online during the summer when I essentially implemented a prepared series of online modules, made myself available via face-to-face and virtual office hours, and graded. When assigned to teach it the following fall, I decided to radically alter the course. I was disappointed in the messages the course sent about the history of the subject as a static rather than evolving discipline and about who does mathematics. I thought it was too focused on historical events rather than the larger context of how the history of the subject has led up to where it is today. Furthermore, the course catalog description includes this phrase, "Philosophical and cultural aspects will be integrated with the structure, theorems, and applications of mathematics," which I thought had been only superficially addressed.

In particular, the original course was problematic for future teachers by reinforcing stereotypes of mathematicians as having lived "in the past" and as primarily White and male. I worried not only that this is false but that if unchallenged, these future teachers would not think of mathematics as an alive discipline and they might not see future students who did not fit stereotypical images of mathematicians as mathematically capable, or might not explicitly encourage them to consider futures in STEM. Although the discipline still struggles to recruit and retain women and students of color, there is far more variety in the people I know and work with daily than would be suggested by the representations in the history course as it had been taught. Our department is among the state's most racially and gender-diverse mathematics departments, and the university as a whole is designated as a Hispanic-serving institution, meaning that at least 25% of our student body is Hispanic. I wanted the course to better reflect our faculty, the undergraduates, and the larger population of students these preservice teachers might go on to teach.

One of the many changes I made was to alter a mathematician biography that had been in the original version of the course. I tweaked that assignment in two ways: a) I asked them to select an "underrepresented" mathematician—living or not—and b) in addition to the biography, they should also comment on the mathematical contributions that person had made. The first change was intended to address the diversity of mathematicians and, thus, to challenge the dominant image, and the second was intended to gather more in the paper than a life biography—I wanted it to specifically address the person's mathematical work. This was important to me because although I recognize that much of mathematicians' scholarly work might be too advanced to unpack, I was looking for a bit more of their subjects' academic lives. For example, it would be too ambitious to fully understand some of Emmy Noether's theorems, but some acknowledgment of her impact on Modern Algebra and special relativity was within reach. In addition, I was concerned that by making the change requesting an underrepresented mathematician that the papers might frame the mathematician through a deficit lens, highlighting hardships and struggles. I also wanted my preservice teachers to address the academic successes and contributions to the mathematical community.

THE DILEMMA

When I first distributed the syllabus, the biography assignment did not elicit any comments. But as the due date approached, a preservice teacher asked in class "who counted as a mathematician?" for the purposes of this paper. Many others soon chimed in that they too were confused. It had not (naively) occurred to me that what was meant by "underrepresented" was unclear, and I also wondered if, for some of the preservice teachers, they were truly confused or were deliberately pushing against my intentions to seek out nondominant representations of mathematicians. I interpreted some preservice teachers as being truly confused, some as being hyper-aware of grades and grading criteria and therefore wanting to know exactly what I wanted, and some as hinting at "reverse discrimination." Those in the latter category were mostly from a group of male preservice 8–12 certification students who sat and studied together and were some of the strongest mathematically in the class. By that I mean they made excellent grades, could write proofs very well, and seemed already familiar with a good deal of the mathematical ideas we discussed. This group questioned whether certain largely unknown and mostly European male mathematicians counted as underrepresented. This group worried me the most because I feared that because they had had success in their education that they might have a narrower view of what and who counts in mathematics.

The day the issue first arose I gave some examples of websites the preservice teachers might seek out like the Biographies of Women Mathematicians website, but it was not until that evening that I fully considered my response. First, I examined why I had wanted them to complete this assignment in this way and how I had hoped it would help them as teachers.

Early in my graduate career I had read Picker and Berry (2000), who had asked middle-grades students to draw images of mathematicians. Many of the themes they identified worried me: The mathematicians were usually male (and many looked like Einstein), and the images were often violent with mathematicians shown as teachers intimidating or harming K–12 students. When asked to provide a scenario where a mathematician might be needed, the students in their study frequently suggested that mathematicians were useful for calculating complicated arithmetic. This suggested to me that middle-grades students do not know what sort of work mathematicians do or who does it. I realized that this article had inspired a desire to widen this image not only for young students, but also for any persons teaching K–12 students mathematics. I had no evidence that this class of preservice teachers thought similarly, but I worried that they might or that they might not have readily available examples of mathematicians to challenge these images if they came across future students who did.

Furthermore, I was worried about the issue of who counts as a mathematician. Working as a mathematics educator in a mathematics department brings this issue to my mind often. Am I a mathematician? Do others see me that way? Although I personally do not actively claim that label, would my "pure" mathematician colleagues challenge me if I did? What about people working in applied fields? Do you need a PhD in mathematics to count? Do you need to publish research in mathematics? If so, what counts as research? And how does this look different across history and location? For example, a current stereotype assumes that Asian students are good at mathematics, but in classic history courses, Asian mathematicians are underrepresented.

Because these sorts of questions arise for me—what I think of as constituting the boundary of mathematics and mathematicians—I worried that in my attempts to broaden my students' perspectives I might unintentionally be drawing different boundaries for them. I did not want to declare that they could only select either female or Hispanic or African American mathematicians, for example. I work with a Kenyan mathematician who is not African American, yet I wanted his story available to them as well. As soon as I drew some line around who counted for this assignment, I faced the dilemma of unintentionally leaving some other story out. I did not want my preservice teachers to draw lines around who counts in mathematics, although my assignment was forcing this issue.

MY RESPONSE

The next class I began by asking them to write out associations they had about mathematicians. They were hesitant to do this—I suspect because they were unsure what I intended to do with what they wrote. Finally, I collected a few responses to share. Here is what one preservice teacher wrote: "I always pictured mathematicians to be old men in libraries wearing fancy robes slaving over piles of books and scrolls with endless lines of mathematics." And another: "Whenever people think of mathematicians the stereotype is a nerdy man with glasses who loves mathematics." Once these were shared many others began to agree, but a few challenged this: For example, one preservice teacher pointed out that her mother was a mathematics teacher so she saw mathematicians as female, too. Then a preservice teacher asked me whether I counted as a mathematician, and I tried answering in ways that I have already mentioned—that to some people I am and to others I am not—that who counts is not so clear cut, but that generally, their mathematics professors counted among many others. The idea that people they knew and took classes from were mathematicians seemed quite new to them, which reinforced to me that not only did they have a narrow image but that it did not seem to include living people.

I next projected images from the Picker and Berry (2000) article, eliciting giggles and agreement. We discussed these images as not reflective of who mathematicians are or can be, and that is why I had created the assignment as I had. I wanted them to be advocates for their future students and to be living role models of mathematicians as well as a resource for connecting students to other models of mathematicians outside their immediate experiences. I next moved to providing them with a more structured writing prompt than I had before:

> For this paper select a mathematician from a culture, ethnicity, and/or gender, etc. typically not well-known in American education for mathematics. Generally speaking a woman, a Latina/o, an African-American, Persian, Hindu, etc. are good choices, but there are definitely examples of European men who might be a good fit for this paper—it depends on your angle. Typically we hear lots of their stories already. When selecting someone consider a hypothetical student who is interested in mathematics but might not see anyone "like them" doing math—who might you teach them about to introduce them to such a role model? Your paper needs to include some basic biographical information as well as a detailed account of their mathematical contribution. This could be someone from "history" or a person still living.

Looking back over it now, I cringe at my use of "etc." as an attempt to open possibilities for their paper. I waffled between providing solid examples and remaining open. I also attempted to ground their choice in an imagined future experience with a student so they might see the assignment as serving their needs as teachers.

I had already pointed them towards the Biographies of Women Mathematicians website, and we also quickly found the Mathematicians of the African Diaspora website as well. After looking at these examples, I suggested that they try the basic structure "<adjective describing mathematician of interest> mathematician" and see what they could find.

THE PRESERVICE TEACHERS' PAPERS

Overall, I was impressed with the papers they submitted. Most preservice teachers elected to focus on a woman, an African American, or a Hispanic mathematician. One of these was particularly interesting to me, and I made sure to share the story with the class. He had selected Ruth Gonzalez, the first U.S. born Latina to earn a PhD in mathematics—in 1986. I had no idea it took so long for a U.S. Latina to achieve this distinction. It was wonderful and awful. Furthermore, she is an applied mathematician currently working in industry and gives motivational talks to K–12 and college students about her experiences and work. Another included Frank Elbert Cox, the first African American to earn a PhD in mathematics, and that paper included specific details on Cox's research and how institutional racism limited his opportunity to publish his work. A few preservice teachers wrote about more famous (in math circles) White male Europeans but in a way, nevertheless, that fit my expectations of addressing some aspect of that person that was different. For example, one preservice teacher selected Tartaglia because he had a speech impediment that the student shared; another wrote about Alan Turing because of the harassment he faced due to his sexuality. In both cases the preservice teachers defended their choices by appealing to the aspect of the revised prompt that suggested they had considered a future student who might need support.

Two papers puzzled me as to how the preservice teachers had framed the underrepresented angle. One was about R. H. Bing, a graduate of Southwest Texas State Teachers' College (Texas State University's old name), who had been a high-school mathematics teacher and coach. The preservice teacher suggested that because Bing had been a local student from this very university and had begun his career as a teacher and coach who went on to a PhD later, that his story was underrepresented. Another student wrote about Andre Bloch, a mathematician who did the majority of his work while incarcerated in an asylum after murdering his family. Whether to accept these two papers presented a second dilemma for me. The first worried me because, as a past student myself of Texas K–12 schools, I had had several male coach mathematics teachers. True, none had also returned for PhDs, but the context was quite familiar to me and did not seem to push at the possibility of who could do mathematics except by suggesting that teachers (and athletes as well), who are often considered less intellectual than their nonteaching mathematics peers, are entirely capable of earning PhDs if their career goals change. I do think that is a valuable message, but the preservice teacher who submitted it did not directly argue this point—I was rationalizing it for him. The second paper reinforced the violent images that Picker and Berry (2000) found middle-grades students drawing as representations of mathematicians. The preservice teacher who submitted that paper supported her choice by pointing out that Bloch hid both his asylum incarceration as well as his Jewish identity via pseudonyms and ambiguous return addresses, because his mathematics papers were being submitted in German-occupied France during World War II. Thus, he feared both the academic stigma of his incarceration and Nazi capture. In the end, I elected to accept the papers and made comments in the text pushing on these choices for the reasons articulated.

EPILOGUE

I was happy with the assignment and the course as a whole. Looking back, I should have anticipated this dilemma and provided more direction for this assignment in the original syllabus but also think the conversation we had when the issue arose was beneficial in a way that even a more fleshed-out syllabus could not communicate. I have been asking to teach the course again, but so far it has not been on my schedule. If it were, I would adjust it as two separate papers, one focusing on a "past" mathematician and a second one, later in the semester, focusing on a 20th or 21st century, preferably living mathematician. I would plan to include a fuller description in the syllabus and a day for discussing it—not just as a reaction to their questions, but as an intended topic to discuss.

I continue to struggle with who counts as a mathematician for the purposes of this assignment (and in general) and am torn between the extremes of providing a set list of mathematicians or exact ways it is acceptable for the paper to "be different" and between letting them freely choose whomever on the grounds that everyone's story can be seen through a particular lens of diversity on some measure. The Andre Bloch example exemplifies this dilemma for me. He perpetuates a "crazy" stereotype for mathematicians and, furthermore, justified the murder of his family as a mathematically rational decision. Yet he was Jewish at a time when Jews in his country and across Europe were heavily persecuted, and it cannot be ignored that his Jewish identity directly influenced his inability to communicate easily with the mathematical community. But he was also Jewish at a time when Jewish scholars and mathematicians were not underrepresented. On the other hand, he went to school as a young man but did not complete any official training in mathematics and thus represents the ways that certifications and degrees can act as gatekeepers to participation. Overall, I do not want to give a specified list of acceptable mathematicians. I do want to hear my preservice teachers' voices and let them teach me. But, maybe next time I will request a quick email on their paper topic in advance.

REFERENCES

Picker, S. H., & Berry, J. S. (2000). Investigating pupils' images of mathematicians. *Educational Studies in Mathematics, 43*, 65–94.

RESOURCES FOR BIOGRAPHIES

Mathematicians of the African Diaspora.
    http://www.math.buffalo.edu/mad/00.INDEXmad.html
Biographies of Women Mathematicians
    http://www.agnesscott.edu/lriddle/women/women.htm

COMMENTARY 1

BUTTONS AND MATHEMATICIANS

A Commentary on Strickland's Case

Zandra de Araujo
University of Missouri

MY POSITIONALITY—*CAN YOU GIVE ME THE BUTTON?*

This seems like a straightforward question to me. My father asked me this very question a great number of times growing up, and I knew exactly what he meant. The first time I asked my husband this question he gave me a confused look and asked, "What button?" I looked at him in a manner that questioned his sanity and replied, "The one in your hand." "This," he said waving the device at me, "is a remote control." Okay, so he was correct, in some circles the device is known as a remote control, but not in my parents' house. We have since come to an understanding that both words can be used interchangeably. So what do buttons and underrepresented mathematicians have in common? Both terms are open, as is any other term, to varying interpretations.

My sister and I were among the first generation of our family born in the United States. This meant that my sister and I frequently had common

Cases for Mathematics Teacher Educators, pages 319–323
Copyright © 2016 by Information Age Publishing
All rights of reproduction in any form reserved.

meanings for particular words, like *button*, that our friends did not share. It did not occur to us that we had different conceptions for particular terms until we encountered situations such as the one with my husband above. In my work with English language learners, the need to attend carefully to the meaning of words and come to a shared understanding is perhaps more obvious than in monolingual English classrooms because differing understandings are expected. Therefore, in classrooms where everyone speaks English, the importance of developing a shared meaning may not be as obvious.

Different interpretations of a term may go unnoticed because the term itself may not be key to a particular interaction. For example, I can pose the following problem.

> There were five birds on a tree. My dog barked and three of the birds flew away. How many birds are left on the tree?

In this case the particulars of what is meant by birds and a tree do not necessarily raise an issue. In creating a visual image of this problem, one person might imagine toucans on a banana plant while someone else might imagine sparrows on an oak tree. Even though some people may argue that banana plants are not really trees, it would, presumably, make little difference to the person solving this problem. However, in Sharon's case, I believe much of the dilemma stemmed from the class's differing conceptions of two terms that were instrumental in completing the assignment: *mathematicians* and *underrepresented*.

INTERPRETING THE DILEMMA

The heart of the given dilemma seemed to be the class's different interpretations of what it means to be a mathematician. Who counts as a mathematician is an interesting question to consider. If you ask five people, as I did in thinking about this case, you may well get five different answers. At a local elementary school I frequently visit, the teachers greet and position their students as mathematicians exclaiming, "Good morning, mathematicians!" The teachers talk with the children, describing the practices of mathematicians and then connecting these practices to the children's work. As I walk around the room, the children refer to themselves as mathematicians. Indeed, in drawing what a mathematician looks like, the diverse field of mathematicians depicted in the children's drawings is wonderful to see. With that said, are they truly mathematicians?

It seems no one can agree on who is deserving of the title mathematician. I do not consider myself a mathematician although I teach mathematics courses for prospective elementary teachers and have completed many graduate-level mathematics courses in addition to my undergraduate degree in mathematics. Searching various websites I came up with the following descriptions of mathematicians.

> A mathematician is a person with extensive knowledge of mathematics who uses this knowledge in their work, typically to solve mathematical problems. ("Mathematician," 2015b, para. 1)

> Mathematicians are people of all ages and from all over the world who enjoy the challenge of a problem, who see the beauty in a pattern, a shape, a proof, a concept. (American Mathematical Society, 2015, para. 8)

> A person who is an expert in mathematics. ("Mathematician," 2015a, para. 1)

RESPONDING TO THE DILEMMA

This diversity of meanings led me to consider how I would approach this dilemma should I give my students a similar assignment. I frame my response in a situated and sociocultural perspective on learning (Moschkovich, 2002). That is to say that learning is a discursive activity that is "inherently social and cultural... that participants bring multiple views to a situation... these multiple meanings for representations and inscriptions are negotiated through conversations" (Moschkovich, 2002, p. 197). Thus, it is necessary to anticipate the learners' multitude of meanings for the term mathematician. Because meaning is "negotiated through conversations," I would have engaged the class in a discussion regarding how we should characterize the term mathematician when introducing the assignment. The negotiation I describe is similar to the negotiation of sociomathematical norms described by Yackel and Cobb (1996). The process allows "subjective ideas [to become] compatible with culture and with intersubjective knowledge" (Voigt, 1996, p. 30) and permits students to bring in their cultural resources and knowledge and contribute those resources to the class (Gay, 2002; Ladson-Billings, 1995). In fostering this communication, the teacher is in turn "building community among divers learners," which is an "essential element of culturally responsive teaching" (Gay, 2002, p. 110). Further, the teacher is not serving as the authority on what constitutes a mathematician or underrepresented. She is instead positioning her students to take ownership for the meaning developed.

In facilitating discussion around what constitutes a mathematician, I would have students consider their own thoughts and then work in small groups to identify common features among their individual insights. Then, we could collaboratively create a list of inclusion and exclusion criteria for what the term mathematician would mean for this particular assignment.

Incorporating both inclusion and exclusion criteria is important in negotiating meaning with others because it more clearly bounds the concept. Thus, students may be better situated to consider whether someone fits the criteria for mathematician. Following this discussion, it would then seem logical to consider what constitutes an *underrepresented* mathematician. The class could similarly negotiate what it means for a mathematician to be underrepresented for this assignment. In valuing the students' contributions to the discussion and allowing them to be brought to the group for consideration, we keep our own views broader and avoid the potential narrowing that worried Sharon in her dilemma.

The final consideration I would add in rethinking this assignment is to have students use the negotiated criteria to develop a rationale for selecting their particular mathematician. This allows the teacher educator to be privy to their thoughts and holds students accountable to the negotiated meaning. In rationalizing their selection, the students must deeply consider the motivations behind their choice, and the teacher educator is then able to evaluate the students' responses using the students' lens. In a sense, the students are providing their subjectivities for the teacher to consider.

Broadening prospective teachers' perspectives and countering stereotypes are important goals for teacher educators. In selecting this assignment, Sharon unintentionally uncovered her students' "buttons." This created an opportunity for her to challenge her subjectivities through a close reflection on her goals and choice of assignment. This process of reflection and revision provides a model for other teacher educators who might encounter similar situations as they seek to push students to renegotiate their interpretation of who and what counts in mathematics.

REFERENCES

American Mathematical Society. (2015). *What do mathematicians do?* Retrieved from http://www.ams.org/profession/career-info/math-work/math-work

Gay, G. (2002). Preparing for culturally responsive teaching. *Journal of Teacher Education, 53*, 106–116.

Ladson-Billings, G. (1995). But that's just good teaching! The case for culturally relevant pedagogy. *Theory Into Practice, 34*, 159–165.

Mathematician. (2015a). In *Merriam-Webster*. Retrieved from http://www.merriam-webster.com/dictionary/mathematician

Mathematician. (2015b). In *Wikipedia*. Retrieved from http://en.wikipedia.org/wiki/Mathematician

Moschkovich, J. (2002). A situated and sociocultural perspective on bilingual mathematics learners. *Mathematical Thinking and Learning, 4*, 189–212.

Voigt, J. (1996). Negotiation of mathematical meaning in classroom processes: Social interaction and learning mathematics. In L. P. Steffe, P. Nesher, P. Cobb,

B. Sriraman, & B. Greer (Eds.), *Theories of mathematical learning* (pp. 21–50). New York, NY: Routledge.

Yackel, E., & Cobb, P. (1996). Sociomathematical norms, argumentation, and autonomy in mathematics. *Journal for Research in Mathematics Education, 27,* 458–477.

COMMENTARY 2

BROADENING PERSPECTIVES THROUGH PURPOSEFUL REFLECTION

A Commentary on Strickland's Case

Jennifer A. Eli
The University of Arizona

MY POSITIONALITY

I am a mathematics educator in the Mathematics Department at The University of Arizona. I hold a BS and MA in mathematics, and my PhD is in mathematics education. I have taught mathematics courses for pure and applied undergraduate mathematics majors as well as mathematics content and methods courses for preservice elementary and secondary mathematics teachers.

This case reminds me of my own experiences as a woman of color interested in pursuing a career in mathematics. Issues of access, opportunity, support, and stereotypical societal messages of who is capable of doing mathematics had an impact on my journey into the discipline. In my youth, societal messages about those who pursued careers in STEM were directed

Cases for Mathematics Teacher Educators, pages 325–328
Copyright © 2016 by Information Age Publishing
All rights of reproduction in any form reserved.

towards kids, typically male, that came from upper middle-class White families. These kids generally had access, opportunity, and financial resources to participate in extracurricular activities that enhanced their knowledge of mathematics. This, in turn, gave them opportunities to take more advanced mathematics while those of us who came from marginalized underrepresented groups had to work twice as hard.

As a high-school student, I had to double up on mathematics courses my sophomore and junior year just to have the option of taking a calculus course my senior year. I remember vividly a high-school teacher discouraging me from taking more mathematics classes because, as she said, "it would be too hard for someone like you" (which, based on past interactions, I took to mean someone who is a mixed-race female from a poor family with an ESL mother). So from my perspective, the case author's concern that "they [future teachers] might not see future students who did not fit stereotypical images of mathematicians as mathematically capable, or might not explicitly encourage them to consider futures in STEM" resonates all too well.

INTERPRETING THE DILEMMA

I see two major issues in this case. The first issue is constructing clear guidelines and expectations for the mathematician biography, and the second is deciding who counts as a mathematician. The way the author reframed the assignment through purposeful reflection did resolve some of the dilemmas that initially surfaced. However, it was apparent that for some students the revised version was still too vague with regards to who counts as underrepresented mathematician because the idea of what is means to be *underrepresented* was not clearly defined. Her main goal through this mathematician-biography task was to broaden students' perspective on who counts as a mathematician by having students research underrepresented people, but in particular, females and minorities. However, there was still concern that students would select boundary cases, that is, choose individuals that the author would feel are not an acceptable target for the assignment. This was evident in students' work on Tartaglia, Alan Turing, R. H. Bing, and Andre Bloch, where it is not so clear if these individuals meet the criteria of what it means to be an underrepresented mathematician.

RESPONDING TO THE DILEMMA

I appreciate the author's work to broaden future teachers' perspectives on underrepresented mathematicians who have made and continue to make significant contributions to STEM. To address some of the concerns that arose, I would start the course off by spending time in class exploring what it means to be a mathematician. This would open up the discussion on underrepresented individuals in a way that would have prospective teachers thinking more broadly about who fits the category as underrepresented. This would then provide an opportunity to confront grey areas head on and provide example situations, such as those identified in the case, where some people would identify the person as an underrepresented mathematician and others would not. From these discussions, I would select and establish clear criteria for the assignment. In fact, I would recommend creating two separate assignments. The first assignment would focus on the contributions of female mathematicians and mathematicians from underrepresented minority groups. Just as the case author did, I would direct the prospective teachers to various websites and resources for ideas. For the second assignment, I would deal with the boundary cases. For the boundary-case assignment, I would provide a list of mathematicians who are arguably underrepresented (e.g., Tartaglia, Touring, Bloch, and Bing). I would then have the students make an argument for why these individuals are or are not mathematicians and whether they are or are not underrepresented. Through the boundary-case assignment, the prospective teachers have another opportunity to think critically and make a convincing argument as to what it means to be a mathematician and how it relates to the challenges of being underrepresented, thereby helping them be prepared to better address these issues in their future practice. By creating two different assignments with clear guidelines and detailed criteria, the case author alleviates the concern that she would receive submissions that she would consider unacceptable.

I see this case as having implications for not just a history-of-mathematics course but for other content courses for prospective teachers. As mathematics teacher educators, we have the responsibility to educate our future teachers by broadening their perspectives on who is capable of doing mathematics. In the content courses I teach for future elementary teachers, I use *Complex Instruction* (Cohen, 1994; Featherstone, Crespo, Jilk, Osland, Parks, & Wood, 2011) as a means to facilitate group collaboration on the intellectual work of solving mathematics problems. Through this process, the prospective teachers begin to see themselves as capable of doing mathematics and begin to appreciate the contributions of their fellow classmates who come from diverse backgrounds. In my own practice, I could challenge and broaden prospective teachers' perspectives of who is capable of doing mathematics by adopting and modifying the case author's assignment on who counts as a mathematician.

REFERENCES

Cohen, E. (1994). *Designing groupwork: Strategies for the heterogeneous classrooms.* New York, NY: Teachers College Press.

Featherstone, H., Crespo, S., Jilk, L., Osland, J., Parks, A., & Wood, M. (2011). *Smarter together! Collaboration and equity in the elementary math classroom.* Reston, VA: National Council of Teachers of Mathematics.

DOING MATHEMATICS AND BEING A MATHEMATICIAN, THESE MAY BE DIFFERENT

A Commentary on Strickland's Case

Tod Shockey
University of Toledo

MY POSITIONALITY

I am a mathematics educator who works in the scholarship of ethnomathe-matics. Since entering higher education in 1999, I have had experiences as faculty in mathematics departments and education departments. I am cur-rently an associate professor of mathematics education with a joint appoint-ment in the Department of Mathematics and Statistics and the Department of Curriculum and Instruction.

As a student I participated in courses taught by the American mathemati-cal historian, Dr. Karen Parshall, at the University of Virginia. At UVA The History of the Calculus investigates mathematics up to the 17th century and how future mathematicians carried on these ideas and improved them (University of Virginia, 2015). The History of Mathematics course explores

Cases for Mathematics Teacher Educators, pages 329–332
Copyright © 2016 by Information Age Publishing

329

mathematics' development from antiquity through the 19th century (University of Virginia, 2015). It has been many years since I took these courses, and I do not recall writing about individual mathematicians. I do recall the proof writing and searching primary sources—It was a wonderful experience.

INTERPRETING THE DILEMMA

Strickland presents an interesting dilemma, as the description seems to attend to two potential courses, a course in Mathematics History or a course in Mathematicians' History. Since joining higher education, many of my students have experienced a course described by Strickland, but what lacks is an attention to culture, to the best of my recollection. I think it is fascinating to consider the lives of mathematicians, both from their contribution to mathematics and their time in history. If we consider the mathematicians' time in history, it may provide insights for students that include: Who was "doing" mathematics and why? How was mathematical knowledge being transmitted, and what role did women play in mathematical work? Who were the scholars in history responsible for translations of mathematics between languages? This is not a course that was ever available for me to teach. I recall a mathematician telling me that such a course could not be mathematically rigorous enough, so he vehemently opposed any discussion of inclusion in the university curriculum. To this day this particular university does not offer such a course. He would certainly have to reconsider his definition of *rigor* if he were to look at the offerings of Dr. Parshall.

RESPONDING TO THE DILEMMA

If I were to teach a course like the one in the case described I would consider expanding it to include the following. First, for me a mathematician has solved an open mathematical problem with a solution that is accepted in the mathematics community. I am not a mathematician; I am a mathematics educator. This may be too narrow of a view, but one I subscribe to. I would not initially share this with my students. I would first ask for their definitions to determine if common ideas emerge. If we, as a class, could agree on a working definition of *mathematician*, the next task would be to explore what is *culture?* My 16 years in higher education have been devoted to ethnomathematics, with a particular lens on improving the mathematics education opportunities for Native Americans. An element of culture that is important is language. Through this single lens I would suggest that mathematicians constitute a culture. Again, working with and negotiating with my students, we would strive to arrive at a working definition of culture.

Arriving at a working definition of culture may necessitate some reading for us as a class, so this would be supported with papers from a variety of disciplines in the attempt to potentially broaden students' meaning of culture (see Bishop, 1991; Gilsdorf, 2012).

There was value in learning that although Strickland had an interpretation of underrepresented, this was different for her students. It reminds us that what "we" in education take for granted might not always be a shared understanding. Thinking about my understanding of underrepresented, I realize how this has been influenced by STEM literature. I also consider remarks that I have heard and wonder about their validity. Was Pythagoras's wife a member of his society? Were female mathematicians writing mathematical manuscripts for publication under pen names? What were the mathematical influences from the African continent in antiquity? These questions and others would tell me what I would want to focus my students' attention initially on for their first biography assignment.

Mason (1998) has stated that "the role of a teacher is to create conditions in which students experience a corresponding shift in the structure of their attention, in which they become aware of acts and facts of which they were previously unaware" (p. 244). Construction of working definitions of mathematician and culture might a good stepping-off point to focus students' attention for whom they choose to write their mathematical biography about. I would purposely craft a list of mathematicians from which students had to choose their subject to write about. Emerging theory from Herrington (2014) suggests that Native American secondary-level students are not realizing "examples of Native Americans in the school curriculum and their representation as natural scientists and engineers" (p. 153). An initial biography would intentionally focus students' attention on underrepresented groups in mathematics. For example, if Robert Megginson were on a list of mathematicians that students could explore, they would learn that Dr. Megginson is Oglala Sioux and one of very few Native Americans holding a PhD in mathematics. If the students were asked to read Murray's (2001) book *Women Becoming Mathematicians: Creating a Professional Identity in Post-World War II America,* they would learn about the contributions made by this group Murray studied as well as the challenges put forth by society and the academy.

With working definitions of mathematician and culture as well as focused writing on underrepresented groups in mathematics, I would want the second assignment for the term to be more open for students to explore. Suggestions for a second paper could include attention to mathematics of a particular time period, from a particular country, or even a continent. Strickland's students were creative in finding mathematicians to learn and write about. I can only hope if I were teaching this course that my students would engage their curiosity and creativity to explore mathematics history through the person, culture, and societal influences.

REFERENCES

Bishop, A. J. (1991). *Mathematical enculturation.* Dordrecht, The Netherlands: Kluwer Academic.

Gilsdorf, T. E. (2012). *Introduction to cultural mathematics: With case studies in the Otomies and Incas.* Hoboken, NJ: John Wiley & Sons.

Herrington, J. B. (2014). *Investigating the factors that motivate and engage Native American students in math and science on the Duck Valley Indian Reservation following participation in the NASA Summer of Innovation program* (Unpublished doctoral dissertation). University of Idaho, Moscow, ID.

Mason, J. (1998). Enabling teachers to be real teachers: Necessary levels of awareness and structure of attention. *Journal of Mathematics Teacher Education, 1,* 243–267.

Murray, M. A. M. (2001). *Women becoming mathematicians: Creating a professional identity in post-world war II America.* Cambridge, MA: MIT Press.

University of Virginia. (2015). *Department of Mathematics course descriptions.* Retrieved from http://www.math.virginia.edu/courses?page=4

PART III

CONVERSATIONS ABOUT INEQUITIES IN GRADUATE AND PROFESSIONAL-DEVELOPMENT CONTEXTS

Joi A. Spencer
University of San Diego

Dorothy Y. White
University of Georgia

The six cases and adjoining 16 commentaries in this section concern themselves with the unique challenges of designing and facilitating learning experiences for teachers centered on mathematics and equity. Like many of our readers, the case writers and commentators in this section encounter practicing teachers in a variety of contexts including graduate-level courses, university-school partnerships, and as participants in their own research projects. The cases are unique, yet several themes emerge including teaching mathematics for social justice, culturally relevant instruction, and addressing teacher beliefs about students who have been marginalized in

Cases for Mathematics Teacher Educators, pages 333–337
Copyright © 2016 by Information Age Publishing
All rights of reproduction in any form reserved.

mathematics. Moreover, the pedagogical challenges related to presenting these ideas to seasoned practitioners versus prospective teachers are of primary concern.

The commentators have done an exemplary job of serving as critical friends. Their analyses and respectful critiques will help the case writers to re-see their work and perhaps approach it in the future with fresh eyes. As we edited this series of manuscripts, it became evident that our field could benefit from more of this kind of peer critique, not solely in the manuscript review process, but also in the crafting and implementation of our instruction and professional-development work. Just as we urge the teachers whom we work with to interrogate their instruction, we must continue to interrogate our own.

In Bartell, Id-Deen, Parker, and Novak's case, "Are These Two Sides of The Same Coin? Teachers' Commitment to Culturally Relevant Teaching While Holding Deficit Views of Poor Communities," practicing secondary mathematics teachers wrestle with Lewis's (1966) treatise on the culture of poverty. One course assignment was designed to help teachers consider the causes of poverty and evaluate the dominant stories that they have been told about poor and low-SES students. However, as the assignment unfolded, "nearly all of the teachers" affirmed congruence with Lewis's problematic ideas on the poor. As Bartell and her colleagues express, "the course seemed to support teachers in recognizing the role of culture in mathematics teaching . . . yet it may have served to perpetuate problematic myths about economically disadvantaged students."

Foote's case, "How Do I Learn to Like This Child So I Can Teach Him Mathematics: The Case of Rebecca," revisits a growing practice in teacher education—student case studies. Teachers participating in Foote's professional-development study conducted home visits, shadowed their case student in nonclassroom contexts, and met with their families. But, in the case of Rebecca (a teacher participating in Foote's study) the project, which was designed to undo stereotypes about students, served to reinforce and even worsen them.

Goffney's case, "Challenging Deficit Language," centers on practicing teachers' use of deficit language when speaking about their K–12 students. "As an instructor," she writes, "I am trying to find a balance between creating a safe environment and correcting this harmful behavior." Teacher educators—particularly those with a constructivist stance—will resonate with the pedagogical conundrum that Goffney faces.

Like many of the contributors throughout this volume, Herbel-Eisenmann works with predominantly White students and teachers. In her case, "Moving From Addressing One's Target Identity to Addressing One's Nontarget Identities," a Latino student enrolls in her master's-level course. When a reading related to systemic racism is discussed, she is surprised by this student's ideas. In wrestling with his response, Herbel-Eisenmann asks, "How can I, as a White professor, help a student of color make sense of patterns of systemic racism?"

Like so much of her work, Rubel's case, "Learning About Students and Communities Using Data Maps," endeavors to "support teachers in developing critical orientations towards data." Just how these critical orientations are developed and the instructional enigmas therein are at the heart of her case.

Finally, in Wager's case, "Let Me Be Your Cultural Resource: Facilitating Safe Spaces in Professional Development," an inexperienced mathematics teacher educator runs a professional-development project attended by experienced teachers. One teacher in particular, a teacher of color, has a long history of supporting marginalized students of color. Her expertise and how it gets taken up and used as a "resource" within the professional development comes into question.

There is a lot to supporting and preparing teachers to teach in ways that invoke mathematics success for marginalized and minoritized students. So that we might have a better appreciation for the vastness of our task, here is a sampling of what mathematics teacher educators focused on equity might attend to:

- intercepting deficit views of children, adolescents, and the communities they come from,
- supporting teachers to instruct in culturally responsive ways across a broad range of student cultures and practices,
- helping teachers support the mathematics identities of students who have previously been marginalized or experienced little success in the subject
- assisting teachers in supporting their students in navigating the postsecondary mathematics pipeline
- helping teachers learn about the mathematics thinking of children and adolescents
- prompting teachers to reconsider mathematics ability-grouping practices and adopt teaching practices that attend to a broad range of student understandings and skills
- encouraging mathematics instruction that helps young people see the relevance of mathematics in their lives

In the context of supporting practicing teachers—as is the focus of this collection of cases and commentaries—our work takes on particular challenges. We are tasked to support teachers, including those who have produced high rates of mathematics achievement for certain segments of their students, to reevaluate their mathematics practices. This requires helping teachers to see their work with fresh eyes, new tools, and new lenses. An activity that asked teachers to look through a critical-race lens, for example, would push teachers to reevaluate their successes through the experiences of their least successful students instead of their most successful. Having teachers interview these unsuccessful students about their experiences with learning mathematics would be revelatory indeed!

Nieto (2006) has written that "teaching is inherently political work" (p. 1). We take this to mean that *our* work as mathematics teacher educators is political. Extending Nieto's words, we would argue that the entire educational enterprise and hence the mathematics education enterprise is political work (Apple, 1992; Martin, 2015). Politics are by definition about interest—a community's interest, a corporation's interest, an individual's interest. Thus, our work is not simply about relaying ideas and changing pedagogical approaches. It is about supporting and preparing educators to teach in the interests of historically disenfranchised people.

Our work as mathematics educators often occurs in small spaces, such as elementary classrooms or seminar rooms on university campuses. However, our work is not small. Its outcomes could mean a sea change in the current arrangements that we now wrestle under, perhaps producing new winners, perhaps rewriting the system entirely. As you read the cases in the section, we offer the following questions to guide your thinking and to facilitate conversations with your own students and teachers.

1. How do the cases and commentaries attend to larger issues of power, privilege, and opportunity?
2. How do the authors attend to teachers' deficit language and practices?
3. How can the cases and commentaries shape and support teacher and principal professional-development efforts?
4. How have the cases inspired you to engage in research or professional-development activities to continue the conversation of inequities in mathematics classrooms with colleagues, teachers, administrators, and parents?
5. Which aspect of the cases resonates with your personal experiences working with teachers and graduate students on equity in mathematics education?

REFERENCES

Apple, M. W. (1992). Do the standards go far enough? Power, policy and practice in mathematics education. *Journal for Research in Mathematics Education, 23,* 412–431.

Lewis, O. (1966). The culture of poverty. *Scientific American, 215*(4), 19–25.

Martin, D. B. (2015). The collective black and principles to actions. *Journal of Urban Mathematics Education, 8*(1), 17–23.

Nieto, S. (2006). *Teaching as political work: Learning from courageous and caring teachers.* Bronxville, NY: Sarah Lawrence College Child Development Institute Occasional Paper Series.

CHAPTER 15

ARE THESE TWO SIDES OF THE SAME COIN?

Teachers' Commitment to Culturally Relevant Teaching While Holding Deficit Views of Poor Communities

Tonya Bartell
Michigan State University

Lateefah Id-Deen
University of Louisville

Frieda Parker and Jodie Novak
University of Northern Colorado

Inequitable distribution of opportunities to learn mathematics along lines of culture, race, and class is a pervasive problem in mathematics education (DiME, 2007). Efforts to address this problem require recognition that mathematics education is a cultural system. Mathematics teacher educators (MTEs) have a responsibility to support teachers in understanding and responding to the role of culture in mathematics teaching and learning.

Cases for Mathematics Teacher Educators, pages 339–346
Copyright © 2016 by Information Age Publishing
339

This case features a dilemma that occurred in a required master's-level course for secondary mathematics teachers, which aimed to support teachers in understanding intersections between culture, mathematics, teaching, and learning and in examining culturally relevant pedagogy in the context of a mathematics classroom. The overarching goal for the course was for teachers to develop the knowledge, skills, and motivation to implement culturally responsive pedagogy in their mathematics classrooms to help students become intrinsically motivated and, ultimately, become more successful math learners. The course projects aimed to support teachers in expanding their awareness of the role of culture in people's worldviews and to consider how to use this knowledge to improve opportunities for student learning.

One component of the course included teachers reading and reflecting on an article focused on myths of the culture of poverty (Gorski, 2008). In the article, Gorski discussed myths associated with classism. These myths reflect deficit perspectives that define students by perceived characteristics and blame individuals as opposed to systemic conditions that exist. Gorski (2008) argued in opposition of Lewis (1961), who had argued that impoverished communities share certain attributes that stood in their way of academic achievement. Some characteristics of Lewis's hypothesized *culture of poverty* included a lack of educational values, poor work ethic, and abuse of drugs and alcohol. Gorski provided research evidence demonstrating that such characteristics of a supposed culture of poverty were myths and established that economically disadvantaged students do not share predictable beliefs and behaviors. Gorski also explained that many teachers blame poverty for perceived deficiencies instead of implementing interventions to help dispel these myths.

This case focuses on the dilemma that although the course supported teachers in recognizing both the role of culture in mathematics education and the effectiveness of culturally responsive teaching, it may have also served to perpetuate problematic myths about economically disadvantaged students. Next, we provide context about the course and details about the dilemma, share how we responded, and consider what we might do differently in order to leverage teachers' interest in and commitment to culturally relevant teaching in efforts to confront myths about the culture of poverty.

CONTEXT

This course was the only course in the teachers' program at this institution that directly dealt with issues of culture or equity. Many teachers noted that this was the first time they had ever considered culture in a mathematics classroom. There were 16 secondary mathematics teachers enrolled in the course. All of the teachers self-identified as White, and all have bachelor's degrees. Teachers' years of teaching experience ranged from 2 to 19 years, with an average of 7 years and a mode of 3 years. The course instructor did not inquire about the teachers' socioeconomic status. The course was taught online by the first author, a White, middle-class female; the instructor team working on the course included two White, middle-class females.

The course was organized into three modules. The first module introduced teachers to what culture is and how to look for and understand the role of culture in mathematics teaching and learning. In the second module, teachers engaged with theories of culturally responsive teaching (e.g., Ladson-Billings, 1994), theories about motivation, and the link between the two (e.g., Wlodkowski & Ginsberg, 1995). The final module engaged teachers in learning about additional practices, such as complex instruction (Cohen, Lotan, Abram, Scarloss, & Schultz, 2002) that could support them in developing equitable mathematical teaching practices. The discussion about the myth of the culture of poverty occurred at the end of Module 1.

A DILEMMA

Although the course supported teachers' orientations around culture in important ways, it did not fully disrupt teachers' deficit beliefs about a fictional culture of poverty. To elaborate, the teachers came to see mathematics education as a cultural activity, a new realization for most of the teachers; the teachers broadened their understanding of culture to include one's values, beliefs, and views of mathematics; and the teachers expressed a commitment toward culturally relevant teaching in their practice (Bartell, Id-Deen, Novak, & Parker, 2015). Further, deficit views that students' or students' families' cultural views were problems that needed addressing arose for only 2 of the 16 teachers early in the course and, in a postcourse survey about the meaning of culture and culturally relevant pedagogy and culture more generally, deficit views did not surface for any teacher (Bartell et al., 2015). On the other hand, teachers' discussions and reflections immediately after reading The Myth of the Culture of Poverty (Gorski, 2008) unearthed deficit views from nearly all of the teachers. Thus, a dilemma: the course seemed to support teachers in recognizing the role of culture in mathematics teaching and the effectiveness of culturally responsive mathematics teaching, yet it may have also served to perpetuate problematic myths about economically disadvantaged students.

The most common theme in teachers' responses considering the causes of poverty was a belief that poverty was caused in large part "due to the mismanagement of funds"[1] where people "choose to spend their money on objects that keep them going like big cars and fancy things and forget about the basics like food and housing," or "we have all seen the family at Foot

Locker buying $200 worth of merchandise on a weekend and then show up to school without a single supply." Another theme that resonated with many teachers was the idea that poor people "may be perfectly happy with their situation," "almost choosing to live in poverty" because they "feel extremely wealthy and happy with the quality of life that they have for themselves and their family...why would they want to change?" Teachers were also wrestling with these ideas, at times labeling their responses as stereotypes and admitting that they still held to these myths. Some teachers pushed back on the problematic views of their peers:

> You [responding to another teacher] state that "the majority of people who lack education also lack motivation, skills, and work ethic." I realize you did not say "ALL," but even saying "the majority" might not be correct. The cycle of poverty (I think) is a difficult one to break. When a person is born into poverty, it can be very difficult to get an education and break the cycle, even IF they have a strong work ethic.

Additionally, in the postcourse survey, when teachers were asked to state the biggest challenges to teaching mathematics to economically disadvantaged students, half of the teachers expressed a myth about a "culture of poverty" that students and students' families do not value education and that there is a lack of support outside of school. As one teacher reflected at the end of the course,

> The biggest challenge for teaching economically disadvantaged students is the lack of support outside of school. Parents are unlikely to help significantly if their child doesn't understand and are also unlikely to be able to hire a tutor.

For half of the teachers, then, it seemed that they did not see that their beliefs about economically disadvantaged students are in contrast with their commitment to and belief in culturally relevant pedagogy. One might even venture to say that the teachers would not see culturally relevant pedagogy as having any effect for "these" students that some teachers described as not valuing education, unmotivated, or lacking work ethic. As Gorski (2008) warned, these myths "lead the most well intentioned of us...into low expectations for low-income students" and "worst of all, it diverts attention from what people in poverty *do* have in common: inequitable access to basic human rights" (p. 33). The course simultaneously supported teachers in recognizing that "knowledge of students' cultural backgrounds is imperative to good teaching," and identifying and confronting the cultural assumptions that may have implications for mathematics teaching and learning, while also not seemingly supporting them in recognizing and challenging the stereotypical beliefs and myths they held about poor students and families.

CRITICAL ISSUES AND A RESPONSE

Here, we share our critical reflections on this dilemma, including issues at play and how we responded. These reflections are not meant to suggest definitive answers but rather to foster continued conversations aimed at transforming the work of MTEs in supporting teachers' disruption of deficit perspectives.

Metanarratives at Play

It is important to remember that teachers are both products of and actors within inequitable systems (Beauboeuf-Lafontant, 2002). The stories told about the poor often relate to *metanarratives,* or broad stories, passed on through the media and perpetuating ways of framing people of poverty, despite evidence and statistics to the contrary (e.g., Gorski, 2008). These narratives structure ways of talking about groups of people, in this case people in poverty, and frame our individual actions (DiME, 2007).

People Can Change

This case reflects one slice of teachers', as well as our own, growth. We should not freeze teachers in time but rather be open to the possibility of change and transformation. Mistakes will be made, and oppressive misconceptions will be voiced, but these teachers have also expressed a commitment to their students, to improving their teaching, and to culturally responsive pedagogy, positioning us in this struggle together and allowing us to work together to "move through and beyond whatever mistake was committed" (Tran, 2013, para. 12).

Culture, Generally Speaking

We defined *culture,* drawing on Hammer (Watson, 2010 as cited in Hammer, 2012), as a dynamic social system, containing the values, beliefs, behaviors, and norms of a specific group, organization, society, or other collectivity that are learned, shared, internalized, and changeable by all members of society. It is important to note that a decision we made when designing this course was to approach culture from a general perspective rather than focusing on particular cultures or cultural groups. We were concerned that if we delved into specific cultures, we could lead the teachers into essentialism, causing the teachers to assign cultural characteristics to students based on students' race or ethnicity. It is possible that maintaining this general focus masked teachers' beliefs about specific groups, allowing a mythical "culture of poverty"—based on controversial studies of Mexican communities (Lewis, 1961) and poor Black families (Moynihan, 1965)—to perpetuate these racist and classist legacies.

A Response

I, the first author of this case and instructor for this course, chose to respond to this dilemma by posting a long, written instructor reflection on our course site for teachers to read. As this course was being taught in the context of a research study, I did not feel that I had the flexibility to alter the course plans to more explicitly take up teachers' beliefs about poor students. But I also knew that I wanted to raise questions about what I had heard to push back on teachers' claims—in fact, I responded directly to teachers' posts, as shown below.

I began my instructor reflection by reiterating the idea that teachers are "products of larger structures and systems that serve to perpetuate inequity (and myths)" and that "a first step toward assessing and correcting our own biases is examining beliefs around the so-called culture of poverty." I began with these points as I wanted teachers to know that I thought the work we were doing was important and necessary and that it was also situated within broader structures of privilege and oppression, thus allowing us to consider simultaneously our roles in perpetuating inequity and ourselves as victims within this same system. I hoped that in writing this, I would not shut down the virtual conversation we had begun.

Next, I expressed some wonderings I had, such as "whether our statements are still reflective of myths about poor people, in that they reflect a deficit view of individuals, which are so *very hard* to break." I also deliberated "which comes first, poverty or 'lack of education'" and about how education was defined, suggesting perhaps "lack of opportunity to learn" was at play. I expressed these wonderings as questions in an effort to potentially position myself as part of the conversation. It was only after these two moves that I directly countered teacher statements.

In the next section of my posted instructor reflection, I challenged whether "we have evidence to make the following statements" and whether I could "think of a counter-example." I pulled quotes from teachers' posts, such as one where the teacher stated that "poverty is caused by mismanagement of funds," and then I wrote, "I know many non-poor people that mismanage funds, including myself. I imagine that there are many poor people who manage funds much better than I do, because they have to, and the fact that I can mismanage my funds and be 'okay' suggests a privilege I have." Similarly, one teacher's idea that "poverty is caused by a compulsion to spend money" led me to question "why is this unique to people that are poor?" Finally, I thanked teachers for their honesty, again referring to their thinking and work being an important step in one's growth and noting the systemic issues they identified, and concluded with a recommendation for a book they might read if they have time and interest: Barbara Ehrenreich's (2001), *Nickel and Dimed: On (Not) Getting by in America*, and I provided a description of the book. Although not perfect, I felt that this instructor response was a place to start challenging teachers' beliefs.

IF WE COULD DO IT AGAIN

We, the authors of this case, then began thinking about what we might do differently in the future. How might MTEs better leverage teachers' interest and commitment to culturally relevant teaching in our efforts to dismantle teachers' long-held myths about poor students? We do not mean to suggest that the ideas we present here are "the" ways forward. Rather, we share some ideas here for consideration and discussion.

First, to more fully insert the instructor reflection into the conversation, we might ask teachers to incorporate a response to the instructor reflection in their next discussion board post. Second, in retrospect, the instructor's targeted statements seem more focused on individual counterexamples (e.g., I am not poor and I mismanage funds) as opposed to upon systems of privilege and oppression. We would work to incorporate instructor reflections targeted at ideas of classism (Gorski, 2008) and to share research about the systemic condition of inequitable access to high-quality schooling, such as that which documents that poor students are more likely to attend schools that have less funding, larger class sizes, and less-experienced teachers (e.g., Oakes, 1985). We also wondered whether the deficit perspectives voiced suggested that teachers may have held deficit perspectives about students' families but not about the students themselves. We further wondered whether teachers' perceptions of race as immutable versus poverty as changeable by individuals may have been at play. In the future, we would ask additional questions to better understand teachers' perspectives. Understanding more about teachers' perspectives would help us better select activities to engage those particular understandings.

Finally, drawing on Gorski's (2008) article, we might have teachers identify a possible, personal action step they would be willing to take on from the list he provided (e.g., invite colleagues to observe our teaching for signs of class bias, respond when colleagues stereotype poor students or parents). We could build in a course project that aimed to support teachers in beginning to implement and reflect upon their chosen action(s). More work is needed in mathematics education that not only identifies possible pedagogical moves MTEs might employ, as in this book, but also provides research-based evidence as to the effects of these moves.

NOTE

1. These are direct quotes from different teachers reflecting the themes that arose for a majority of teachers in the course.

REFERENCES

Bartell, T. G., Id-Deen, L., Novak, J. D., & Parker, F. (2015). *"More than a side conversation": Secondary mathematics teachers consider culture in the mathematics classroom.* Manuscript submitted for publication.

Beauboeuf-Lafontant, T. (2002). A womanist experience of caring: Understanding the pedagogy of exemplary black women teachers. *Urban Review, 34,* 71–86.

Cohen, E. G., Lotan, R. A., Abram, P. L., Scarloss, B. A., & Schultz, S. E. (2002). Can groups learn? *Teachers College Record, 104,* 1045–1068.

DiME. (2007). Culture, race, power, and mathematics education. In F. Lester (Ed.), *Second handbook of research on mathematics teaching and learning* (pp. 405–434). Charlotte, NC: Information Age.

Ehrenreich, B. (2001). *Nickel and dimed: On (not) getting by in America.* New York, NY: Metropolitan Books.

Gorski, P. (2008). The myth of the culture of poverty. *Educational Leadership, 65*(7), 32–36.

Hammer, M. (2012, November). *The intercultural development continuum.* Presentation at the IDI Qualifying Seminar, Baltimore, MD.

Ladson-Billings, G. (1994). *The dreamkeepers: Successful teachers of African American students.* San Francisco, CA: Jossey-Bass.

Lewis, O. (1961). *The children of Sanchez: Autobiography of a Mexican family.* New York, NY: Random House.

Moynihan, D. (1965). *The Negro family: The case for national action.* Washington, DC: United States Department of Labor.

Oakes, J. (1985). *Keeping track: How schools structure inequality.* New Haven, CT: Yale University Press.

Tran, N. L. (2013, December 18). Calling in: A less disposable way of holding each other accountable. [Web log post]. Retrieved from http://www.blackgirldangerous.org/2013/12/calling-less-disposable-way-holding-accountable/

Wlodkowski, R. J., & Ginsberg, M. B. (1995). A framework for culturally responsive teaching. *Educational Leadership, 53*(1), 17–21.

COMMENTARY 1

RESPONDING TO MATHEMATICS TEACHERS' DEFICIT PERSPECTIVES ABOUT ECONOMICALLY DISADVANTAGED STUDENTS AND THEIR FAMILIES

A Commentary on Bartell et al.'s Case

Richard Kitchen
University of Denver

"Are These Two Sides of the Same Coin? Teachers' Commitment to Culturally Relevant Teaching While Holding Deficit Views of Poor Communities" is a captivating case that challenges the reader to consider complexities involved in work with teachers centered on culture and equity. Specifically, the 16 White secondary mathematics teachers enrolled in the author's course were "compelled" by culturally relevant pedagogy (CRP), but they

Cases for Mathematics Teacher Educators, pages 347–351
Copyright © 2016 by Information Age Publishing
All rights of reproduction in any form reserved.

continued to hold deficit perspectives about economically disadvantaged students and their families. The case implicitly draws attention to teachers' beliefs about the poor and the question of how to deal with teacher resistance to change. In my commentary to this case, I start by introducing myself and my positionality in this work. I will then provide an interpretation of the case. I close by responding to the dilemma outlined in the case.

MY POSITIONALITY

I am a middle-class, White male. Although I am multiply defined, I also recognize that having white skin in the richest country in the world has granted me innumerable privileges and access to countless opportunities. For instance, I benefit from white privilege in simple activities such as being able to visit White middle-class neighborhoods without fear and in my professional life (e.g., some may trust me as knowledgeable in mathematics simply because I have white skin). As the father of three children whose biological father is African American, I have experienced racism firsthand. Our family has been treated poorly in commercial outlets such as stores and has even been denied service in several restaurants. To be clear, whiteness affords benefits to Whites, even those who actively resist it as a social construction designed for the benefit of White people. I started my career in education as a Teaching Assistant in inner-city elementary and middle schools. I have taught at the elementary, middle, high-school, community-college, and university levels and have extensive experience working with diverse student populations in schools located in the U.S. and abroad. I also have 5 years of administrative experience, having worked as the principal of a middle school that almost exclusively served immigrant families from Mexico. Forming positive relationships with the families and community we served was a primary goal for me, and I was devoted to the school being a place where students could intellectually, emotionally, socially, and spiritually grow and thrive.

I believe that at this point in history, the mathematics education research community needs to do more to highlight the economic and racial injustices that continue to plague the U.S. and hinder public schools from providing the sorts of educational opportunities in urban and highly rural schools that serve poor students and students of color that have long been offered to economically advantaged students (Kitchen, DePree, Celedón-Pattichis, & Brinkerhoff, 2007). An important part of this work must include coming to terms with white privilege and the structural classism and racism in which our society is rooted. White privilege has and continues to provide differential access to economic and educational opportunities based upon class and race (Battey, 2013). As a White male who has been involved in work related to diversity and equity in mathematics education for some time, I am committed to contributing to improved access in mathematics for students who have historically been denied such access, students living in poverty and students of color.

INTERPRETING THE DILEMMA

In this case, the authors describe a dilemma in which the teachers enrolled in the first author's class embraced culturally relevant pedagogy, yet the teachers' deficit perspectives about economically disadvantaged students endured. In my opinion, the author approached the dilemma well by initially posting her "wonderings" before countering the teachers' statements. I also like the author's confession about how she is privileged to be able to "mismanage my funds," just like poor people. Lastly, the author is quite thoughtful in her appraisal about how all of us are victims in an oppressive and racist society, yet people can change and learn from their misconceptions. We all need support to confront difficult things like privilege and power. Moreover, having the teachers read Ehrenreich's (2001) book to learn more about the poor and then to have discussions about the book should help get differing viewpoints out in the open.

To the credit of the author and her instructional team, teachers challenged one another about their various perspectives. It is also a tribute to the author and her team that the teachers in the course were willing to share their views about students and their families. Clearly, high levels of trust had been established in the course, which is essential in work focused on examining classism. In my experience, teachers may not be willing to share their perspectives about class out of fear of being judged adversely by peers or the instructor.

I believe the teachers could have been challenged to clarify what specifically they were referring to when they expressed that there is "a lack of support outside of school" for their economically disadvantaged students. Were they expressing the belief that there is a lack of academic support provided by families in the home? By high school, many parents, not just parents of economically disadvantaged students, may in fact struggle to provide academic support in mathematics for their children (Kitchen et al., 2007). I would also want the teachers to elucidate what they meant by the statement "students' families do not value education." I think it would be important to have all the teachers, not just those who expressed this idea, discuss their past experiences with families of economically disadvantaged students as a means to more deeply examine why some of the teachers were advancing the notion that these families do not value education.

RESPONDING TO THE DILEMMA

I applaud and agree with the author and her team's insights concerning how they could do things differently if given the opportunity. Specifically, I agree that there is a need to more explicitly focus on "systems of privilege and oppression" in future classes, rather than just on individual counternarratives as well as for the need to bring in more literature on classism (as well as on structural racism). For instance, enrolled teachers could explore the "geography of opportunity" (Tate, 2008, p. 397), that educational opportunities and access to them are influenced by where one lives. Brynes and Miller (2007) found that SES has direct effects on mathematics achievement because of privileges afforded students in high-SES communities (e.g., access to well-trained teachers) and indirect effects on opportunities students have to enroll in advanced mathematics classes in high school (i.e., students in low-SES communities may be less inclined to enroll in more advanced courses).

I also agree about the need to "ask additional questions to further understand teachers' perspectives as they relate to students and their families." An appropriate response would be for the author and her team to continue to engage with the teachers in intentional dialogue about the various perspectives that have been expressed. A goal here should be to provide a forum where everyone involved in the class can share their perspectives with one another and then explore these diverse perspectives without fear of judgment.

It also seems reasonable to me for the instructor and her team to continue to follow up by not only posting to the teachers why they believe some previous postings were problematic, but also asking if others agree with their particular takes on these postings. For example, how do others in the group view the notion that by high school many students may not have academic support in mathematics at home? Do the teachers view this as a deficit perspective? Perhaps a next step is to dialogue about actual strategies that the teachers are using or could use to meet the academic needs of their economically disadvantaged students.

Finally, I am drawn to the recommendation made near the end of the case about inviting colleagues to make classroom observations to look for signs of class bias, and recommend that race be examined as well. In addition to poverty, student access to a challenging standards-based mathematics education is influenced by race, ethnicity, and English language proficiency (Gutiérrez, 2008; Martin, 2013). Low academic expectations have historically been the norm at schools that serve both low-income communities and students of color (see, for example, DiME, 2007; Kitchen et al., 2007; Payne & Biddle, 1999). Given the research that demonstrates how both class and race influence the expectations that mathematics teachers hold for their students, it would be worthwhile for teachers to begin to systematically work together to explore how classism and racism are influencing their instruction.

REFERENCES

Battey, D. (2013). Access to mathematics: A possessive investment in Whiteness. *Curriculum Inquiry, 43*, 332–359.

Brynes, J. P., & Miller, D. C. (2007). The relative importance of predictors of math and science achievement: An opportunity-propensity analysis. *Contemporary Educational Psychology, 32*, 599–629.

DiME. (2007). Culture, race, power and mathematics education. In F. K. Lester (Ed.), *Second handbook of research on mathematics teaching and learning* (pp. 405–433). Charlotte, NC: Information Age.

Ehrenreich, B. (2001). *Nickel and dimed: On (not) getting by in America*. New York, NY: Metropolitan Books.

Gutiérrez, R. (2008). A "gap gazing" fetish in mathematics education? Problematizing research on the achievement gap. *Journal for Research in Mathematics Education, 39*, 357–364.

Kitchen, R. S., DePree, J., Celedón-Pattichis, S., & Brinkerhoff, J. (2007). *Mathematics education at highly effective schools that serve the poor: Strategies for change*. Mahwah, NJ: Erlbaum.

Martin, D. B. (2013). Race, racial projects, and mathematics education. *Journal for Research in Mathematics Education, 44*, 316–333.

Payne, K. J., & Biddle, B. J. (1999). Poor school funding, child poverty, and mathematics achievement. *Educational Researcher, 28*(6), 4–13.

Tate, W. F. (2008). "Geography of opportunity": Poverty, place, and educational outcomes. *Educational Researcher, 37*, 397–411.

COMMENTARY 2

TEACHING PRIVILEGE ABOUT EQUITY

A Commentary on Bartell et al.'s Case

Brian R. Lawler
Kennesaw State University

The Culture in a Mathematics Classroom (CIMC) course served high-school mathematics teachers who lacked knowledge of the role of culture in a maths classroom and of culturally relevant pedagogy (CRP). The case authors' dilemma seems to be that at the end of the course, the teacher-learners continued to express problematic beliefs about underachieving students. It appears they failed to embrace the view that mathematics education is a social system and as a result did not question the system, mathematics, or their role within it.

An alternative view of the dilemma may be that the teacher-learners were disinclined to confront their powered and privileged participation in "mathematics *knowing* as a cultural activity" and "mathematics *learning* as a cultural enterprise" (Nasir, Hand, & Taylor, 2008, p. 194, italics added). I suggest that rather than viewing the results of the course as promulgating the deficit viewpoints of the teacher-learners, the course failed to disrupt the

Cases for Mathematics Teacher Educators, pages 353–357
Copyright © 2016 by Information Age Publishing
All rights of reproduction in any form reserved.

teacher-learners' consideration for their own privileged role within mathematics education. It may also be that the instructors failed analogously.

POSITIONALITY

I am not a race, poverty, or culture expert. I grew up the eldest male of a White, Midwestern family. I am tall, heterosexual, and work among the academic and intellectual elite. In short, I lurk amid the summit of culturally assigned privilege. Other than living inside that privilege, my present positionality has also been informed by early family life as activists at Milwaukee's Open Housing marches, hosts of civil-rights rallies, and members of the Open Door Society; as a youth I was often in settings not as a member of the racial majority.

What *do* I possess that may qualify me to respond? I have graduate degrees in mathematics education and mathematics—that version of mathematics taught to students in the White schools of the United States; I also taught this mathematics in high school. For 20 years I have conducted professional development (PD) for high-school mathematics teachers, PD that begins with a premise that all children do interesting mathematical things, and that a schooling could and should allow for each to express one's own proclivities (Dewey, 1902). I have worked closely with development and implementation of the Interactive Mathematics (IMP) curriculum and am a contributing author. My research is on power and privilege in maths education (Lawler, 2010), grounded in a radical constructivist theory of learning and a poststructural and postepistemological status for knowledge. Such an orientation to knowledge defies the possibility for one's own knowledge to be more true than another's, disrupting hierarchical relations of knowledge and thus of persons based upon knowing.

I have an extreme aversion to authoritarian relations. I value liberty, equality, and solidarity (Mueller, 2012). I am deconstructive[1] in my experiencing of the world, seeking to disrupt structures at work in text and experience, not to discard but to reinscribe. I see the goal for education to be social equality (Mueller, 2012), not merely a rally cry in support of the subaltern but "an emancipating precondition for *all* to actualize themselves fully" (p. 18). In such an orientation to social justice, individuals recognize dependence on others' attainment of full humanity so that they themselves can pursue their own limits of intellectual, emotional, and relational capacities (Lawler, 2012).

DECONSTRUCTING THE CASE

In this commentary, I ignore elements of the course structure that likely contributed to the dilemma, particularly the lack of personalization and social responsibility present in face-to-face interactions. Instead, I respond more broadly, having experienced similar results when addressing issues of social equality with high-school mathematics teachers.

I do not foreground race and poverty in a traditional manner; rather I bring forth the privilege of mathematics, the identities it creates in its knowers, priests, teachers, sages.[2] Race and poverty may be easy to recognize and easily excused signifiers for underlying power structures, yet I contend that the more insidious beast is the intellectual classism the system of mathematics education (Nasir et al., 2008) has established and is being propagated here. Although the CIMC course may have appropriately troubled the oft-ignored role of culture in the mathematics classroom, evidence supports that it did little to disrupt both the privileged status of mathematics as a way of knowing, and the colonizing power of mathematics education. *I wonder if by avoiding a deep examination of the teacher-learners' own privilege—as teachers of mathematics as well as by more traditional, demographic metrics such as class—the discussion of culture that did emerge necessarily perpetuated problematic myths of the Other.*

Following on this wondering, as a second consideration I again sidestep the foregrounding of demographics and instead interrogate how the teacher-learners view their work, as teaching mathematics or teaching children? Said less diametrically, do the teacher-learners listen hermeneutically (Davis, 1997) alongside children? Do they view their work as participants in the activity of students? Do both students and teachers revise their mathematics in the social coupling of the high-school classroom? Is authority displaced, shared, collective? Do their children see themselves in the curriculum (Alper, Fendel, Fraser, & Resek, 1997)? *I contend that when the teacher assumes a role to educe mathematical activity from kids, the necessity to listen and co-construct a mathematics will supersede a culturally relevant pedagogical theory.* Further, the need to learn and embed into pedagogy each child's cultural mores will be dissolved as *the* quick fix to complex issues of power and privilege fabricated by hierarchal views of knowing.

Such hierarchy underscores a third deconstructive move, which is to notice the deficit orientation toward the teacher-learners as they are constructed in the case itself. The teacher-learners are identified as deficient from knowing about intersections of culture, mathematics, and their work as teachers. What funds of knowledge they may possess seems neither recognized nor built upon. Practices of CRP seem not to have been applied in the pedagogical design for interactions with the teacher-learners of the CIMC course.

Instead, I contend that the course is structured with an orientation toward delivery of knowledge. Not only are the teacher-learners' own cultures *and* ways of knowing disregarded, but there also is *a* knowledge to be *acquired* rather than *constructed,* or *co-constructed.* As a result, the pedagogical design appears to hope for "discovery" of that particular knowledge, which

I contend is pernicious and *iniquitable* (Lawler, 2012). A pedagogy of discovery serves to sediment and enhance inequitable power relations, maintaining an "I know and you don't" positionality. Discovery pedagogy may be impossible to avoid when an *end*, whether knowledge or belief, is held as the instructional goal. Such an end is a teacher's ideal, a spook (Stirner, 1845/2005), and success toward that end is judged by that teacher.

It may be possible that a Funds of Knowledge or CRP pedagogical orientation by the authors would help to redesign the course and its objectives, shape differently the instructors' decisions made during the course, and ultimately create a more profound interrogation and revision for knowing mathematics education for both the teacher-learners and the course instructors.

ALTERNATIVE DESIGN

To close, I suggest two possible course activities. First, as course instructor I would attempt to teach with the premises of CRP, learning and then utilizing the teacher-learners' funds of knowledge. Begin the course with a sincere effort to engage the learning community with one another. This might be an initial journal entry for instructor and teacher-learners alike to describe their own culture and beliefs about economically disadvantaged students.

Second, I suggest reorienting the action-plan task toward one defined and directed by the teacher-learners of the course. Rather than require the teacher-learners to take an "action step," provoke self-directed inquiry (Cochran-Smith, 1994). After the coupled study of cultures' role in mathematics teaching and CRP, (a) ask teacher-learners to wonder about ways in which their own expectations for student success may be steeped in deficit views of students, families, or culture and privileged perspectives on maths or knowledge that make themselves as teachers power-full, and position others as novice, docile, apprentice, or otherwise unequal. Next, teacher-learners (b) design an empirical study to attempt to capture data regarding this wondering. Third, teacher-learners (c) design a tool to collect and a method to analyze this data. Teacher-learners must define how the tool and analysis is constructed to overcome the bias their privilege brings to the study. Of course, this action research project would conclude with (d) a report of findings.

To close, I too now must deconstruct my comments from my owned privileged point of view. I point toward one disciplining problem pervasive in my response, that of my hypervaluation of knowledge and knowing. My commentary may be another, or the same, metanarrative at play; that the learner is dysfunctional and the knower can repair. *Might a refiguration of the teacher/ student narrative destabilize our privileged status as mathematics teacher educators?*

NOTES

1. Deconstruction may be "a persistent critique of what one cannot not want" (Spivak, 2013, p. 28).
2. I suggest that *race* and *poverty* are replacement terms that work to redirect more sinister issues of status, privilege, and power.

REFERENCES

Alper L., Fendel, D., Fraser, S., & Resek, D. (1997). Designing a high school mathematics curriculum for all students. *American Journal of Education, 106,* 148–178.

Cochran-Smith, M. (1994). The power of teacher research in teacher education. In S. Hollingsworth & H. Sockett (Eds.), *Teacher research and educational reform: Ninety-third yearbook of the National Society for the Study of Education* (pp. 142–165). Chicago, IL: University of Chicago Press.

Davis, B. (1997). Listening for differences: An evolving conception of mathematics teaching. *Journal for Research in Mathematics Education, 28,* 355–376.

Dewey, J. (1902). *The child and the curriculum.* Chicago, IL: University of Chicago Press.

Lawler, B. R. (2010). Fabrication of knowledge: A framework for mathematical education for social justice. In U. Gellert, E. Jablonka, & C. Morgan (Eds.), *Proceedings of the Sixth International Mathematics Education and Society Conference* (pp. 330–335). Berlin, Germany: Freie Universität Berlin.

Lawler, B. R. (2012). The fabrication of knowledge in mathematics education: A postmodern ethic toward social justice. In A. Cotton (Ed.), *Towards an education for social justice: Ethics applied to education* (pp. 163–190). Oxford, England: Peter Lang.

Mueller, J. (2012). Anarchism, the state, and the role of education. In R. H. Haworth (Ed.), *Anarchist pedagogies* (pp. 14–31). Oakland, CA: PM Press.

Nasir, N. S., Hand, V., & Taylor, E. V. (2008). Culture and mathematics in school: Boundaries between "cultural" and "domain" knowledge in the mathematics classroom and beyond. *Review of Research in Education, 32,* 187–240.

Spivak, G. (2013). Bonding in difference, interview with Alfred Arteaga. In D. Landry & G MacLean (Eds.), *The Spivak reader: Selected works of Gayati Chakravorty Spivak,* (pp. 15–28). London, England: Routledge.

Stirner, M. (2005). *The ego and his own: The case of the individual against authority.* New York, NY: Dover. (Original work published 1845)

WHAT ARE WE DOING WHEN UNDERSTANDING CULTURE IS NOT ENOUGH?

A Commentary on Bartell et al.'s Case

Crystal H. Morton
Indiana University Purdue University Indianapolis

MY POSITIONALITY

My upbringing in a low-income household, a household where our economic position was not viewed as a deficit, strongly impacted my lens for analyzing this case. For my family and members within my community, failure in school was not an option. My parents, family, and community members provided me with assets that continue to support my professional and personal growth today.

Because of the support I received from my parents, family, and community members, I was able to successfully pursue my education to become a high-school mathematics teacher and now mathematics teacher educator. As a mathematics/teacher educator and former high-school mathematics teacher, I can relate to the specific issues raised in this case. In my

Cases for Mathematics Teacher Educators, pages 359–363
Copyright © 2016 by Information Age Publishing

experiences, I have encountered and worked to challenge the deficit views held by preservice and inservice teachers about students from low-income backgrounds, students like me. Challenging these views was often an arduous task because it was extremely difficult to shift someone's mindset. This shifting is especially difficult when teachers do not realize or do not want to admit to having deficit views of children.

As an African American female student, I personally experienced the benefits of the incorporation of culture into the teaching and learning of mathematics. Mathematics became more enjoyable to learn when I saw myself in the curriculum and likewise to teach when my students saw themselves in the curriculum. As an educator, I believe that "if the curriculum we use to teach our children does not connect in positive ways to the culture the young person brings to school it is doomed to fail" (Delpit, 2012, p. 21). Incorporating a socially transformative approach to curriculum is also essential. A socially transformative approach to curriculum addresses students' culture as well as the social and historical conditions of the students' community and how students can use the curriculum as a tool for social change (Mutegi, 2011). For example in a mathematics course, students may explore the concept of *big data* from a socially transformative perspective. This perspective would position students to address five areas of mastery, content (What is big data?), context (How is big data relevant to human beings?), currency (How is big data relevant to people of African descent?), critique (How is big data contributing to the maintenance of systemic racism?), and conduct (Given what we know about big data, what should we do?) (Mutegi, 2011).

INTERPRETING THE DILEMMA

The case, "Are These Two Sides of the Same Coin? Teachers' Commitment to Culturally Relevant Teaching While Holding Deficit Views of Poor Communities" speaks to the challenge of unearthing deeply entrenched deficit views of students of poverty. In my teaching experiences, the terms students from low-income backgrounds, students from poverty, and the like often served as code words for Black and Brown children. As such, the case raises several issues about what we are doing and not doing in mathematics education courses to prepare teachers for diverse populations.

Milner and Laughter (2015) found that inservice and preservice teachers' anxiety about teaching students who live in poverty pale in comparison to their concerns about teaching Black and Brown children. What are we doing to prepare mathematics teachers for situations in which they will teach Black and Brown children living in poverty? What are we doing to support teachers in developing a critical consciousness and becoming self-reflecting teachers who can recognize and address race and poverty both individually and in their intersections?

Other dilemmas present in this case were that the Culture in a Mathematics Classroom (CIMC) course was the only course where teachers explored the role of culture in the teaching and learning of mathematics. We live in a rapidly diversifying society, and it is imperative that we prepare all teachers to teach meaningful, relevant, and transformative mathematics to ALL learners. Secondly, all students enrolled in the course were White. This demonstrates clearly that the diversity of our teaching force does not reflect the diversifying student population (Gay & Kirkland, 2003).

When Understanding Culture Is Not Enough

Reading this case reminded me of how difficult it is to separate issues of race from issues of poverty; their intersections are too embedded. For example, there are a disproportionate number of Black and Brown children living in poverty. Living in poverty increases the likelihood of not having access to quality schools and teachers and other societal benefits afforded to the economically privileged (Milner & Laughter, 2015). In order to understand the cycle of poverty, one most explore and understand the systems of structural racism prevalent in the fibers of our society.

The CIMC course was effective in helping teachers value the role of culture in the teaching and learning of mathematics, but the course did not unearth teachers' deficit views of students of poverty. The CIMC course provided the space to explore issues related to culture in general, but the instructors acknowledged "this general focus masked teachers' beliefs about specific groups, allowing a mythical "culture of poverty."

RESPONDING TO THE DILEMMA

If I were the instructor represented in this case, I would have responded similarly. I believe it is important to post an overall reflection to teachers' comments reminding them that they are "products of the larger structure" but at the same time helping them to critically analyze and deconstruct those structures. Teachers must examine their own stances on the issues of poverty and educational achievement (Gorski, 2013). In my response to students, I would ask specific questions to challenge their views. For example, in response to the quote on page 5 that ends by saying, "Parents are unlikely to help if their child doesn't understand and are also unlikely to be able to hire a tutor," I would first ask teachers to define what is meant by "a lack of support" and ask teachers to consider what systemic structures contribute to limiting parents' financial ability to hire a tutor. I would challenge students to consider parents that seek out and advocate for free or reduced-rate tutoring support for their children. My hope is that this questioning would push teachers to see counterexamples to their deficit views.

Additionally, I would share my personal stories and accounts from members of my immediate and extended family as counterexamples to the dominant deficit discourse around students from poverty. In these examples, I would highlight the many assets passed on to me by my parents and members of my community, assets that greatly impacted my education and my overall well-being.

I would end my reflective response with a list of supplemental readings, such as *The Brilliance of Black Children in Mathematics: Beyond the Numbers and Toward New Discourse* (Leonard & Martin, 2013), to help encourage and empower teachers to view students from a point of strength and not deficiency. Even though there is little to no flexibility to change course content, I would set up online sessions (via Google Hangout, Adode connect, etc.) to meet with students in smaller groups in order to provide an alternative space for them to talk through their views and the rationale behind those views with other students and their professor. These online conversations could create a space for students to deconstruct their deficit views (Caruthers & Friend, 2014).

Since I have more flexibility in my current context, I would respond to the current dilemma by doing the following:

- conduct a postsurvey of teacher beliefs that aligns with the current instructor's recommendation on page 9,
- integrate real-time activities that challenge the teachers' notions of classism, race, and poverty early on in the semester, and
- continue to focus on the role of culture but also integrate aspects of socially transformative curriculum espoused by Mutegi (2011).

Although it is important that we as mathematics educators and classroom teachers continue to focus on the role of culture in mathematics teaching and learning, it is equally important for us to focus on the role of race and poverty.

REFERENCES

Caruthers, L., & Friend, J. (2014). Critical pedagogy in online environments as thirdspace: A narrative analysis of voices of candidates in educational preparatory programs. *Educational Studies, 50,* 8–35.

Delpit, L. D. (2012). *Multiplication is for White people: Raising expectations for other people's children.* New York, NY: The New Press.

Gay, G., & Kirkland K. (2003). Developing cultural critical consciousness and self-reflection in preservice teacher education. *Theory Into Practice, 42,* 181–187.

Gorski, P. (2013). *Reaching and teaching students in poverty: Strategies for erasing the opportunity gap.* New York, NY: Teachers College Press.

Leonard, J., & Martin, D. B. (Eds.). (2013). *The brilliance of black children in mathematics: Beyond the numbers and toward new discourse.* Charlotte, NC: Information Age.

Milner, R., & Laughter, J. (2015). But good intentions are not enough: Preparing teachers to center race and poverty. *The Urban Review, 47,* 341–363.

Mutegi, J. (2011). The inadequacies of "science for all" and the necessity and nature of a socially transformative curriculum approach for African American science education. *Journal of Research in Science Teaching, 48,* 301–316.

CHAPTER 16

HOW DO I LEARN TO LIKE THIS CHILD SO I CAN TEACH HIM MATHEMATICS

The Case of Rebecca

Mary Q. Foote
Queens College—City University of New York

INTRODUCTION—CONTEXT

Over a span of several years and as part of a university-school district collaborative project in a moderately sized city in the Midwest (approximately 230,000 residents and approximately 24,000 students in grades K–12) teachers participated in professional development (PD) focused on mathematics and equity. One of these PDs was a semester-long teacher study group comprised of six elementary school teachers from the same school. This particular school is a large one for the district with more than 650 students and 42 classrooms in grades K–5. Similarly to the year of the PD, during the 2014–2015 school year, 73% of students were eligible for free or reduced-priced lunch, 40.3% received English-as-a-second-language services, and 12.9% special-education services. The composition of the student body was 37.1% Hispanic/

Cases for Mathematics Teacher Educators, pages 365–371
Copyright © 2016 by Information Age Publishing
All rights of reproduction in any form reserved.

Latino, 30% Black or African American, 23.2% White, 7% Multiracial, and 2.7% Asian. (All statistics from school district website, purposefully not cited for anonymity.) A district class-reduction initiative meant that in 22 of the 32 district elementary schools (including the one in this study), the number of students in classrooms in grades K–3 was reduced to 15.

I facilitated the PD being reported on in this case, which was also a research site. The purpose of the study group was for teachers to examine how learning about a student's mathematical thinking and their in- and out-of-school competencies as well as how examining cultural differences between them and their students might support their teaching of students who were different from themselves. To accomplish this, each teacher conducted a case study of a student from her classroom who struggled with mathematics. In order to minimize issues of essentializing based on comparisons across cultural groups of students, the teachers were all asked to choose an African American learner. The teachers and I were all White.

The activities of the PD included teachers observing and interacting with the case-study student during the course of their teaching, shadowing the student for one day during the semester, and meeting with parents, which in all cases was the mother. These activities were discussed in weekly PD sessions. Each teacher was asked to have informal discussions with their case-study student in order to learn from them about activities outside of school that they participated in and enjoyed. They were also asked to notice, within the classroom setting, the activities that the student chose to participate in when given a choice. The shadowing provided the opportunity for the teacher to observe the child in situations to which a teacher does not usually have access such as lunch and recess, pull-out programs, and special classes such as art and gym, as well as to observe the child in the classroom itself, but from the position of an observer rather than a participant in the activity of the classroom. The meetings with the mother supported the teachers to learn more about the student's out-of-school interests, competencies, and funds of knowledge (their cultural, linguistic, home, and community knowledge) by reviewing photos that the mother had taken of her child. The mothers had been asked to photograph their child when they were (a) engaged in an activity that was particularly interesting to them, (b) engaged in an activity at which they were particularly competent, (c) engaged in a household routine such as cooking or grocery shopping, and (d) engaged in an activity that involved mathematics or attention to number. In order to take field notes for research purposes, I accompanied the teachers to the meetings with the mothers. Along with work around the PD just described, I also spent one or more periods per week in the classrooms of the teachers in the study group in order to support them in their teaching. Due to this, I also had many casual conversations with them about their teaching and their case-study student.

CASE DESCRIPTION

The case that follows is that of Rebecca (a pseudonym), one of the teachers in the study group. She taught a combined 2nd/3rd-grade class of 15 students and had 9 years of teaching experience. Rebecca began and ended the PD struggling to understand the mathematical thinking of her case-study student, an eight year old in the third grade whom I will call Robert, although she spoke on several occasions about the need to find and work within children's "zones of proximal development." In an early session, Rebecca reported that she found Robert easily distractible in the classroom setting, somewhat unmotivated, a bit uncoordinated, and immature socially. Robert did present as an immature third grader, playful, and distractible. Robert was a child who was tall for his age; since taller children are sometimes expected by the teacher to be more mature than their shorter classmates are, this may have influenced Rebecca's opinion of his level of maturity. She found his behavior to be more like the second graders in her 2nd/3rd-grade class than the third graders. For example, Rebecca felt that Robert was more interested in using base-ten blocks to build with than using them as a problem-solving tool. She did, however, recognize that this interest in manipulating three-dimensional objects extended to making symmetrical designs with pattern blocks, an activity he completed successfully; yet this interest and show of competence in this area of mathematics was not built on by Rebecca. She also seemed at the beginning of the year to have some understanding of his mathematical thinking and performance in the area of number and operations. She reported in an early PD session that he relied on direct-modeling strategies when the problems contained larger numbers (numbers between 20 and 100), and she recognized that his explanations of his solutions didn't always match the problem. For example, when he solved a problem in which someone had 22 cats and was given 8 more, he arrived at a correct solution of 30 but then explained his thinking by saying that $5 \times 6 = 30$. Her initial concern was to support him in representing his thinking on paper and verbalizing that thinking. In terms of knowing about the student's interests outside of school, even at an early point in the year, she recognized that Robert had an interest in video games because he talked about them frequently. So we see that Rebecca did have some early knowledge of Robert's mathematical thinking and his interests and competencies. Yet her understanding of his abilities progressed little throughout the course of the PD.

Later in the semester, Rebecca, referring to her understanding of where Robert was functioning in mathematics, said, "I'm still figuring out the puzzle. I don't think he's progressing because I still haven't found out what his zone of proximal development is." By this, Rebecca led me to believe that she meant that she hadn't done well in finding that line between what

Robert could do mathematically and what appropriate next steps to take with him would be. As a nascent researcher, I perhaps did less than I might have to provide feedback to teachers about their interactions with students so as not influence the data. And yet, I did provide feedback to Rebecca following visits I made to her classroom. She was fairly resistant to my suggestions. She was frustrated by Robert and resisted taking responsibility for her continuing inability to "like" the child.

In the meeting with Robert's mother to discuss the photographs that she had been taken, three major interests that emerged were video games, sports, and cooking. Robert's interest in cooking was confirmed by his mother, who reported he enjoyed both real and pretend cooking and owned an extensive set of play dishes and pretend food. None of these three interests of Robert's, all of which might conceivably have led to rich mathematical discussions or activities within the classroom, were taken up by Rebecca. These were missed opportunities for connecting out-of-school knowledge to in-school mathematics. Furthermore, during the conversation with Robert's mother at the meeting, Rebecca was very negative about his involvement with video games, telling the mother that she (Rebecca) was not going to allow her young child the same access to video games when he was older. In our conversation after the meeting, Rebecca also expressed negativity about Robert's interest in pretend cooking. She indicated to me that she thought it was a babyish pursuit for a child his age. Her difficulty in connecting to any of Robert's interests continued throughout the semester. In her reflective writing after the penultimate PD session, Rebecca asked, "How can I incorporate his love for video games and sports, even though I have no interest in either?" The question points to Rebecca's struggle to try to connect with the student's interests. This problem of not being interested in activities that interest a particular child is undoubtedly a problem not unique to Rebecca but rather one that may easily be shared by many teachers. In this case, however, no resolution was reached to this question. No inroad was found.

In the later part of the semester, Rebecca noted, "I think that in some ways he and I have a personality conflict. I just haven't figured out how to reach him with my own personality. . . . The more I push him the more he's going to lapse into his imaginative world, and not have a true sense of wanting to learn." She struggled to build a relationship with the child that for a variety of reasons did not come easily. She found it difficult to contend with what she saw as Robert's lack of commitment to his own learning. She also was frustrated by his tendency to give any answer, to guess wildly in order to present an answer to a question. In the middle of the semester, Rebecca explained it this way, "He's so concerned about getting it wrong. . . . He's just waiting for someone else to tell him it's right or wrong. There's no internal motivation to feel good about problem solving." Rebecca seemed to base her analysis of the child having no internal motivation on her view of his mother as overly dominating, as someone who provided Robert with external motivation in many cases. As the study group progressed, Rebecca grew to understand that her demeanor toward the child might be feeding into this dynamic and damaging the learning environment. She made small steps toward taking responsibility for her part in squelching Robert's ability to explain his thinking and problem-solving strategies. At the final session of the study group, in her brief update to the group about Robert she said, "In math I've been more patient with him . . . instead of being so authoritarian." She was hoping that with other tactics he might be less anxious and more able to express his thinking so that she could better understand his mathematical thinking and problem solving.

THE DILEMMA

Rebecca was one of the two teachers in the PD who struggled with making a connection to and building a relationship with both her case-study student and the student's mother. She found both of them difficult and frustrating to work with, and this impacted her willingness or ability to focus on the student's strengths and interests, leading to lost opportunities both for making potentially fruitful connections to meaningful contexts for mathematical problem solving and to building a solid working relationship with the child. Furthermore, Rebecca often voiced negative perceptions about the mother (such as the comment that she [Rebecca] did not approve of video games) that, although not explicitly racist, seemed not to take into account parenting practices that might not align with her own middle-class, White parenting practices. Comments by Rebecca to the mother that she (Rebecca) did not approve of the child's engagement with video games or her comments to me that she thought of Robert's interest in pretend cooking as babyish were not respectful of decisions that Robert's mother had made about his out-of-school activities. She was quick to judge the situation instead of trying to find a way into understanding why things transpired the way they did in this household.

What was difficult for me was the extent to which Rebecca appeared to dislike and be dismissive of the child and his mother. I could see that there was a sort of clash of personalities between the mother and Rebecca, at least from Rebecca's perspective. Rebecca wished it were otherwise but seemed to indicate that it was outside of her ability to make connections with either mother or child. Furthermore, she was resistant to comments and suggestions from other teachers in the PD and myself as to steps she might take to reorient her perspective toward the case-study student and his mother so as to support the student in his learning of mathematics.

For example, although Rebecca knew that her target student had an avid interest in cooking that had begun at an early age as well as an interest in sports and video games, she never discussed the idea of building on her student's competency in cooking, even when other participants and I probed the issue during study-group sessions. In one dramatic instance, Rebecca turned her chair around and sat so that her back was facing into the circle at the study group, cutting off eye contact between her and the rest of the group. She struggled with the fact that she personally had no interest in either sports or video games, and she said she believed this made it difficult for her to connect with those interests of the child. In addition to the formal PD meetings, Rebecca and I had opportunities when I was visiting her classroom over the course of this semester to talk about the situation with this student. She most often presented any thinking around the student and his mother in a negative light. Rebecca had grave difficulty relating to and even liking the student and yet was resistant to suggestions as to how to ameliorate the situation. I was frustrated by the way she related to the child and his mother. This was the core of the dilemma for me. Unlike Rebecca, I had considerable sympathy for Robert's mother. Robert's mother seemed to me to be actively and assertively trying to support her young Black male child's academic success; and yet this went unappreciated by Rebecca, who read the mother's actions as pushy.

REFLECTION ON THE DILEMMA

I understand that teachers' receptivity to learning can be constrained by many factors. Personal situations can leave them less available for examining their practice or their relationships with students. Whatever the reason, it was very difficult for Rebecca to reorient herself toward the student. I was uncomfortable pushing her further during PD sessions or in personal communications because of a fear of alienating her. It's a delicate balance to walk with a teacher. My frustration with Rebecca could conceivably be read as a situation parallel to her frustration with Robert and his mother. Do I need to accept some responsibility for Rebecca's inability to connect with this student and his mother? Did I in some way not meet her where she was, not attend to her needs? If I had had the opportunity to re-engage with this situation, I might have tried to find a time to interact with Rebecca more informally or perhaps to meet with her and a teacher friend she worked closely with. This teacher had a classroom next door to Rebecca and knew Robert well. It might have been the case that the teacher friend would have had some positive insights into Robert and that Rebecca might have been more receptive to hearing from her.

Rebecca began to take a few steps toward the end of the semester, realizing that there was something in the way SHE was interacting with the student that may have been contributing to his anxiety around mathematics and his behavior in the classroom more generally. Perhaps this indicates that she needed more time to process the situation and to be able to accept responsibility for her own complicity in the way the relationship with the student had developed. Maybe it was unrealistic of me to have been disappointed that Rebecca was not able to reflect and reorient toward this child on a schedule that coincided with our work in PD.

COMMENTARY 1

EXAMINING INTEREST CONVERGENCE AND IDENTITY

A Commentary on Foote's Case

Robert Q. Berry III
University of Virginia

MY POSITIONALITY

My roles as a researcher, a mathematics teacher educator, and a former elementary and middle-grades teacher of mathematics are the impetus for me to provide perspectives on how Robert and his mother are positioned by Rebecca. As a mathematics teacher educator, I have asked teachers to engage in activities requiring them to know and understand the experiential, familial, cultural, and communal resources students bring to classrooms to understand the strengths and motivations of students. My experiences as a teacher and teacher educator influence how I frame Rebecca's teaching and interactions with others throughout the professional development. I cannot discount the fact that being a Black man influences ways in which I make sense of the complexities of negotiating identities. I have

Cases for Mathematics Teacher Educators, pages 373–377
Copyright © 2016 by Information Age Publishing
All rights of reproduction in any form reserved.

done research and developed academic and summer programs focused on the schooling and mathematical experiences of Black boys. This work has allowed me to position myself as close as possible to understanding the perspectives of Black boys who are developmentally evolving and in need to support and mentorship. Rather than minimize my roles and identities, I use these to provide a perspective for framing this case around the constructs of interest convergence and identity.

INTERPRETING THE DILEMMA

The activities described in the professional development provide a context in which teachers can learn and develop an understanding of the resources that students bring with them to classrooms. Students' out-of-school interests, competencies, and funds of knowledge represent the kinds of resources students bring to classrooms and in many cases represent students' motivations. Rebecca engaged in activities, such as observations, shadowing, and meeting with Robert's mother, that provided her with opportunities to learn about Robert's resources and motivations. These activities provided Rebecca the context for understanding and interpreting who Robert is, how Robert and his mother see him, how other teachers and students see him, and how Robert acts as a result of these understandings and interpretations. Rebecca's observations and comments suggest that she did not understand that her interactions and dispositions towards Robert had a tremendous effect on the ways she accessed his knowledge and understandings, and the ways Robert participated in class. She appeared to be egocentric and focused on interest convergence as evidenced by asking, "How can I incorporate his love for video games and sports, even though I have no interest in either?" and stating, "I think in some ways he and I have a personality conflict. I just haven't figured out how to reach him with my own personality." These statements suggest that Rebecca is less interested in knowing and understanding Robert's motivations and more interested in how Robert's interests do not converge with her interests, values, and personality. Because of the incongruence, she positions Robert and his mother as deficit by ignoring Robert's interests and admonishing Robert's mother for allowing him access to video games. Further, she positioned pretend cooking as "babyish," which contributed to her view of Robert as being immature.

It is not clear whether Rebecca had an opportunity to reflect on shadowing Robert in a way that would allow her to see how or whether Robert used his resources and interests in other school settings. I wonder if other educators in the school had an appreciation for knowing what it is like being a third-grade Black boy who is tall for his age having interest in video games, sports, and cooking. Because of Robert's size, I am concerned that

Rebecca did not see his interests as age-appropriate and developmentally evolving. Seeing Robert as a young Black boy who is developmentally evolving and in need of support provides a perspective of his vulnerability. Imagine, Robert going through the school day where there are no opportunities to connect to his interests and resources. I would wonder how Robert would participate in school and what would motivate Robert to participate given that there are few connections to who he is and how he sees the world. Now, imagine a school day in which Robert's interests and resources are used to motivate his participation in mathematics. Again, I would wonder how Robert would participate in school. But, I can see a student who might be highly engaged, one who has a connection to the learning, and one who uses his connections to deepen his understanding of mathematics. I wonder if this thought process would appeal to Rebecca in ways that would allow her to see the extremes in the range of experiences Robert could be receiving in school. Reflecting on the shadowing experience by considering the range of Robert's experiences may be beneficial for Rebecca. The challenge with Rebecca is to help her see that Robert's interests can be leveraged to support positive teaching and learning and not position Robert's resources as deficits that teaching and learning must overcome.

RESPONDING TO THE DILEMMA

Understanding the strengths and motivations that serve to develop students' identities and sense of agency should be embedded in the daily work of all teachers (Aguirre, Mayfield-Ingram, & Martin, 2013). Agency is students' identity in action and their presentation of their identity to the world (Murrell, 2007). Mathematical agency is about participating in mathematics in personally and socially meaningful ways (Aguirre et al., 2013). As a mathematics teacher educator, one goal I have when working with teachers is to help them understand that making connections to students' identities impacts their sense of agency. That is, mathematics teaching involves more than helping students develop mathematical skills and mathematical understanding but also empowering students to seeing themselves as capable of participating in and being doers of mathematics. This understanding of identity and agency gives teachers insights to how and why some students might make positive connections with mathematics and others do not (Aguirre et al., 2013).

Because it appears that Rebecca is egocentric or unable to connect with Robert, it may be helpful to have her reflect on positive mathematics teaching (or another subject if mathematics is not positive) and learning from her own K–12 experiences as a student that connected with her resources and interests. I would ask her to reflect on one or two experiences to

remember how she participated in mathematics and the ways she was motivated to do well or persevere with tasks or problems. My hopes are that her
reflections would provide a context focusing on active participation, asking
questions, reasoning, and motivation. During the reflection, I would ask
Rebecca to consider the actions of teachers that supported questioning,
reasoning, and motivation. The kind of teaching that supports questioning,
reasoning, and motivation values students' thinking and uses pedagogical
practices, such as differentiated tasks and publicly praising contributions
and perseverance, to cultivate and affirm participation and behaviors (National Council of Teachers of Mathematics, 2014). After this reflection, I
would ask Rebecca to consider how Rebecca as learner would participate
and learn in her classroom. I would push her to consider how her teaching,
actions, and decisions would support Rebecca as a learner of mathematics. By asking Rebecca to reflect on herself as a learner in her own classroom, my goal is to bring forward the role of leveraging students' interests
and identities as motivating factors for participation and learning. Finally,
I would ask Rebecca to transfer her reflections as a learner in her own classroom to contrast it with Robert's experiences as a learner in her class. My
hope is that the contrast in reflections would highlight that teachers who
engage in activities that affirm students' identities impact the ways students
participate and learn in mathematics. I may receive "push back" from Rebecca because she may focus on interest convergence as a factor that supports positive teaching and learning. In my work with teachers common
push backs are often stated as not having common interests, backgrounds,
or connections. These push backs often situate students and families in a
deficit position because some teachers perceive the lack of convergence as
a problem for the student and families. If interest convergence is a factor, I
would ask Rebecca to consider who should make alterations in their interests: third-grade students or the teacher. I would challenge Rebecca on the
role of the teacher to adjust and meet the needs of students and to embed
some parts of Robert's resources and interests to see if it had any impact on
his motivation and participation.

One major lesson mathematics teacher educators can take from this
case is working with teachers like Rebecca should focus on helping teachers develop entry points for examining the relationship between students'
resources, identities, and supporting students' sense of agency. One way to
do this is to have teachers examine their own stories and experiences then
contrast them with the experiences they are providing for their students.
For teachers like Rebecca, this can provide her with an experience that
may be highly personal and may need to be unpacked personally prior to
engaging in a larger group. It is hard to imagine that teachers do not want
to provide the best opportunities for their students to learn mathematics.
For teachers like Rebecca, the challenge is to help her understand that her

actions, motives, and decisions impact students' identities, sense of agency, and ability to learn mathematics.

REFERENCES

Aguirre, J. M., Mayfield-Ingram, K., & Martin, D. B. (2013). *The impact of identity in K–8 mathematics learning and teaching: Rethinking equity-based practices.* Reston, VA: National Council of Teachers of Mathematics.

Murrell, P. C. (2007). *Race, culture, and schooling: Identities of achievement in multicultural urban schools.* New York, NY: Erlbaum.

National Council of Teachers of Mathematics. (2014). *Principles to actions: Ensuring mathematical success for all.* Reston, VA: Author.

COMMENTARY 2

SUPPORTING A TEACHER'S SHIFT FROM DEFICITS TO FUNDS OF KNOWLEDGE

A Commentary on Foote's Case

Maura Varley Gutiérrez
Elsie Whitlow Stokes Community Freedom
Public Charter School

MY POSITIONALITY

I read this case with my role as a teacher educator in the forefront of my mind. In my day to day as the Director of Teaching and Learning at an elementary school in Washington, DC, I spend time thinking about how to best support teachers so that they can spend their time creating learning environments that will best support children. In addition, I think about ways to support teachers in their learning. Whenever I plan for professional development, coach teachers, or make decisions about curriculum, teacher change and support are squarely at the center of my mind.

In addition to my current role as a teacher educator, I have multiple identities (bilingual, White, middle-class, and female, I grew up outside of

Cases for Mathematics Teacher Educators, pages 379–382
Copyright © 2016 by Information Age Publishing
All rights of reproduction in any form reserved.

the United States) that work together with the critical and feminist lenses I bring to my work and, therefore, to my reading of this case. I am undoubtedly an outsider to the communities in which I often live and work, which are predominantly comprised of low-income, marginalized people of color. My childhood living overseas and my study of critical and feminist theories have shaped the way I view situations, relationships, and society. I think often about how power relations operate and how the dominant narrative works to impose upon and color my perspective.

INTERPRETING THE DILEMMA

While reading this case, my interpretation of the dilemma shifted. Is the dilemma how a teacher should manage a student she does not "click with," as the title of the case suggests? Or is the dilemma about how to best prepare teachers to facilitate learning considering the lives of their students? Having made a few missteps myself in terms of how I have prepared (or underprepared) teachers for change, I would argue that the dilemma lies in how to best facilitate teacher change and handle resistance, which in this case is toward seeing and maximizing the strengths of a particular student, Robert. Rebecca seems to be struggling, not with a particular student, but perhaps with what the child represents: a child from a background that is different than her own and who values things that she does not. If a teacher comes to work with biases, preconceived notions, and judgments about a child (and do we not all bring these to our work?), how do we best support that teacher to understand that child (and others that will come after him) in a different way?

In reflecting on the case, I noticed the same parallel that Foote describes in her own reflection on the dilemma: "My frustration with Rebecca could conceivably be read as a situation parallel to her frustration with Robert and his mother. Do I need to accept some responsibility for Rebecca's inability to connect with this student and his mother? Did I in some way not meet her where she was, not attend to her needs?" The intention of the professional development was "for teachers to examine how learning about a student's mathematical thinking and their in- and out-of-school competencies as well as how examining cultural differences between them and their student might support their teaching of students who were different from themselves." Given this, it seems that the notion of the author not having met her student's (Rebecca's) needs is central.

Teachers deserve the same approach to learning that we ask them to bring to their students. The central activities of the professional development (meeting with Robert's mother, getting to know Robert's interests, and shadowing the student) are all powerful activities capable of shifting a teacher's perspective and enriching their practice. However, they can also serve to reinforce preconceived beliefs. Having prepared teachers to conduct home visits for the past 5 years, I have learned that if you do not set teachers up to look for assets and opportunities within a home, they may just see deficits. For example, one time a teacher returned from a home visit, and their description of the environment was that a television was taking up the main living area, always on, and that there was no quiet, separate space for the student to do homework. She had overlooked what could have been potential assets or funds of knowledge (González, Moll, & Amanti, 2005) from which to build upon in the classroom. The framing and structure of these interactions is as important as the activities themselves. If we set up a student to do a mathematical task without activating prior knowledge, ensuring they understand the goals and outcome of the activity, and adjusting course during the activity if that student is struggling, we will not maximize learning. In addition, we do not expect our students to develop at the same rate, which also holds true for teachers as learners.

RESPONDING TO THE DILEMMA

Having spent the last 4 years improving my practice as an advocate and supporter of teachers, I have learned about ways to best support teacher change that shape how I would approach the dilemma in this case. If I were the teacher educator in this case, I would first focus on my relationship with the teacher. Are there ways I can engage her in reflecting on her practice so that she will be open to my suggestions or even come to some of the conclusions on her own? What kinds of experiences could facilitate her learning about utilizing student strengths as a tool for their learning? I do not think there is one "correct" way to do this with a teacher, but a trusting relationship is essential. Perhaps Rebecca needed some modeling on how to engage Robert's strengths in her classroom. What if Foote had volunteered to model mathematical interviews or engaged Robert in problem posing within the classroom? Perhaps Rebecca needed more guidance on what to discuss with Robert's mom and how to interpret her findings.

I can turn to family engagement work that I do at my school to think about another way to approach the family component of the professional-development experience of this case. At the school in which I work, teachers go on home visits with their students' families. Did Rebecca meet Robert's mom at school? How would that activity have been different if Rebecca had been to Robert's home? It is entirely different for a teacher to enter the space of her student than it is for a parent to meet with them at school, which is really the teacher's space. Perhaps Rebecca would have been more open to learn from Robert's mom. Relationships between parents and teachers are fraught with power, entangled with issues of race, class, status, parents' own schooling experiences, and so forth (Valencia & Black, 2002). Schools often make this even more pronounced. Perhaps she needed to step out of the school in order to approach the learning experience in a different way.

Rather than being disappointed with the impasse that Rebecca seemed to have with Robert, I would focus on the possibilities that remain and the small steps that Rebecca did make. Foote ended the case by saying, "Rebecca began to take a few steps toward the end of the semester, realizing that there was something in the way SHE was interacting with the student that may have been contributing to his anxiety around mathematics and his behavior." Rebecca *did* recognize that her relationship with Robert played a role, possibly hindering his mathematical learning. Coming to this conclusion is a first step in wanting to make a change. This seems like the ideal time to engage Rebecca in moves that would begin to break down that wall. One example would be to conduct a home visit as described, prefaced by a discussion of funds of knowledge. A further example would be a mathematics interview. I find that having teachers conduct mathematics interviews in order to determine a student's level of *understanding* of a concept (emphasizing understanding in order to counter the notion of a student "not knowing" a concept) often opens a teacher to the learning possibilities of a student, versus their deficits. It appears that Rebecca, like many students, needed more time. Where has Rebecca taken this now that the professional development is over?

As educators, we have at least two ideas to reflect upon from this case: (1) Just as in our teaching of children, we need to value relationships as essential to learning, and (2) students grow at different rates and we need to meet them where they are. A final question remains, which I continuously consider in my work: How do we create learning environments and experiences that address power relations, whether between a teacher and student, teacher educator and teacher, or between families and schools?

REFERENCES

González, N., Moll, L., & Amanti, C. (Eds.). (2005). *Funds of knowledge: Theorizing practices in households, communities, and classrooms.* Mahwah, NJ: Erlbaum.

Valencia, R., & Black, M. (2002). "Mexican Americans don't value education!" On the basis of myth, mythmaking, and debunking. *Journal of Latinos and Education, 1*, 81–103.

COMMENTARY 3

A COMMENTARY ON FOOTE'S CASE

Nora G. Ramírez
Tempe, Arizona

MY POSITIONALITY

I currently work as a mathematics professional-development consultant with school districts or schools that have large numbers of students who are traditionally underserved in mathematics. My work often involves a district-wide approach to professional development in mathematics for teachers, coaches, and administrators. In this work I focus primarily on mathematics content and standards implementation—intertwining equity into that content. I also work with grade-level teams on an ongoing basis focusing on in-depth understanding of the mathematics standards, effective instructional strategies, teaching English language learners (ELLs), understanding student thinking, and planning and modeling lessons. We engage in difficult conversations about important issues such as low expectations and deficit views often held about their student population.

Although I am not an ELL, I am particularly interested in fostering high-quality mathematics instruction for English language learners and students with disabilities. More than 15 years ago, I was fortunate to have participated

Cases for Mathematics Teacher Educators, pages 383–387
Copyright © 2016 by Information Age Publishing
All rights of reproduction in any form reserved.

in the Equity and Mathematics Education Leadership Institute (EMELI). Through the Institute's activities, I increased my commitment to equity in mathematics education. I cried when I heard about the young child who tried to wash the brown color off his skin. I reflected on my own teaching when I read a story from Maryann Wickett, a teacher who believed in equity but through her own research found "unexpected patterns of bias" in her facilitation of classroom discussions. I now use stories to give educators opportunities to begin to understand the experiences and emotions of Black, Latina/o, and low-SES students.

While attending EMELI, I realized that all individuals have biases. Professional development should provide educators with opportunities to identify and reflect on their biases. I learned that the work is not easy but necessary.

INTERPRETING THE DILEMMA

Several issues surfaced as I read *The Case of Rebecca*. These concerns relate to Rebecca's mathematical content and pedagogical knowledge, her possible biases and deficit approach to teaching and learning including her relationships with Robert and his mother, and the role of the facilitator in preparing and supporting Rebecca through this process

Rebecca's Mathematical Content Knowledge

I am aware that a lack of mathematics content knowledge can result in teachers raising barriers because they feel mathematically inadequate. I noted that the author mentioned that Rebecca was "resistant to my suggestions." Making connections between mathematics, cooking, sports, and video games should not be a difficult task for a teacher whether she was interested in these activities or not. Rebecca also did not build on Robert's interest in pattern blocks, which again makes me wonder about her facility with mathematical concepts and connections.

Rebecca's Mathematical Pedagogical Knowledge

I wonder if Rebecca has the instructional skills to create a positive classroom culture in which students are not concerned about being wrong, where they use manipulatives as tools and not toys, and they understand what is expected of them. These things often need to be explicitly taught and practiced in classrooms. I have no evidence that Rebecca is aware of and able to ask appropriate questions and facilitate activities and experiences that will move students from direct modeling to using more sophisticated mathematical strategies and tools.

Rebecca's Relationships

Being aware of the cultures of the students in your school is a first step in developing a respectful attitude and a classroom culture that celebrates differences rather than sees them as inferiorities. I wonder if Rebecca were more aware of her Black students' culture would she would have been able to build a better relationship with Robert's mother.

Because the school composition is 30% Black, I am assuming that there are other Black students in the class. I am very interested in Rebecca's relationship with her students—in particular other Black students. Observing Rebecca's class might give insight as to whether Rebecca "liked" other non-Black students with or without behaviors similar to those of Robert. It is difficult to totally understand without observation.

The Role of the Facilitator

I do not know what kind of commitment Rebecca made before she began participating in this project, but I know that clear commitments and defined expectations are imperative for a successful project. I also am aware that when working with issues of equity one needs to begin with thoughtful experiences, discussions, and reflections; this is foundational in developing a culture of open and honest discussion. Because having discussions related to equity can be emotional, facilitators need to set an appropriate tone in the beginning, recognize and respond when teacher-participants are struggling, and ask questions that may raise disturbing feelings.

RESPONDING TO THE DILEMMA

If I had been working with Rebecca, I would have had some of the same concerns as the facilitator. I might have tried to support her in attempting to understand Robert's mathematical thinking. Rebecca expressed that she wanted to figure out the puzzle of where Robert was functioning in mathematics. She wanted to do something about his learning of mathematics. Working on the mathematics issue should not be as emotionally charged for Rebecca as working on the cultural and acceptance issues. I might have coached Rebecca in selecting strategies and tools that could help Robert explain his thinking and enable him to transition from a direct-modeling approach to more abstract thinking. For example, asking Robert to use base-10 blocks to explain his answer for 22 + 8, then helping him record his thinking with numerals to move him to an abstract model of the situation. If Rebecca could find some success with Robert in mathematics, she may become more open to him and his mother, and may begin to like him. As a teacher, Rebecca knows that she is responsible for Robert's learning but might not be as committed to understanding his culture or knowing about his interests. I feel that having a positive experience with Robert in mathematics first would open the doors to address Rebecca's dislike for Robert and her unacceptance of Robert's mother.

Rebecca may need support to continue taking steps in "taking responsibility for her part in squelching Robert's ability to explain his thinking and problem-solving strategies." I might coach Rebecca to identify the steps she has taken and their effect on Robert and also begin to think of additional steps she should take with Robert. I would also celebrate these small steps with Rebecca, taking the time to connect this success to Robert's motivation and the role that Rebecca has in motivating students.

I wonder if I might have had to work hard at liking Rebecca. I do know that just as one cannot accept Rebecca putting all the blame on Robert for his current situation, one cannot put all the blame on Rebecca for her current situation. So, prior to giving the assignment of investigating cultural differences, I might have had the teachers read and discuss classroom-related cases that focus on this topic. I suggest cases because the dilemmas and discussions are not as personal to the reader, thus leading to more open dialogue. I would use the case discussions to establish open relationships and develop a risk-free environment. I would use structures for discussion and reflection that I learned in EMELI, such as a dyad. A dyad consists of two people who take turns talking, then listening attentively to each other for a specific period of time allowing each to explore their thoughts and feelings.

In mathematics classrooms, we attempt to have risk-free environments. Risk-free environments are also important in professional-development settings, where participants are engaged in often difficult dialogue about racial and cultural differences. From experience, I know that neglecting to develop a supportive environment can result in feelings of personal attack and defensiveness as was exhibited when Rebecca turned her chair away from the group.

In addition to those professional dialogue tools mentioned earlier, I have found position statements from the National Council of Teachers of Mathematics (NCTM) and the National Council of Supervisors of Mathematics (NCSM) to be useful. I also have used *Classroom Practices That Support Equity Based Mathematics Teaching* (Chao, Murray, & Gutiérrez, 2014) and rely on articles and the accompanying reflection questions in the journal articles such as Gutierrez's (2009) article, "Framing Equity: Helping Students 'Play the Game' and 'Change the Game.'" Having discussed and reflected on equity may have been helpful in addressing Rebecca's deficit thinking. I truly believe that educators need to examine and reflect on our own biases if we are to make a difference in the lives of all students.

REFERENCES

Chao, T., Murray, E., & Gutiérrez, R. (2014). *NCTM equity pedagogy research clip: What are classroom practices that support equity-based mathematics teaching? A research brief.* Reston, VA: National Council of Teachers of Mathematics.

Gutierrez, R. (2009). Framing equity: Helping students "play the game" and "change the game." *Teaching for Excellence and Equity in Mathematics, 1*(1), 4–8.

CHAPTER 17

CHALLENGING
DEFICIT LANGUAGE

Imani Masters Goffney
University of Maryland–College Park

CONTEXT OF THE CASE

This case draws upon an online graduate level course entitled "Education in a Multicultural Society." Students enrolled in this course follow a cohort model—all taking the same courses at the same time. These courses meet synchronously on alternating weeks throughout a traditional semester. During the "off" week, students complete projects in their schools and submit them online. During the "on" week, we meet one day, face-to-face via webcam for 3 hours. Each of the students enrolled in this course is currently employed as a full-time mathematics or science teacher with at least 3 years of classroom experience; yet more than half of the students have at least 7 years of experience. More than 75% of the teachers in this program are White, and they teach in many different school settings including urban, rural, charter, public, and private.

The course explores a broad array of issues pertaining to equity and diversity in mathematics and science education including teaching for understanding, culturally relevant teaching, complex instruction, as well as

Cases for Mathematics Teacher Educators, pages 389–394
Copyright © 2016 by Information Age Publishing
All rights of reproduction in any form reserved.

equitable and ambitious teaching (Goffney, 2014). Readings, discussion questions, reflection assignments, and projects focused on these topics and issues are designed to build and deepen their understanding of broader issues of social justice and equity as it relates to mathematics and science education. Readings are clustered into common themes for synchronous class meetings. For example, one theme is "culture and race" so students read articles about the over-identification of African American students in special education (Blanchett, 2006; Ladson-Billings, 2011) and microaggressions that African American students face in schools, as well as articles about the *educational debt* (Ladson-Billings, 2006), the dangers of using deficit language (Ladson-Billings, 2007), and articles about *geography of opportunity* (Tate, 2008). Students focus on guiding questions to help them explore big ideas, unpack the theoretical constructs, or examine the research contributions for each article. In addition to broadening their understanding of these issues, another goal of this course is to connect what they are learning from readings, discussions, and projects to strategies to improve their classroom instruction. For example, students complete a "studying practice project" for which they have to study their own practice with a goal of improving their teaching practice along equitable dimensions. Teaching this type of course raises many dilemmas as the structure and the content often challenge core beliefs and perspectives of how enrolled teachers view themselves and the world around them. This case focuses on one dilemma in particular—the use of deficit language. Specifically, it focuses on the use of this language when referring to marginalized students (e.g., economically disadvantaged students, English-language-learning students, and African American and Latino students) and my responsibility as the instructor in addressing the deficit language.

DESCRIPTION OF EQUITY-RELATED DILEMMA AND CASE

Deficit language is particularly problematic as it is generally drawn from deficit perspectives. Teachers using this language often believe that the problem of low achievement rests with students and view diversity in student lifestyle, language, and ways of learning as problematic (Ladson-Billings, 2006, 2007). This case is important because language is a primary medium of instruction and is the central mechanism for communication between teachers and students as well as students and their peers.

The use of deficit language has many consequences. One such consequence results when deficit language is used around students. When this occurs, students who are the subjects of the deficit language are often positioned in the classroom as having lower ability and being lower skilled as compared with their peers (Boaler & Staples, 2008; Cohen, Lotan, Abram,

Scarloss, & Schultz, 2002; Ladson-Billings, 2007). These students also interpret the deficit language used in talking to or about them, to indicate lower expectations for their academic achievement. This often constrains student engagement and participation, thus constraining their opportunities for learning (Esmonde & Langer-Osuna, 2010). These constraints often result in not only lost academic opportunities but also behavioral problems (Blanchett, 2006; Ladson-Billings, 2011). Similarly, teachers sometimes share their deficit language and labels with their colleagues. When this occurs, these dilemmas follow students throughout their day. For example, a student may struggle with math but may be a gifted reader who is very athletic and charismatic. Therefore, the student may experience low status in the classroom, especially during mathematics, yet move to high status during reading, language arts, or physical education (Boaler & Staples, 2008). When deficit language is used in classrooms about particular students and then shared with colleagues who may teach or supervise students in other classes, students' opportunities to move flexibly through status levels throughout their day are reduced (Featherstone, Crespo, Jilk, Oslund, Parks, & Wood, 2011).

DILEMMA FACED BY THE AUTHOR

The primary dilemma I have faced when teaching this course is deciding how to respond when teachers use deficit language about students. As an instructor, I am trying to find a balance between creating a safe environment for class discussions and correcting this harmful behavior. Because this course is designed to broaden teachers' understandings of equity issues in middle-school mathematics and science classrooms, readings and discussion prompts are designed to challenge their perspectives and views about issues of race, class, proficiency with English, discipline, and other social factors that influence teaching and learning. Said another way, features of this course are designed to interrogate marginalizing perspectives about students and communities. The course discussions are designed around key prompts that connect ideas from the readings to their instructional practices. Class discussions are a public venue for exchanging ideas and perspectives about the prompts both between the teachers as peers and in my responses as their professor about the accuracy and depth of the conversations. I expect teachers to talk about students using respectful, ability-oriented language; yet in our opening discussions and throughout the first several weeks, many teachers have used negative and marginalizing language to describe the students in their schools. This dilemma is compounded when teachers feed off of each other in a discussion and borrow each other's terms, using them repeatedly when they build on each other's comments. One such example was when students were described in terms of categories, such as "6 SPEDs[1] and 4 ELLs[2] make it my hardest class," or in terms of perceived ability, "this is my low, low class, they don't know how to do anything and it's so hard to teach them—why do I even try?" Another example is when a teacher stated that he worked "in a low-income area and most of his students don't care and don't try" so why should he? In numerous incidences, many teachers mention "apathy," "lack of parental involvement," and "lack of student motivation" as reasons for their students not mastering course content. In other discussions, the teachers use a range of deficit descriptions about students, such as "culturally deprived," "culturally deficient," "low," "language deficient," "struggling," "at-risk," "wild," "rough," "illegal," and "undesirable" as descriptors of the students in their classrooms and in their schools.

AUTHOR REFLECTION OF THE CASE

Teachers' use of this kind of language and these terms for describing their students and their schools is problematic, in my opinion, for several reasons. First, use of deficit language impacts students' experiences in schools. Students to whom such language is directed (e.g., African American and Latino students, low-income students, and those for whom English is a second language) must grapple with the consequences of teachers' deficit perspectives. These perspectives often result in fewer opportunities for ambitious learning, separation into groups or classes that offer less content, less analytical thinking, and weaker tasks and projects that do not provide the types of learning opportunities that would help them develop proficiency in mathematics. The beliefs and perspectives that influence these comments are often rooted in racism and privilege. As such, they are dangerous and morally problematic. Especially problematic is that many of these terms and comments are based on characteristics for which students have no agency. Students do not choose to live in poverty and have no say in society's portrayal of members of their community or their parent's occupation, nor do they select the community in which they live. As a teacher educator, I find it troubling that teachers are often adopting this language from the schools that they teach in, where principals, counselors, other teachers, and support staff use this same deficit language.

I wrestle with how to address this use of deficit and marginalizing language. First, I recognize that the topics, ideas, and perspectives in this course are very new to my students. Based on reflections written in courses previously taken to earn this degree and on responses to survey questions before the class begins, it is evident that many of the teachers enrolled in this class have not yet interrogated privilege; the intersection of race, poverty, and education; or considered that their views and teaching practices may not be ideal.

I also know from experience that teachers often shut down and become unwilling to explore these topics if they perceive that the goal of the activities is to make them feel guilty versus empowering them as social justice advocates for their students or to learn new strategies for teaching in more equitable ways. To deepen and broaden each teacher's understanding and views of equity, diversity, race, poverty, language, and their relationships in mathematics and science teaching and learning, I have to engage them in productive class discussions about the underlying concepts that support the need and usefulness of knowledge about these issues. This means everyone talks, everyone shares, everyone reflects. Yet, how do I simultaneously encourage free expression and create a space where their perspectives can be challenged and their boundaries pushed? Likewise, how can I hold my students accountable for their frequent use of deficit language? If I discourage them in sharing their ideas from their own perspective, then I can't know where their thinking is and will have limited opportunities to push their ideas forward. If I allow them to freely use deficit and marginalizing language without interrupting and questioning their word choice, then in some ways I am perpetuating the type of behavior this class is designed to interrupt. Many mathematics teacher educators who seek to engage in these discussions face similar dilemmas, and each decision path has consequences.

AUTHOR NOTE

The research reported in this chapter was developed as a part of the iSMART program in the College of Education at the University of Houston. I would like to acknowledge the work of the iSMART program under the leadership of Dr. Jennifer Chauvot. This program is generously funded through the Greater Texas Foundation (http://greatertexasfoundation.org/). The opinions expressed in this report are those of the author and do not necessarily reflect the views of the Greater Texas Foundation.

NOTES

1. SPEDs is an abbreviation of the terms "special education students."
2. ELLs is an abbreviation for the terms, "English Language Learners."

REFERENCES

Blanchett, W. (2006). Disproportionate representation of African American students in special education: Acknowledging the role of White privilege and racism. *Educational Researcher, 35*(6), 24–28.

Boaler, J., & Staples, M. (2008). Creating mathematical futures through an equitable teaching approach: The case of Railside School. *Teachers College Record, 110,* 608–645.

Cohen, E. G., Lotan, R. A., Abram, P. L., Scarloss, B. A., & Schultz, S. (2002). Can groups learn? *Teachers College Record, 104,* 1045–1068.

Esmonde, I., & Langer-Osuna, J. M. (2010). Power in numbers: Student participation in mathematical discussions in heterogeneous spaces. *Journal for Research in Mathematics Education, 44,* 288–315.

Featherstone, H., Crespo, S., Jilk, L. M., Oslund, J. A., Parks, A. N., & Wood, M. B. (2011). *Smarter Together! Collaboration and equity in the elementary math classroom.* Reston, VA: National Council of Teachers of Mathematics.

Goffney, I. M. (2014). Toward a theory of mathematical knowledge for equitable teaching. In C. W. Lewis, *Proceedings of international conference on urban education* (pp. 334–337). Montego Bay, Jamaica: University of North Carolina-Charlotte.

Ladson-Billings, G. (2006). From the achievement gap to the education debt: Understanding achievement in U.S. Schools. *Educational Researcher, 35*(7), 3–12.

Ladson-Billings, G. (2007). Pushing past the achievement gap: An essay on the language of deficit. *The Journal of Negro Education, 76*(3), 316–323.

Ladson-Billings, G. (2011). Boys to men? Teaching to restore Black boys' childhood. *Race Ethnicity and Education, 14*(1), 7–15.

Tate, W. F. (2008). "Geography of opportunity": Poverty, place, and educational outcomes. *Educational Researcher, 37,* 397–411.

COMMENTARY 1

ADJUSTING PERSPECTIVES

A Commentary on Goffney's Case

Joel Amidon
University of Mississippi

INTRODUCTION

When I look in the mirror I see a stereotypical mathematics teacher and most of the math teachers of my youth. White, male, glasses, it's all there (see Picker & Berry, 2000). Through most of my mathematics education career I did not recognize the power and privilege that came with my particular demographics. I did not recognize that as a White male being taught by White males there could have been an unearned assumption of mathematical competence by both my teachers and my peers. Now, as a teacher of mathematics methods and general education courses at a university in the Deep South, I cannot help but recognize how power and privilege play into education... including my own.

In my role as a teacher educator, I make it my responsibility to provide opportunities for both preservice and inservice teachers to consider their background, to consider how power and privilege factored into their education, and to consider how those experiences shape how they teach. It

Cases for Mathematics Teacher Educators, pages 395–399
Copyright © 2016 by Information Age Publishing
All rights of reproduction in any form reserved.

is with this responsibility and the above recognition that I write this commentary on the use of deficit language in a hybrid classroom environment.

INTERPRETATION OF THE DILEMMA

Goffney describes the central dilemma in her case as finding "a balance between creating a safe environment for class discussions and correcting this harmful behavior" of using deficit language in talking about students. I empathize with the author and, at the same time, I offer an alternative interpretation. My interpretation of the dilemma is that there is a need to improve the collective relationship between the teachers and the students they are responsible to teach.

I consider the classroom a community of practice (Lave & Wenger, 1991) and the teacher as an insider who is demonstrating how to become a more central participant within the community. The teacher is in a relationship with the students and mathematics (Lampert, 2001) with the ultimate purpose being to put processes and products (Udvari-Solner, Villa, & Thousand, 2005) into play within the classroom in order to facilitate a desired relationship between the student and mathematics. Returning to my interpretation of the dilemma, the use of deficit language suggests that the relationship between the students and the teacher is damaged or not perceived as a means for creating better processes and products for implementing in the classroom, thus hampering the teaching and learning that is happening in the classroom (Milner, 2012).

For example, Emdin (2008) has described a classroom where the teacher is enacting what is thought to be an engaging lesson, but when Emdin looked at the students he noticed disengagement and boredom. Then music from a passing car could be heard through the open windows of the classroom. As the familiar sounds and lyrics were heard throughout the class, smiles were exchanged between the students along with synchronous bobbing of heads as lyrics were being silently spoken across the room. As the sounds from the car faded into the sounds of the city, the faces of the students returned to the reality of the lesson and the previous looks of disengagement. Rather than perceive students as "disinterested," "apathetic," or "uncaring," Emdin noticed a means for focusing their attention on the content through the cultural practices of hip-hop. Instead of a deficit he saw an asset.

My interpretation of the dilemma is considering how to shift the perspective of the teachers from a deficit perspective of students to an asset perspective of students and do it in a way that honors and maintains their engagement within the course. What I present below is how I would respond to this dilemma, both in the moment and in the design of the course.

RESPONSE TO THE DILEMMA

My initial response to teachers' use of deficit language would be to share my perspective on how our relationships with our students, just like our relationship with the content (Hill, Rowan, & Ball, 2005) shape the processes and products, which facilitate learning (Milner, 2012). In addition, I would share my experiences and how I confronted my own deficit perspectives of students. Because I am a part of this community of practice of teachers, I need to share how I confronted similar issues and that we need to overcome this issue to be better teachers.

For example I would share my encounter the first time I had a student fall asleep in my class. This student was labeled as "at risk" and was "obviously" trying to disrespect me by laying his head down on his desk and ignoring my carefully prepared lesson. After engaging the rest of the students in a lesson, I woke the offending student and had him follow me into the hallway so I could properly reprimand him. In the hall, he apologized for his inability to keep his head up. He then revealed that he had spent the previous night under the pool table at a local establishment waiting for his parent to take him home after enjoying a few beverages. He could have easily (and understandably) stayed home to sleep; instead he came to school because he didn't want to miss my class. For me, the "offense" went from a sign of disrespect to an example of perseverance. An additional example came from a different student whom I caught cheating in my class, resulting in another return to the hallway for a carefully planned speech about honesty and responsibility. The conversation led to the student revealing that he wanted to be successful, wanted to match his parents' expectations, but did not know how to ask or seek help in an acceptable way. Both of these examples started with me assuming the worst about a student, with deficit language filling my thoughts, but in the end the situations led me to a better relationship with the students and tangible ways to help them learn mathematics.

As a mathematics teacher educator the use of deficit language shows me, at a minimum, that the mathematics teacher is paying attention to the student and has some curiosity about the student and how best to teach them. A way to proactively confront deficit language is to design experiences and assignments that will reveal deficit language as being counterproductive to the practice of teaching. One such experience would be to have teachers share some of their personal stories where someone had a faulty assumption about them that hindered their performance. Building off of that experience could be the sharing of teaching stories where they assumed below a student's means of participation (similar to what I shared above).

Finally, I would respond to teachers' comments in the moment by providing them with language to use in place of their deficit language. Instead of describing a "problem" or "deficiency," I would offer the word "puzzlement" as a way to describe the "problems, concerns, or student behaviors or attitudes that teachers do not understand" (Jacob, Johnson, Finley, Gurski, & Lavine, 1996, p. 30). This language honors the teachers' contributions to the course and what they have noticed in the classroom but has them reposition it as something to investigate. Jacob and colleagues (1996) provided a structure, the *cultural inquiry process*, for systematically examining such puzzlements and considering what assets are being revealed. The ultimate goal of the process being to improve the learning opportunities provided for the student in question. This structure could be used as the foundation for designing a course assignment (see Jacob, 2004) or series of assignments to directly address deficit language in a proactive way.

CONCLUSION

The mathematics teacher educator needs to recognize that language is an artifact of the relationship between a teacher and their student. Thus, this language needs to be interrupted, even challenged, but in a way that doesn't shut down engagement. Teachers need to be given alternative language and shown how growth experiences occur through the use of story. Finally assignments and experiences can be developed to productively investigate puzzlements associated with students to produce greater means of teaching mathematics.

REFERENCES

Emdin, C. (2008). The three C's for urban science education. *Phi Delta Kappan, 89,* 772–775.

Hill, H. C., Rowan, B., & Ball, D. L. (2005). Effects of teachers' mathematical knowledge for teaching on student achievement. *American Educational Research Journal, 42,* 371–406.

Jacob, E. (2004). *Cultural inquiry process.* Retrieved from http://cehdclass.gmu.edu/cip/index.htm

Jacob, E., Johnson, B. K., Finley, J., Gurski, J. C., & Lavine, R. S. (1996). One student at a time: The cultural inquiry process. *Middle School Journal, 27,* 29–35.

Lampert, M. (2001). *Teaching problems and the problems of teaching.* New Haven, CT: Yale University Press.

Lave, J., & Wenger, E. (1991). *Situated learning: Legitimate peripheral participation.* Cambridge, England: Cambridge University Press.

Milner, H. R. (2012). Beyond a test score: Explaining opportunity gaps in educational practice. *Journal of Black Studies, 43,* 693–718. doi:10.1177/0021934712442539

Picker, S. H., & Berry, J. S. (2000). Investigating pupils' images of mathematicians. *Educational Studies in Mathematics, 43,* 65–94.

Udvari-Solner, A., Villa, R. A., & Thousand, J. S. (2005). Access to the general education curriculum for all: The universal design process. In R. Villa & J. S. Thousand (Eds.), *Creating an inclusive school* (2nd ed., pp. 134–155). Alexandria, VA: Association for Supervision and Curriculum Development.

SUPPORTING STRENGTH-BASED PERSPECTIVES AND UNDERSTANDINGS

A Commentary on Goffney's Case

Amy Roth McDuffie
Washington State University Tri-Cities

MY CONTEXT

As a mathematics teacher educator and researcher, my teaching and re-search interests overlap in that I focus on mathematics teachers' profes-sional learning and development. Within the broader arena of teachers' professional development, I study teachers' learning in and from practice with attention to equity and teachers' use of curriculum resources. I teach a K–8 mathematics methods course for undergraduate students and a range of graduate courses in mathematics education (including a course on lan-guage and culture in mathematics education), and facilitate professional development for practicing teachers. Currently, I serve as a co-principal in-vestigator on a National Science Foundation-funded project, Teachers Em-powered to Advance Change in Mathematics (TEACH Math), collaborating

Cases for Mathematics Teacher Educators, pages 401–405
Copyright © 2016 by Information Age Publishing
401

with Corey Drake, Julia Aguirre, Tonya Bartell, Mary Foote, and Erin Turner. In my teaching and through my work on the TEACH Math project, I have experienced the dilemma Goffney described, and so this case resonated well with my own experiences, challenges, and wonderings. Indeed, along with my TEACH Math colleagues, we have discussed issues relevant to this case in recent publications (Bartell, Foote, Drake, Roth McDuffie, Turner, & Aguirre, 2013; Foote, Roth McDuffie, Turner, Aguirre, Bartell, & Drake, 2013; Roth McDuffie, Foote, Bolson, et al., 2014).

MY INTERPRETATION OF THE DILEMMA

The dilemma Goffney described centers on how teacher educators respond to teachers' (both prospective teachers' and practicing teachers') use of deficit language when discussing students and families, as well as underlying deficit perspectives. Deficit language can take many forms such as labeling and categorizing students in ways that feature deficits (e.g., "SPED students," "ELLs," "low students," "from [a neighborhood known for poverty]," "THESE student need to be shown how to do math"), describing deficits (e.g., "parents do not speak English," "student is behind in math"), and blaming the child or family (e.g., "student does not try," "parents are never home," "parents do not care"). As Goffney discusses, the consequence of this deficit language and these perspectives can be severe for students' sense of identity, positioning in the classroom, and academic opportunities (Boaler & Staples, 2008; Hand, 2009; Ladson-Billlings, 2006; Nieto, 2004; Villegas & Lucas, 2002). As mathematics teacher educators (MTEs), it is imperative that we support teachers in developing supportive and strength-based perspectives for students and their families (e.g., "student asks questions that push others' thinking," "student jumps in and tries different approaches," "grandparents are involved in raising the student" [considering grandparents as a positive influence]). In doing so, as Goffney describes, we face tensions in finding a balance between creating a safe environment for teachers' learning with open discussions and addressing language and perspectives that can be detrimental to students.

RESPONDING TO THE DILEMMA

In responding to the dilemma, I have learned to not expect deeply held beliefs and perspectives to shift in a short period of time (e.g., a semester-long course). Yet it is critical that teachers have opportunities to understand, experience, and discuss the consequences of deficit language and perspectives, as well as the affordances for students' learning that result from teaching with strength-based perspectives and actions. Although my TEACH Math colleagues and I continue to grapple with ways to provide these opportunities and negotiate the tensions, I have found that the following approaches can be helpful for teachers.

- Providing readings and facilitating discussions that highlight issues related to deficit language and teachers' decisions based on deficit perspectives. If these are planned early in a course or workshop to increase teachers' awareness, teachers might be better prepared to monitor and resist statements that reflect a deficit perspective. Moreover, teachers have more opportunities to engage in discussions that support a strength-based perspective as the course progresses. Readings include Boaler (2008, especially Chapter 5, "Stuck in the Slow Lane: How American Grouping Systems Perpetuate Low Achievement"), Featherstone, Crespo, Jilk, Oslund, Parks, and Wood (2011, on complex instruction, status, and competence), and Varley Gutierrez (2009–2010, on empowering students and featuring strength-based perspectives in action).
- Encouraging teachers to share their own experiences as students or as parents when they were labeled, classified, or assumed that they were not capable or deficient. Discuss ways for teachers to understand and learn about students and families that could contribute to a more productive stance and relationship. These personal stories can be powerful for teachers to understand consequences and shift deficit perspectives.
- Establishing and posting norms for ways to talk about students and families. Teachers are often surrounded by colleagues for whom deficit discourse is common. As MTEs, we need to acknowledge that deficit language is common in some schools, but it should not be accepted. These norms could be developed from the readings and discussions described above. The norms serve as a reminder and a referent for teachers to support each other in monitoring and practicing strength-based language and shifting what is normal and acceptable talk. Featherstone et al. (2011) provided suggestions for establishing norms, as well as other norms that support productive and respectful discourse in mathematics teaching and learning.
- Providing experiences for teachers to engage in routine teaching practices with a focus on incorporating strength-based perspectives. For example, teachers analyze and adapt curriculum materials by connecting to and building on students' understandings and experiences (Drake et al., 2015). As another example, teachers view classroom videos from lenses that focus on students' thinking about mathematics (highlighting strengths), students' power and par-

ticipation in learning mathematics, students' resources (e.g., family-, cultural-, linguistic-, and community-based experiences) that students bring to and leverage in a lesson, and teaching decisions and moves associated with teaching from a strength-based perspective (Aguirre, Turner, Bartell, Drake, Foote, & Roth McDuffie, 2012; Roth McDuffie, Foote, Bolson, et al., 2014; Roth McDuffie, Foote, Drake, et al., 2014).

Perhaps the most important idea is not in any one of these approaches, but instead is to continually connect to issues and approaches related to equitable pedagogy throughout a course or workshop. As MTEs, we need to be patient but persistent in supporting teachers as they work to develop new understandings and perspectives for students and their families. In the TEACH Math project, we have worked to integrate discussions of equity throughout the methods course and have found that with multiple and varied opportunities to engage in and revisit issues and perspectives related to equitable instructional practices teacher can shift from deficit toward more productive and supportive perspectives.

REFERENCES

Aguirre, J., Turner, E., Bartell, T. G., Drake, C., Foote, M. Q., & Roth McDuffie, A. (2012). Analyzing effective mathematics lessons for English learners: A multiple mathematical lens approach. In S. Celedón-Pattichis & N. Ramirez (Eds.), *Beyond good teaching: Advancing mathematics education for ELLs* (pp. 207–221) Reston, VA: National Council of Teachers of Mathematics.

Bartell, T. G., Foote, M. Q., Drake, C., Roth McDuffie, A., Turner, E. E., & Aguirre, J. M. (2013). Developing teachers of Black children: (Re)orienting thinking in an elementary mathematics methods course. In J. Leonard & D. B. Martin (Eds.), *The brilliance of Black children in mathematics: Beyond the numbers and toward a new discourse* (pp. 343–367). Charlotte, NC: Information Age.

Boaler, J. (2008). *What's math got to do with it?* New York, NY: Viking.

Boaler, J., & Staples, M. (2008). Creating mathematical futures through an equitable teaching approach: The case of Railside School. *Teachers College Record, 110,* 608–645.

Drake, C., Land, T., Bartell, T. G., Aguirre, J. M., Foote, M. Q., Roth McDuffie, A., & Turner,

E. (2015). Three strategies for opening curriculum spaces: Building on children's multiple mathematical knowledge bases while using curriculum materials. *Teaching Children Mathematics, 21,* 346–352.

Featherstone, H., Crespo, S., Jilk, L., Oslund, J., Parks, A., & Wood, M. (2011). *Smarter together: Collaboration and equity in the elementary mathematics classroom.* Reston, VA: National Council of Teachers of Mathematics.

Foote, M. Q., Roth McDuffie, A., Turner, E. E., Aguirre, J. M., Bartell, T. G., & Drake, C. (2013). Orientations of prospective teachers towards students' families and communities. *Teaching and Teacher Education, 35,* 126–136.

Hand, V. (2009). Constructing competence: An analysis of student participation in the activity systems of mathematics classrooms. *Educational Studies in Mathematics, 70,* 49–70.

Ladson-Billings, G. (2006). From the achievement gap to the education debt: Understanding achievement in US Schools. *Educational Researcher, 35*(7), 3–12.

Nieto, S. (2004). *Affirming diversity* (4th ed.). Boston, MA: Pearson.

Roth McDuffie, A., Foote, M. Q., Bolson, C., Turner, E. E., Aguirre, J. M., Bartell, T. G., Drake, C., & Land, T. (2014). Using video analysis to support prospective K-8 teachers' noticing of students' multiple mathematical knowledge bases. *Journal of Mathematics Teacher Education, 17,* 245–270. doi: 10.1007/s10857-013-9257-0

Roth McDuffie, A., Foote, M. Q., Drake, C., Turner, E., Aguirre, J. M., Bartell, T. G., & Bolson, C. (2014). Mathematics teacher educators' use of video analysis to support prospective K-8 teachers' noticing. *Mathematics Teacher Educator, 2,* 108–138.

Varley Gutierrez, M. (2009–2010, Winter). I thought this U.S. place was supposed to be about freedom. *Rethinking Schools,* 36–39.

Villegas, A., & Lucas, T. (2002). *Educating culturally responsive teachers: A coherent approach.* Albany, NY: State University of New York Press.

CHALLENGING MATHEMATICS TEACHERS' DEFICIT-LANGUAGE USE

A Commentary on Goffney's Case

Eugenia Vomvoridi-Ivanovlc
University of South Florida

MY POSITIONALITY

I am a bilingual mother of two young trilingual children. Although English is not my first language (my K–14 education was in Greece and in Greek), this is not typically apparent because I "sound American." Furthermore, although I have witnessed other parents being reprimanded by strangers for talking to their children in their non-English home language, I am regularly praised for doing so. This is perhaps because, unlike the parents I am referring to, I have an "American" accent. Talking to my children in Greek, then, is viewed as an asset (i.e., teaching them an additional language) as opposed to a deficit. I am also a U.S. citizen, White, heterosexual, abled, from a middle-class background, and have reaped many privileges associated with being all these things. As a student, I was considered smart and

Cases for Mathematics Teacher Educators, pages 407–411
Copyright © 2016 by Information Age Publishing

did very well in school, especially in mathematics. I remember being very young when I realized that high mathematics achievement was associated with high intellect and that girls were not considered as strong as boys in this subject. This realization along with my mathematics teachers' comments such as "are you as good as your brother is in math?" and "you are the strongest female mathematics student in this class" pushed me to excel in mathematics. However, through this process, I also developed and overcame mathematics and testing anxiety, I constantly questioned my abilities, and in my early adolescence developed a behavioral disorder, trichotillomania, that is still difficult to control under acute academic pressure. Even though I had taken on the positive label of a "smart student" who is "strong in mathematics," I was also a girl and navigated the negative label of not being as good as my male counterparts.

INTERPRETING THE DILEMMA

During the few years that I taught mathematics to adolescents and nearly a decade of working with preservice teachers (PSTs), I too have encountered teachers using the type of deficit language and labels to refer to students that Goffney describes. As a mathematics teacher educator who strives to privilege equity in her work, I am especially tuned in to how educators talk about students, their families, and their communities. I have heard teachers use the exact terms that Goffney lists. Every time this occurs I face the same dilemma as Goffney: How do I help teachers become aware of the negative effects their use of deficit language has on students, while maintaining a learning environment where all teachers feel safe and express their thoughts freely? Creating and maintaining a learning environment where the teachers I work with can express their thoughts freely is vital. Such an environment helps me gain insight into teachers' knowledge and beliefs, informs my instruction, and allows for teachers to become aware of their own knowledge and beliefs. However, expressing ones' thoughts freely may occur through the use of deficit language. Like Goffney, I am worried that in my attempt to correct this type of problematic behavior, some teachers may keep quiet—and then I will never know what they are thinking. Others may regulate their way of talking in the context of our university class but continue using the same language at their schools.

Goffney describes the impact teachers' use of deficit language has on students' school experiences. While reading the case, I reflected on the impact that deficit language has on both teachers and students. In the context of mathematics teacher education, Aguirre (2009) defined equity as ensuring that "all students in light of their humanity...have the opportunity and support to learn rich mathematics" (p. 296) Acknowledging students' humanity is a necessary component for equitable teaching practices and for being pedagogically responsible (van Manen, 1991). The use of deficit language and labels to refer to students, however, has a double effect: It reflects and reinforces deficit ideas and stances teachers have about particular students, and at the same time strips students from their humanity. In this era of high-stakes testing, teachers feel enormous pressure and frustration. Unfortunately, students are at the receiving end of this. Deficit language is present in most schools and is shared by counselors, administrators, and support staff. PSTs and novice teachers enter this type of school climate and oftentimes adopt this language. One of my students shared that, as a result of the pressures to increase their student achievement, many teachers in her school "do not see children when they look at their students; all they see are numbers" that reflect their students' test scores (Bannister, Bartell, Battey, Hand, & Spencer, 2007). Viewing children as numbers certainly dehumanizes them, but it affects teachers as well. As Freire (2003) noted, dehumanization "afflicts both those whose humanity has been stolen" (students) "and those who have stolen it" (teachers) as it "distorts the process of becoming more fully human" (p. 28). One cannot be fully human if s/he views other individuals as less than human. Talking about children as if they are not fully human validates neglecting them and their education. With such an orientation, teachers are certainly not being pedagogically responsible to their students (van Manen, 1991).

RESPONDING TO THE DILEMMA

As I consider the ways in which I would respond to this dilemma, I have to acknowledge the fact that the author of the case is African American and I am Caucasian. This may affect how the teachers we work with perceive us. Even without ever having seen Goffney, one could safely assume that she is African American by her first name, "Imani." Goffney and I face the same challenge in our courses, and both I and people dear to me have been negatively affected by deficit-language use. However, I am at lower risk of my students thinking that I have an agenda or thinking that the issue I have with their use of deficit language is of a personal nature and is therefore not a "real" issue. In Goffney's case, the teachers in her class did not use deficit language to talk about children who look like themselves (or like myself). Instead, they used deficit language to describe "other people's children," children that resemble Goffney's children (Aguirre, 2009).

I wonder if a way to handle this situation is to first identify the reasons behind teachers' use of deficit language. Does their use reflect a lack of awareness about how it may affect students, or does it reflect their negative beliefs about particular students? Are teachers using this language to fit in with other teachers? If other teachers used ability-oriented language, as Goffney describes it, would these teachers also use ability-oriented language? Do unexamined assumptions about students and communities lie beneath the deficit language? Perhaps one strategy to help teachers develop relevant knowledge and awareness would be to assign readings on the negative effects of mathematics-course tracking on students. Tracking is a particularly important entry point because deficit language often accompanies low-tracked mathematics classrooms (Spencer, 2006). Tracking also leads students to take on labels—both in their own minds as well as in the minds of their teachers (Ansalone, 2010). These labels are usually associated with students' pace of learning, such as the "slow" or "fast." Subsequent to track placements, teachers confuse students' pace of learning with their capacity to learn and create different expectations for different groups of students. As Boaler (2005) has argued, students who take on negative labels quickly learn to view themselves as unsuccessful and develop antischool values that lead into general antisocial behavior.

Another strategy that would target teachers' beliefs is to provide them with counterexamples (Solorzano & Yosso, 2002) that illustrate the capabilities and the funds of knowledge of marginalized students (e.g., Terry & McGee, 2012). Finally, a third strategy is to share commonly used deficit language about students and have teachers discuss potential implications.

Upon reading the case, I decided to share parts of it with the middle-grades mathematics PSTs I taught at the time. After listing deficit-oriented terms and phrases from Goffney's case, I asked the PSTs if they have noticed similar language use in their school placements and whether they thought that referring to students in such ways was problematic. The majority of the class expressed that this type of language is common and makes them feel uncomfortable. Several expressed that due to this kind of language they refrain from having lunch in the teachers' lounge because it is a "toxic" environment. Some PSTs, however, admitted to adopting this language when talking with their collaborating teachers. Other PSTs, who had previously used deficit language in classroom discussions, remained silent. The PSTs' comments led me to consider a related dilemma: How do PSTs negotiate and maintain ability-orientated language in deficit-based school settings? Finally, I can't help but think about the PSTs in my classes who come from marginalized backgrounds. How do they feel when their peers talk about students of similar backgrounds in these ways?

REFERENCES

Aguirre, J. M. (2009). Privileging mathematics and equity in teacher education: Framework, counter-resistance strategies, and reflections from a Latina mathematics educator. In B. Greer, S. Mukhopadhyay, A. B. Powell, & S. Nelson-Barber (Eds.), *Culturally responsive mathematics education* (pp. 295–319). New York, NY: Routledge.

Ansalone, G. (2010). Tracking: Educational differentiation or defective strategy. *Educational Research Quarterly, 34*(2), 3–17.

Banister, V., Bartell, T., Battey, D., Hand, V., & Spencer, J. (2007). Culture, race, power, and mathematics education. In F. Lester (Ed.), *Second handbook of research on mathematics teaching and learning* (pp. 405–434). Charlotte, NC: Information Age.

Boaler, J. (2005) The "psychological prisons" from which they never escaped: The role of ability grouping in reproducing social class inequalities. *FORUM, 47,* 135–144.

Freire, P. (2003). *Pedagogy of the oppressed* (30th anniversary ed.). New York, NY: Continuum.

Solorzano, D. G., & Yosso, T. J. (2002). Critical race methodology: Counter-story telling as an analytical framework for education. *Qualitative Inquiry, 8*(1), 23–44.

Spencer, J. A. (2006). *Balancing the equation: African American students' opportunities to learn mathematics with understanding in two Central City middle schools.* ProQuest.

Terry, C. L., Sr., & McGee, E. O. (2012). "I've come too far, I've worked too hard!" Reinforcement of support structures among Black male mathematics students. *Journal of Mathematics Education at Teachers College, 3*(2), 73–85.

van Manen, M. (1991). *The tact of teaching: The meaning of pedagogical thoughtfulness.* Albany: State University of New York Press.

CHAPTER 18

MOVING FROM ADDRESSING ONE'S TARGET IDENTITY TO ADDRESSING ONE'S NONTARGET IDENTITIES

Beth A. Herbel-Eisenmann
Michigan State University

For 5 years starting in 2003, I taught a graduate level course related to algebra in K–12 classrooms that met weekly. The course enrolled a range of students: people returning to school post-BA as part of an alternative teaching certification program, elementary prospective teachers who were earning a mathematics minor, full-time teachers who were earning a master's degree, and full-time master's and PhD students. In the class, we had a series of algebra-related questions upon which we focused. Two of the questions were related to *algebra for all* and to various perspectives on what constitutes "algebra" and the purposes of such a course (e.g., a modeling view, a functions view, a symbolic manipulation view).

In about 2005, I started reading *Rethinking Schools* and came across some articles related to mathematics and teaching mathematics for social justice.

Cases for Mathematics Teacher Educators, pages 413–419
Copyright © 2016 by Information Age Publishing

In 2006 and 2007, I decided to incorporate some of these tasks and readings into the course. In particular, I used a chapter from Rico Gutstein's (2006) book about using mathematics to read and write the world in order to unpack some of the subquestions about algebra for all in relationship to his view on doing mathematics. It was also the first time I had decided to use some of the problems from *Rethinking Mathematics* (Gutstein & Peterson, 2005). I felt that this perspective and types of problems filled important holes in the course by raising significant issues related to equity that the syllabus and readings had not previously addressed. Although we discussed many equity-related readings in my elementary mathematics methods courses, I had not previously used social justice mathematics tasks.

The class typically enrolled about 15 students each fall semester, with fairly equal numbers of male and female students. Similar to most of the undergraduate classes I taught, students who enrolled in this course were White, and most of them were from Iowa or the surrounding states. The final time I taught the course was an exception, however. That year, the students were all White except one male student, Manuel (a pseudonym), who identified as Latino. He had recently relocated to Iowa from another state, where he had grown up and had been an engineer for a short period of time. He was returning to school to get certified to teach mathematics. I later learned that he had also experienced extreme poverty and homelessness during his childhood.

One of the first tasks I decided to use, more specifically in relationship to this case, was a task called "Mortgage Loans—Is Racism a Factor?" (Gutstein, 2006). This task includes an article from the *Chicago Tribune* that analyzed mortgage rejection rates for African American, Latino, and White people in 68 small, medium, and large metropolitan areas. The information is presented in a table, listing percentage of applicants denied a home mortgage in which White applicant rejection rates, Black applicant rejection rates, and a disparity ratio are reported for 10 of the cities from which data were collected. In the article itself, the information is discussed by at least three different representatives: a researcher from Association of Community Organizations for Reform Now (ACORN) who describes it as institutional racism, a representative from Bank One who denies racism is involved, and the president of the Woodstock Institute who addresses the issue as being about the competition for loans. The following information from the ACORN report is also included in the task:

> The disparities in rejection ratios remain even if we compare applicants of the same income. Upper income African Americans [Latinos] (earning more than 120% of the median income) were denied 5.23 [3.02] times more often than upper income Whites. Upper-middle income African Americans [Latinos] (earning between 100%-120% of the median income) were denied 5.22 [2.61] more often than upper-middle income Whites. Moderate-income

African Americans [Latinos] were rejected 4.51 [2.18] times more often than moderate-income Whites while low-income African Americans [Latinos] were rejected 3.26 [1.87] times more often than low-income Whites. (Gutstein, n.d., p. 2)

Similar to the task designed by Gutstein, I asked my students to examine Bank One's mortgage-granting practices by analyzing the data presented. They were asked to decide whether they thought racism was involved in Bank One's decisions and to make an argument using the data to back up their stance. After they completed the assignment, they were asked to bring it to class to discuss, and I assigned the book chapter Gutstein (2005) wrote about using this task with his students. One issue Gutstein raised about using this task was that most students tended to focus on racism at the interpersonal level, rather than recognizing that racism also functions at the institutional and cultural levels. Going into the discussion of the task, I thought that my students might also focus on the interpersonal level and had considered a few ways I might respond to this kind of response.

As with other nights, we tended to start class sessions with questions and discussion about homework problems. I opened the class by asking whether there were any questions on the homework problem. No one responded, so I moved to asking students about how they responded to the question of whether racism was involved and reminded them to use data from the article and the report to back up their responses. I remember looking around the room and that many of the students were shaking their heads yes. Manuel had his hand raised, so I asked him to share. I was caught off guard when he responded by saying that racism was definitely not a factor in the Bank One's mortgage practices.

Although I do not remember for sure what I said in response to Manuel's answer, I remember feeling very uncomfortable because I was not sure how I should respond. Being White and a professor for the course raised all kinds of questions and issues for me because I was speaking as a person in power (institutionally by my position) and as a person who had unearned privilege in relationship to my race. I have a feeling I probably asked how others answered the question to draw attention away from Manuel's response. Afterwards, however, I had many questions about what I could have done differently: How could I, as a White professor, help a student of color make sense of patterns of systemic racism? What are some ways I might engage and challenge deficit perspectives about students of color when working in a context where most students are White and the student denying inequities is a student of color? How could I do this in an authentic way that builds relationships instead of potentially marginalizing someone or replicating an imperialistic (i.e., enabling the powerful to act and speak on behalf of the oppressed) viewpoint? I did not want to make it seem as if I was discounting his experience. It was possible that he may not have wanted others to perceive him as "playing the race card." In some ways, his response may have also felt more personally empowering because he was not constructing himself as someone who was a victim of the system in relationship to being able to get a loan.

AUTHOR'S REFLECTIONS

This particular classroom episode has stayed with me. In my graduate coursework and with colleagues, I had often discussed issues of equity, diversity, and culture. But, this was one of the first times I had actually tried to use more explicit tasks and readings that named race as an interlocking system of privilege and oppression. I have come to understand the idea of *interlocking systems* to mean that the oppression of some people does not exist without systems supporting the unearned privilege of other people and that these systems occur in relationship to many social identities (e.g., race, class, gender, sexuality, ability, age). For example, *racism* does not exist without systems supporting white privilege. That is,

> racism is understood to be widespread and ingrained in society, rather than manifested only in the actions of a few "irrational" people . . . [and] . . . through this perspective, racism is perceived as an entity that affects everyone in society, benefiting some and victimizing others. (Marx, 2006, p. 5)

Previously, I had felt comfortable pushing on issues of gender with my students. For example, we would talk about how teachers often interacted with girls differently than with boys and the fact that girls tended not to pursue mathematics at higher levels in high school and university. We would look for differences in videos for how teachers probed boys and girls, would read chapters written by teachers who were doing action research and learning about some of the ways they undermined equitable practices (e.g., Wickett, 1997), and would talk about the ways in which the media shaped our views about girls and mathematics (e.g., Mendick, 2005). I also tried to share ways in which I tried to push and support the girls in my mathematics classes when I was a teacher.

I had also, to a lesser extent, pushed on issues of race in the context of being a White faculty member in teacher education in the Midwest, where almost all of our students were White students. For example, while serving on hiring committees and admission committees, I often asked questions about who we discounted, what we were losing in terms of possibly new, different, and interesting experiences and perspectives, and the ramifications for diversifying our department and university community. I had not anticipated that a student of color would deny racism was at play in a task where it seemed obvious (to me!) that race was an issue in the decisions being made by mortgage lenders. In a quest to better understand this instance, I talked to many colleagues and friends and have read articles and books to try to understand what might have been happening in this interaction. For example, I am now aware of internalized oppression, which Tappan (2006) explained, has been used to

> describe and explain the experience of those who are members of subordinated, marginalized, or minority groups...; those who are powerless and often victimized, both intentionally and unintentionally, by members of dominant groups; and those who have "adopted the [dominant] group's ideology and accept their subordinate status as deserved, natural, and inevitable" (Griffin, 1997, p. 76). (p. 2117)

Although many scholars have treated this idea as a psychological one (or one that is an internal, hard-to-change characteristic of someone), Tappan (2006) argued that it should be seen as a sociocultural process of appropriation that is mediated by people and cultural artifacts. By coming to see internalized oppression as a sociocultural idea that is mediated by actors and cultural tools instead of only a personal one, it is easier to see how structures that perpetuate systems of privilege and oppression occur at broader levels than the individual. In relationship to using the *Mortgage Task*, for example, I could have been better prepared to talk about the history of things like access to housing and discriminatory mortgage practices that people of color have endured for decades in order to put this particular article into a larger perspective. Such a sociocultural view allows us to consider the individual and broader structures at the same time and can be used thoughtfully in antioppressive education (see Tappan, 2006, for example).

I have also come to recognize the pervasive myth of meritocracy, "that democratic choice is equally available to all" (McIntosh, 1990, p. 36), that operates in our society as a socializing myth. This construct also serves a role in the hidden curriculum in schools as a way of implicitly establishing dominant and subordinate positions in society (McLaren, 1989). As McIntosh (1988) stated, "Keeping most people unaware that freedom of confident action is there for just a small number of people props up those in power, and serves to keep power in the hands of the same groups that have most of it already" (p. 36).

My more recent work also makes me reflect on the important role of positioning (see Wagner & Herbel-Eisenmann, 2009) in the discussions we were having. This class was a situation in which we were talking about algebra in K–12 classrooms, ideas that many people still find to be politically neutral and culture-free (which has been called into question by many – see Nasir, Hand, & Taylor, 2008). I now understand that, even if a student thought racism might be involved, he probably would not voice this viewpoint in a setting where I had not established norms for such discussions. There is also a possibility that, if he had agreed that racism was involved, it could have positioned him as a recipient of such oppression rather than as a person who has agency in the world. This illustrates the complex interactions between positionings and the storylines (or grand narratives) that inform such positionings.

Although I now have theoretical ideas like internalized oppression, the myth of meritocracy, and positioning that help me understand possible interpretations of responses like Manuel's, I must confess that I am not sure exactly what these mean for what I should *do* as a mathematics teacher educator in a similar situation. I also do not know if any of these ideas were at play for Manuel when he responded the way he did, either. And, I am not sure how I would know whether these might be at play for other students of color who respond similarly.

I do think that having prospective and practicing teachers engage in social-justice tasks is important, and I continue to use these kinds of tasks in the courses I teach. I have designed my methods course, for example, to highlight the fact that we teach values in all tasks we offer. For instance, even if we focus on something as mundane as students buying items and spending money and see this as value-free, we are, in fact, supporting consumerism and capitalism. I have tried to further infuse my courses with readings that provide lenses that can prompt discussion about equity issues and have found the idea of positioning as particularly fruitful in these discussions. For example, in some professional-development materials I have been designing with colleagues,[1] we have seen teachers discuss issues of status and competence in the classroom with less discomfort than we had previously and also raise issues about systems of privilege and oppression in relationship to things like tracking or discipline in their schools. We have also seen them use the idea of positioning to take each other to task if someone is positioning students in deficit frames. Finally, I have also come to see how important it is to develop relationships with the practicing and perspective teachers with whom I work because it allows us to engage in harder dialogue, work through issues when we disagree, and continue pushing forward. That said, learning about interlocking systems of privilege and oppression, learning to incorporate those ideas in mathematics education courses in authentic ways, and learning how to engage in hard conversations are all ongoing processes that I will continue to learn and engage in conversations about.

NOTE

1. The professional development materials, *Mathematics Discourse in Secondary Classrooms,* are funded by the National Science Foundation Award #0918117, Herbel-Eisenmann, PI; Cirillo and Steele, co-PIs. Opinions, findings, and con-

clusions or recommendations expressed here are the authors' and do not necessarily reflect the views of NSF.

REFERENCES

Griffin, P. (1997). Introductory module for the single issue courses. In M. Adams, L. A. Bell, & P. Griffin (Eds.), *Teaching for diversity and social justice: A sourcebook* (pp. 61–81). New York, NY: Routledge.

Gutstein, E. (2005). "Home buying while brown or black": Teaching mathematics for social justice. In B. Peterson & E. Gutstein (Eds.), *Rethinking mathematics: Teaching social justice by the numbers* (pp. 47–52). Milwaukee, WI: Rethinking Schools.

Gutstein, E. (2006). *Reading and writing the world with mathematics: Toward a pedagogy for social justice.* New York, NY: Routledge.

Gutstein, E., & Peterson, B. (Eds.). (2005). *Rethinking mathematics: Teaching social justice by the numbers.* Milwaukee, WI: Rethinking Schools.

Gutstein, R. (n.d.). A Reading the World *mathematics project: Mortgage loans—Is racism a factor?* Retrieved from http://www.rethinkingschools.org/static/ publication/math/racismInMortgagesAssgnmnt.pdf

Marx, S. (2006). *Revealing the invisible: Confronting passive racism in teacher education.* New York, NY: Routledge.

McIntosh, P. (1990). White privilege: Unpacking the invisible knapsack. *Independent School, Winter,* 31–36.

McLaren, P. (1989). *Life in schools: An introduction to critical pedagogy in the foundations of education.* Toronto, Canada: Irwin.

Mendick, H. (2005). A beautiful myth? The gendering of being/doing 'good at maths'. *Gender and Education, 17,* 89–105.

Nasir, N. S., Hand, V., & Taylor, E. V. (2008). Culture and mathematics in school: Boundaries between "cultural" and "domain" knowledge in the mathematics classroom and beyond. *Review of Research in Education, 32,* 187–240.

rethinkingschools.org (n.d.). http://www.rethinkingschools.org/static/publica- tion/math/racismInMortgagesAssgnmnt.pdf

Tappan, M. (2006). Reframing internalized oppression and internalized domina- tion: From the psychological to the sociocultural. *The Teachers College Record, 108,* 2115–2144.

Wagner, D., & Herbel-Eisenmann, B. (2009). Re-mythologizing mathematics through attention to classroom positioning. *Educational Studies in Mathemat- ics, 72,* 1–15.

Wickett, M. (1997). Uncovering bias in the classroom—A personal journey. In J. Trentacosta & M. J. Kenney (Eds.), *Multicultural and gender equity in the math- ematics classroom: The gift of diversity* (pp. 102–106). Reston, VA: National Coun- cil of Teachers of Mathematics.

ANTICIPATING THE UNEXPECTED: MANAGING A DILEMMA DURING FACILITATION OF A SOCIAL JUSTICE MATHEMATICS TASK

A Commentary on Herbel-Eisenmann's Case

Lawrence Clark
University of Maryland–College Park

MY POSITIONALITY

As a mathematics teacher educator at a large, predominantly White university, I am familiar with the teaching context described in this case. I have taught an undergraduate mathematics methods course for several years, and the students who enroll are typically White. In the past four sections of the methods course I teach, a course that averages 25 students per section, I recall teaching three African American students and two Latino students

Cases for Mathematics Teacher Educators, pages 421–425
Copyright © 2016 by Information Age Publishing

across all four sections. Furthermore, like Herbel-Eisenmann, I infuse the course with tasks and discussions that require students to consider issues of equity, agency, and participation structures, activities that require students to consider their experiences and the experiences of others through the lenses of race, gender, and socioeconomic status.

INTERPRETING THE DILEMMA

The author describes a scenario in which she is surprised by the reaction of Manuel, a Latino student, during the facilitation of a task exploring patterns of systemic racism in obtaining mortgage loans. Manuel denies racism is at play when interpreting mortgage-rejection-rate data; data that shows that African Americans and Latinos are denied mortgage loans at higher rates than Whites. The author's dilemma is that she, as a White educator, did not feel she was well positioned or possessed the resources to help Manuel, a student of color, to make sense of patterns of systemic racism.

From Herbel-Eisenmann's perspective, Manuel did not hold the expected perspective, namely that racism is evident in the mortgage-rejection-rate data. She anticipated that Manuel would acknowledge that racism was at play for two reasons: (1) the task and the data suggest that *something* inequitable is going on, and (2) as a function of him being a person of color, Manuel is predisposed to seeing inequities when others (i.e., White people) may not.

Herbel-Eisenmann attempts to understand Manuel's response through a number of theoretical perspectives including internalized oppression (adopting the dominant group's ideology and accepting their subordinate status as deserved, natural, and inevitable), the myth of meritocracy, and positioning. She also acknowledges that none of these perspectives may have contributed to his response.

RESPONDING TO THE DILEMMA

All of these perspectives (or some combination of them) may explain Manuel's response, yet I cannot help but wonder why Herbel-Eisenmann did not employ what appears to me to be the first best response to Manuel's statement—asking him to explain his thinking. As opposed to becoming uncomfortable with, ignoring, or trying to interpret his response through theoretical perspectives, Herbel-Eisenmann could have posed a series of questions that may have revealed why he did not think racism was at play and heard him out. Herbel-Eisenmann could then have allowed other students to restate or interpret Manuel's extended explanation, very much like what might take place when facilitating a mathematical discussion

(Walshaw & Anthony, 2008). These questioning and facilitation techniques may have transformed Manuel's unexpected initial response into a set of connected rationales that may have provided opportunities for the class to move forward towards achieving the task's learning goals.

If carefully questioned, Manuel's extended response may have revealed that his perceptions of racism are predominated by images of interpersonal racism (Gutstein & Peterson, 2005; Krieger, 1999), which could be argued are not particularly evident in the data. He may not have a working understanding of other forms of racism at play in the mortgage-loan task, namely structural or institutionalized racism. Therefore, it may be that he does not see the form of racism with which he is most familiar evident in the task. Another potential explanation may be that Manuel has a sophisticated knowledge of the components of obtaining (or being denied) a mortgage, including buyer's credit score, amount of down payment the buyer has access to, and the relationship between the price of the home and the buyer's income, all variables that he may interpret as within the buyer's control and not a function of the buyer's race. He may believe that had more variables than income been held constant and racial disparities still remained, then racism may be at play. Simplistically holding one variable constant (income) and not throwing more into the mix may have not been sufficient for him. However, until his position is explored through asking him why he holds this position, it becomes an exercise in trying to make sense of his response without providing him the opportunity to explain or explore his thinking.

Employing strategic questioning techniques in this case may first appear to be an overly simplistic suggestion. Yet I would argue that Herbel-Eisenmann and other mathematics educators who find themselves in her position possess a unique skill set that they expertly employ in their daily practice in teaching mathematics or teaching preservice and inservice mathematics teachers to teach. This skill set consists of knowing and facilitating a series of thoughtful questions that reveal students' thinking (Walsh & Sattes, 2004; Wilen, 1987).

As mathematics educators, we would certainly do this—engage in a series of strategic questions—if the task and his response were strictly mathematical in nature. Why? Because we have accumulated evidence that formative assessment in general and good questioning in particular are powerful pedagogical tools that positively influence student learning. Furthermore, when the task is mathematical in nature, the mathematics teacher or teacher educator is typically very familiar with the mathematical content associated with the mathematical task and has a relatively clear conception of the learning goals associated with the task. However, it is important to acknowledge that there are instances when the best mathematics educators are at a loss for identifying and facilitating the best set of questions. These instances typically occur when a student's response is unfamiliar, unanticipated, or unexpected.

So how might mathematics teacher educators draw on their existing knowledge base and skill sets to productively facilitate a social-justice task such as the one described in this case? How, in this case, can a White mathematics teacher educator manage an unanticipated response from a student of color when facilitating a social-justice task such as the one described in this case? Here are four interrelated possibilities:

Be Clear on the Learning Goals Associated with the Social-Justice Task

In my review of the materials associated with this task (Gutstein & Peterson, 2005), there are no specific learning goals (mathematical goals or goals associated with social justice issues) listed. This is not to say that specific learning goals do not exist or have not been developed by individuals who have facilitated this task. When planning to facilitate such tasks, I try to think carefully through what I want students to learn when engaged in the task. For example, if the facilitator of this task has established that she wants students to come away with a clear understanding of the difference between institutional and interpersonal racism, Manuel's response may have been more easily explored and responded to.

Be Prepared to Employ a Series of Nonevaluative Questions That Explore Student Thinking Related to the Mathematics of the Task or the Social Issues Associated with the Task

When students interact with social-justice tasks in mathematics classrooms, students' mathematical understandings and perspectives of social justice issues are simultaneously on the table for discussion. Much like students' mathematical understandings are difficult for a mathematics teacher to understand without engaging students in a thoughtful series of questions, students' perspectives of social justice may be equally difficult for a mathematics teacher educator to understand. When engaging students with such tasks, I attempt to provide a substantial amount of time exploring students' perspectives on the issues at hand. I have found good questioning and the encouragement of productive discourse to highly influence the extent to which task learning goals are met.

Anticipate Responses Associated with the Mathematical Thinking and the Thinking around the Social Issues, yet Don't Be Surprised if an Unanticipated Response of Either Kind Emerges

Social justice tasks are not easily facilitated and do not always engender the predicted responses of participants (Brantlinger, 2011; Leonard, Brooks, Barnes-Johnson, & Berry, 2010). Tasks of this nature are predictably challenging to facilitate because the issues at hand are complex and demand that facilitators have a good sense of differing perspectives. It is critical, therefore, for facilitators to anticipate the nature of those perspectives and be ready to productively manage differing perspectives in service of the task's learning goals.

Acknowledge the Heterogeneity of the Experiences of Students of Color

It is my belief that the root of Herbel-Eisenmann's dilemma is that Manuel did not respond to the data in a predictable way and, thus, disrupted her vision of the flow of the classroom discourse. She expected his response and perspective from White students but not from him. It is important for facilitators of social justice tasks in mathematics classrooms to be aware of historical forces that shape the perspectives of communities (racial, geographical, economic) yet leave space for heterogeneity of experiences within these communities (Clark, Johnson, & Chazan, 2009).

REFERENCES

Brantlinger, A. (2011). Critical mathematics in a secondary setting: Promise and problems. In B. Atweh, M. Graven, W. Secada, & P. Valer (Eds.), *Mapping equity and quality in mathematics education* (pp. 543–554). Dordrecht, The Netherlands: Springer.

Clark, L., Johnson, W., & Chazan, D. (2009). Researching African American mathematics teachers of African American students: Conceptual and methodological considerations. In D. Martin (Ed.), *Mathematics teaching, learning, and liberation in the lives of Black children* (pp. 75–102). New York, NY: Routledge.

Gutstein, E., & Peterson, B. (Eds.). (2005). *Rethinking mathematics: Teaching social justice by the numbers.* Milwaukee, WI: Rethinking Schools.

Leonard, J., Brooks, W., Barnes-Johnson, J., & Berry, R. Q. (2010). The nuances and complexities of teaching mathematics for cultural relevance and social justice. *Journal of Teacher Education, 61,* 261–270.

Krieger, N. (1999). Embodying inequality: A review of concepts, measures, and methods for studying health consequences of discrimination. *International Journal of Health Services, 29,* 295–352.

Walsh, J. A., & Sattes, B. D. (2004). *Quality questioning: Research-based practice to engage every learner.* Thousand Oaks, CA: Corwin Press.

Walshaw, M., & Anthony, G. (2008). The teacher's role in classroom discourse: A review of recent research into mathematics classrooms. *Review of Educational Research, 78,* 516–551.

Wilen, W. (1987). Effective questions and questioning: A classroom application. In W. Wilen (Ed.), *Questions, questioning techniques, and effective teaching* (pp. 107–135). Washington, DC: National Education Association.

COMMENTARY 2

CHALLENGING PSTS' VIEWS AND THE INHERENT SUBJECTIVITY WHILE DOING SO

A Commentary on Herbel-Eisenmann's Case

Laura McLeman
University of Michigan–Flint

MY POSITIONALITY

As I considered this case and how I would have responded if I were in this situation, I could not help but identify with the author. I too am a mathematics teacher educator (MTE) who identifies as White and is currently working to support the next generation of secondary mathematics teachers conceive of mathematics teaching through the lenses of equality and justice. In my situation, I completed my undergraduate- and master's-level studies in mathematics and used that knowledge when I taught mathematics in both public and private high schools. I earned my doctorate

Cases for Mathematics Teacher Educators, pages 427–430
Copyright © 2016 by Information Age Publishing
427

in education, focusing specifically on mathematics education and issues of equity and social justice. I now work as an MTE at a public urban institution in the Midwest.

INTERPRETING THE DILEMMA

The dilemma presented in this case is relevant to me as I have also sometimes felt at a loss on how to push people to articulate their thinking about a given issue, especially when their thinking differs from mine or is contrary to what research has shown. Added to the dilemma in this case is that race is a factor. Specifically, the author, who does not identify as a person of color, was unclear on how to challenge a statement made by a student of color, Manuel, who did not respond to a task in an anticipated way. Underlying this predicament of how to challenge Manuel's thinking was the quest to do so in a way that was respectful and honored his lived experiences while at the same time engaged him in critical reflection about systemic structures.

RESPONDING TO THE DILEMMA

Throughout the case, I yearned to know more from Manuel. Why did he think that racism was not at play? What evidence did he have that the others in the class did not? In the author's reflections, she notes many possibilities that might be behind Manuel's response — however we do not know for sure if any of those possibilities are reality. With that said, the author's conscientious and thoughtful reflection about what might be possible reminds me that we as educators have an immense responsibility to take seriously the lives of each of our students by listening attentively to what they are saying (Johnson, 2006). Further, we must value the stories and experiences of our students by prying into the reasons behind their responses, as uncomfortable as that may make us (Swalwell, 2013) and even if the students are not thinking in ways that are expected. By doing so, we can begin to more fully understand our students and authentically build relationships with them. For students of color, in particular, this attentive listening begins to honor their knowledge, skills, and experiences, something that does not often occur in teacher education programs (Kohli, 2009).

One issue that was raised for me throughout this case is the issue of perspective. The author notes that to her it was obvious that racism was a factor in the denial of mortgages. However, data is subjective. Depending on one's outlook, an event can be construed from different points of view — this is seen frequently on the 24-hour news shows where two individuals are arguing opposing points using the same set of data. So, what seems to be obvious to one person may not be obvious to another, as our experiences and biases, among other things, shape how we see and read the world (Gutstein, 2006). Consistent throughout the possible interpretations for Manuel's answer that the author provides is the notion that Manuel's experiences shaped the way he interpreted the data. It is those experiences we should be drawing on within our classes. By entering into a dialogue with Manuel, the author may have received insight into the data she had not previously considered, insight that could only come from Manuel's experiences and knowledge as a person of color. Likewise, Manuel could have seen the data in a different light, thus further developing his own knowledge about systems of racism.

This leads to a broader issue when addressing issues of equity within mathematics education. In a move to push students away from making broad statements based on personal experience, we oftentimes ask them to use evidence to support a particular position or response. In the context of this case, the author asked students to support their responses about institutional racism in lending practices using data about mortgage loans—a laudable and worthy goal. However, when the author qualified to the students that they needed to use the data given in the task to support their responses, it is possible that the message received by the students is that their opinions needed to be supported by something objective. In this sense, they may see the data, and thus by extension mathematics, as the "objective other." However, this message is in conflict with what social-justice educators and others advocate (e.g., Felton, 2010). Although it is understandable there is a desire to help students see that they can use mathematics to understand the world around them, we must be cognizant that we are doing so in ways that are not reinforcing the notion that mathematics is neutral and context-free.

It is unclear from the case what prior work was done to help unpack the ideas related to systemic racism. The author notes that there could have been more framing of the task, specifically through discussions related to the history of practices surrounding access to housing and loans, to support the students' understanding of the broader issues of systemic racism. This is an important component, because if and when you encounter an unanticipated response, you can draw upon the framework as a way to constructively challenge any deficit thinking that may be taking place. However, I have found in my context that students need more than this type of framing to help support their learning. Students also need help in identifying the existence of structural and societal norms and practices that are imposed upon individuals. To this end, I have begun having students read texts such as *Fire in the Ashes* (Kozol, 2012) and underlining any instances where they see evidence of these types of norms and practices. Through these underlined passages, we then deconstruct some of the ways these norms and practices impact individuals' lives. For example, after identifying the passage, it can't be easy for the cops to know when somebody is telling them the truth. If these boys are goin' to dress like gangsters, walk like gangsters, talk like gangsters, and behave like gangsters, and their friends are gangsters, then they *got* to know that they'll attract attention (p. 123)

as an example of societal norms, we discuss how society has normalized what gangsters look, sound, and act like. We further discuss, then, what impact these ideas about gangsters have on the psyches of the individuals being targeted as well as on the actions and perceptions of other individuals.

In addition to focusing on what can be done to support what is happening within a specific class, we also need to consider how the physical environment can serve as a means to challenge our thinking. When we are sitting in a university classroom, for example, we are depriving our students and ourselves a chance to engage with some of the very systems we are trying to understand. My colleagues and I have taken this notion to heart and have sought to forge deeper relationships with the urban schools that surround our institution. As these relationships deepen, we have been able to integrate some of our education classes into one high school's schedule, so that the courses are offered at the school site during the school day where we have regular interaction with students, faculty, and staff. In this way, we are helping to see ourselves as part of the educational structure, which has been important for our development as professionals. When we see ourselves as part of a structure as opposed to separate from it, we gain a more nuanced understanding of how that structure operates and perpetuates deficit views. We can then draw on this knowledge and understanding and challenge *each other* to think more deeply and critically.

REFERENCES

Felton, M. D. (2010). Is math politically neutral? *Teaching Children Mathematics, 17,* 60–63. Retrieved from http://www.jstor.org/stable/41199594

Gutstein, E. (2006). *Reading and writing the world with mathematics: Toward a pedagogy for social justice.* New York, NY: Routledge.

Johnson, A. G. (2006). *Privilege, power, and difference* (2nd ed.). New York, NY: Mc-Graw Hill.

Kohli, R. (2009). Critical race reflections: Valuing the experiences of teachers of color in teacher education. *Race, Ethnicity, and Education, 12,* 235–251.

Kozol, J. (2012). *Fire in the ashes: Twenty-five years among the poorest children in America.* New York, NY: Broadway Books.

Swalwell, K. M. (2013). *Educating activist allies: Social justice pedagogy with the suburban and urban elite.* New York, NY: Routledge.

COMMENTARY 3

ON DENIAL AND THE SEARCH FOR EXPLANATION

A Commentary on Herbel-Eisenmann's Case

José María Menéndez
Pima Community College

MY POSITIONALITY

I was born and raised in Latin America where I got my high-school mathematics teacher degree. I came to the United States at the age of 25 years old to continue my education. After obtaining my doctorate in mathematics, I taught mathematics content courses for prospective elementary school teachers for several semesters while working on a postdoctoral position in mathematics education at the Center for Mathematics Education of Latinas/Latinos (CEMELA) at the University of Arizona. I continued teaching these courses at Radford University, Virginia, whose teacher preparation program included a third-semester mathematics course on mathematics for the public interest, *Elementary and Middle Grades Mathematics for the Social Analysis*. This last course had a strong social-justice orientation; we used

Cases for Mathematics Teacher Educators, pages 431–434
Copyright © 2016 by Information Age Publishing
431

Rethinking Mathematics (Gutstein & Peterson, 2005) as the main reference book, and students had a choice of either Service Learning (mostly tutoring at local schools in afterschool programs) or a research project based on issues that reflected a social-injustice situation and for which they needed to use mathematics to understand the problem or propose a solution. Students in these courses were mostly White females, with very few students of color, and most of the students were from the region, predominantly rural areas. It was very common for students to resist addressing issues of social injustice, and quite often students had negative reactions to discussing racism, poverty (locally and around the world), and militarization, among other topics. For example, in looking into how many hospitals or schools could be built with one year's budget on defense, sometimes students would express that the topic is not appropriate for a mathematics course (where we should be learning mathematical facts, not sociological facts or criticizing the government) or that they were preparing to teach at grade level and these examples were not appropriate for children, thus we would do better to bring examples that would be useful in their future classrooms.

INTERPRETING THE DILEMMA

Herbel-Eisenmann possesses three questions for herself: (1) "How could I, as a White professor, help a student of color make sense of patterns of systemic racism?" (2) "What are some ways I might engage and challenge deficit perspectives about students of color when working in a context where most students are White and the student denying inequities is a student of color?" and (3) "How could I do this in an authentic way that builds relationships instead of potentially marginalizing someone or replicating an imperialistic (i.e., enabling the powerful to act and speak on behalf of the oppressed) viewpoint?" These three questions make me think of two dilemmas: The first one is the immediate response of a teacher thinking on her feet before an unexpected answer, and the second dilemma (which comes after having reflected on the possibility of such an answer) is how to respond in a way that reaches the desired objectives raised by the questions the instructor possesses.

RESPONDING TO THE DILEMMA

For the first situation, the immediate response, unfortunately we do not have much information on exactly what happened after Manuel gave his answer. I think that a good strategy is to stay true to the task and press on the mathematical reasoning for the answer. Whether the answer is right or wrong, the exercise of argumentation is valuable by itself. Depending on the classroom norms, the group might spontaneously question or counter-argue the position, or side with Manuel. Maybe the instructor would need to promote the conversation around Manuel's answer. This takes me to the second dilemma.

The theories presented by Herbel-Eisenmann (internalized oppression, the myth of meritocracy, positioning) are all sound. However, without more information on the case, we do not know exactly where Manuel was coming from, why he was making the moves he was making, how he perceived himself, or what was his understanding of racism. As an international student (both undergraduate and graduate programs), I was oblivious to personal racism towards or around me, especially the former. I was not part of the system. Intellectually I understood systemic racism (and structures of oppression) from the books, and it was easy to see the historical patterns and evolution of the dynamics of exploitation. Yet, I refused to acknowledge that it was happening to me and people like me, right then, right there. We don't know whether Manuel thought of himself as an international student or not, but we do not know what other identities Manuel manages. The point I am trying to make is that even though these theories can shed light on aspects of a complex problem, the way that a student individually processes his or her experience is quite personal and quite obscure for an instructor to know what it is and, even harder, what to do about it. We must humbly acknowledge that we do not know everything.

The second dilemma seems to be centered about the dichotomy of powerful (White, professor) and not so powerful (student, of color). I would argue that most students perceive the role of a professor being in a position of authority regardless of the race or gender of the professor. Moreover, students will react to this perception differently. However, it is not easy for a professor in most U.S. colleges to change this perception because the professor-student relationship is framed, institutionally, in a particular way with the ultimate result of a grade at the end of the semester. As the professor in this case has done, I would first try to understand what is going on and try to be prepared to the best of my ability, which reflects owning the position of authority a professor has. I find it useful to continue with this idea and assume the role of the expert in the subject matter. It is from this position that students could be more receptive to challenging, even threatening, ideas. Showing sensitivity to the power dynamics is important, mostly for the teacher, instructor, or professor but does not necessarily translate into the desired outcome as the way an individual student perceives the situation is, for the most part, beyond our control. That being said, I am hopeful that the environment we can try to create in our classrooms allows for these conversations about controversial topics to occur, that the conversations among classmates, even if orchestrated by the teacher, allow for knowledge to emerge, especially shared, coconstructed knowledge. More importantly, that the professor does not think that she or he has to teach students the patterns of systemic racism, but as Herbel-Eisenmann states, to help them make sense of these patterns or, for starters, to acknowledge or identify them even if they make no sense. We can only try to create situations and the environment that promotes learning, but we cannot force learning to occur. Depending on many things ranging from the student's experiences to the way the professor comes across, being a White, female professor could be an asset not an obstacle. I think that this comes back to how the professor positions herself. As for the specific questions Herbel-Eisenmann asks, that is for each of us to discover, but one must be authentic oneself, that is recognizing what our limitations are in understanding how students perceive specific situations and what motivates them to react in a particular way, what the experiences and ideologies that shape our sense of self are, how we really perceive our students (what narratives do we tell ourselves about them), what we believe our role as educators is (for example, how committed we are about challenging students' perceptions on social issues), and what uniquely we bring to the academic arena (in particular as teachers) given our salient identities.

REFERENCE

Gutstein, E., & Peterson, B. (Eds.). (2005). *Rethinking mathematics: Teaching social justice by the numbers*. Milwaukee, WI: Rethinking Schools.

CHAPTER 19

LEARNING ABOUT STUDENTS AND COMMUNITIES USING DATA AND MAPS

Laurie H. Rubel
Brooklyn College of the City University of New York

INTRODUCTION: WORKING WITH LOCAL DATA ABOUT PEOPLE AND COMMUNITIES

Data about people, places, and practices are increasingly available and can be used by teachers as one way to learn about students, geographical neighborhoods, and cultural communities. Analysis of data can be part of a process for teachers toward critical reflection about data and how it is used to frame people and communities. For instance, what types of data are collected by local, state, or federal agencies? What categories are used in, for example, the census, and how do those categories end up framing our understanding of people and of communities? What kinds of data are available beyond the census?

Many datasets include geospatial components, which allow for representations of data using maps. Representing data on a map invites questions about spatial patterns that can further guide teachers toward critical

Cases for Mathematics Teacher Educators, pages 435–443
Copyright © 2016 by Information Age Publishing
435

orientations about data and to using data to mathematically investigate issues of fairness and equity. For example, the variable of median household income has a geospatial pattern in New York City. Why are those neighborhoods located in that particular spatial arrangement, and what significance might that spatial arrangement have? Spatial arrangements of variables can be compared; for example, how does the spatial arrangement of family income compare to an arrangement of transportation or other resources? Considering why particular patterns coincide can illuminate interactions and relationships. For example, why are there greater rates of asthma in low-income neighborhoods in New York City?

Data that represents a school's neighborhood relative to its surrounding context invite these types of questions and support teachers both in developing critical orientations *about* data and *with* data and, in so doing, in refining their political knowledge (see Gutiérrez, 2012) about students, families, and the local city. Beyond bringing local data into their own classroom activities, teachers can draw upon studies of students' neighborhoods to identify potential pressing issues for students or residents of the school's neighborhood and use what they learn from such analyses to consider how they might be able to better support students.

NEIGHBORHOOD, SCHOOL, AND PROFESSIONAL-DEVELOPMENT CONTEXT

Harwood High School (a pseudonym) is a small high school in New York City of about 400 students. Harwood is in a majority Latin@[1] neighborhood, diverse in terms of its residents' countries of origin and length of time in the United States. At the time of the professional-development project described here (2009–2010), about half of the families in the neighborhood were on federal assistance, and more than half of the families had incomes in the city's lowest two quintiles.

Harwood is located in a building that previously housed a large high school that was closed by the city because of poor performance, and students must pass through metal detectors to enter. In the 2009–2010 school year, about two thirds of Harwood's students identified as "Hispanic or Latino" and the other students as "Black or African American," and 85% of students returned paperwork to qualify for free or reduced lunch. That year, approximately half of the incoming 9th-grade students had not met proficiency levels on their 8th-grade standardized tests in mathematics. In contrast with the way that most data about Harwood represents it as a dangerous school, attended strictly by weak students of color from low-income families, Harwood consistently receives high marks from its students and their parents on city surveys in terms of satisfaction with the school and its teachers.

In 2009–2010, Harwood's mathematics department was comprised of six mathematics teachers: four teachers identified as Black or African American (two men and two women) and two teachers as White (one man and one woman). All of the teachers were early career: The most experienced teacher was an 8th-year teacher, the least experienced was a 3rd-year teacher, and the median number of years of experience was 5. None of the six teachers were raised in this city, and all of the teachers commuted to Harwood from other parts of the city.

The department participated in a 2-year professional-development project about culturally relevant mathematics pedagogy (see Rubel, 2012; Rubel & Chu, 2012), which consisted of two 8-day summer institutes and follow-up meetings once a month at the school. The meeting that is described in what follows took place in an afterschool meeting in the project's first year. I facilitated the meeting. I am a teacher educator and researcher, am about a decade older than the median teacher age in this group, identify as Ashkenazi Jewish, and pass as White. Like the teachers, I relocated to this city as an adult and commuted to Harwood from another part of the city.

Earlier that year, we had conducted "mathematics of community in community" (see Rubel, 2012) field studies by doing investigations about several city neighborhoods, using various protocols, together with school students. The goal of the session that is highlighted here was to share with teachers an additional approach to studying communities, in terms of experiencing the potential of investigating data about communities. Prior to the meeting, I created a collection of data and maps, data tables, and text resources, all centered on the local neighborhood. I organized the teachers into three pairs and gave each pair several datasets, in various representations, pertaining to the school neighborhood and its residents. Teachers were asked to interpret these sources of information with an eye toward learning more about life in this neighborhood, who lives here, and the benefits and challenges that residents might face. After 15 or so minutes of work in pairs, the rest of the hour meeting was spent with teachers presenting their unique data source to the rest of the group. The meeting was audio recorded.

DILEMMAS WITH DATA

The first pair received information that summarized demographics changes in the neighborhood over time. They explained that although the neighborhood has been majority Latin@ for 40 years, these data had allowed them to unpack the census category of "Hispanic" in learning about rates of change in population for various subgroups. They found and reported, for instance, that Puerto Ricans were the largest group in the neighborhood, but with a smaller rate of change in population than newcomers from Mexico, Ecuador, and El Salvador. I pointed out that they might consider the category of "Other" in these Census tables, the rate of change in its frequency, and possible changes in its meaning over time.

The second group presented information about the neighborhood relative to the city totals, in terms of an array of variables included in the census. For instance, the group noted that relative to the rest of the city, this neighborhood has a high unemployment rate, low family income levels, and low numbers of college graduates. I had intended for teachers to interrogate the framing of some census categories—for instance, having multiple generations in one household can be an indicator of poverty but, alternatively, can be seen as an indicator of strong and positive family presence. Similarly, a low number of college graduates could be interpreted in different ways given the context of a majority immigrant population. The teachers in this second group voiced frustration that they had tried to look for what they called "positives" but found this difficult to do with the data tables that they were given. In other words, the census variables seemed fixed according to deficit interpretations for the teachers that afternoon.

My dilemma as to how to support teachers in developing critical orientations about data was further highlighted by the third group's engagement with their datasets. This pair had received a choropleth map showing rates of rodent complaints to the city's 311 call center across city community districts or neighborhoods. The map (an example is shown in Figure 19.1) shows community districts (numbered by borough or county) colored in terms of rodent complaints. Since neighborhoods are vastly different in population, the number of rodent complaints has to be normalized to be able to compare one neighborhood to another. In this case, the rodent complaints were normalized according to the number of residents (and not according to land area), and expressed as a rate per 10,000 residents. For example, in the neighborhood highlighted in Figure 19.1, there were 8.2 rodent complaints per 10,000 residents, which represents 1.4% of the city's total rodent complaints. The school at which this professional development was held was located in Community District 4 (immediately above the area marked in red), which is one of the four neighborhoods in the city with the highest rates of rodent complaints per person.

I gave this group a map of the same area colored according to a different variable, rates of cash assistance, which had the same spatial pattern as the rodent complaint map. I wanted the teachers to notice that in this given year, there were relatively more rodent complaints per person in low-income areas. My goal was for teachers to unpack that relationship. Why might these two patterns coincide? In other words, and I facetiously posed this to the teachers in their small-group focus on these two maps and in their presentation to the rest of the teachers: How does this happen, or how do the rats know where to go? My intention with this absurd question

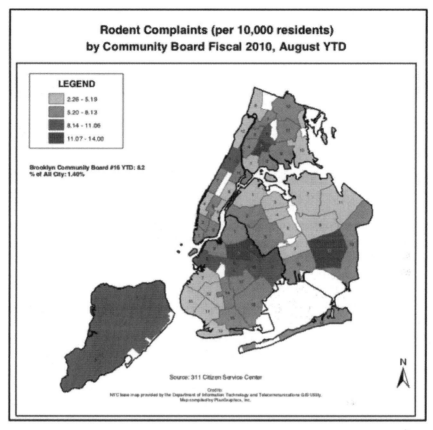

Rodent Complaints (per 10,000 residents)
by Community Board Fiscal 2010, August YTD

LEGEND

2.26 - 5.19

5.20 - 8.13

8.14 - 11.06

11.07 - 14.00

Brooklyn Community Board #16 YTD: 8.2
% of All City: 1.46%

Source: 311 Citizen Service Center

Credits:
NYC base map provided by the Department of Information Technology and Telecommunications GIS Utility.
Map compiled by PlanGraphics, Inc.

N

Data may include duplicate complaints/requests for services and are reported whether or not they are founded.
Complaints/requests for services may also be received through other channels such as NYC.gov and direct contact
with an agency.

Figure 19.1 Rodent complaint map.

was to help the teachers consider whether greater rates of complaints to the city imply that there are more rats, and if so, why there might be more rats in low-income neighborhoods. However, if the map indicates data about complaints, then perhaps more complaints does not imply more rats. Why might there be more rodent complaints in low-income neighborhoods? In other words, analyzing a rodent complaint map might quickly lead away from the rats and toward questions about housing types, owner-occupancy rates, frequency of vacant lots, and complaint rates in general.

Similar to the second group, the third group did not question the rodent complaint map or the data it represented. Instead, this group accepted the greater rates of rat *complaints* to mean that there are *more rats* in the city's low-income neighborhoods. They attributed the supposed presence of more rats to the cleanliness of people who live in the school neighborhood. One of the teachers explained to the whole group, *"Well, I imagine if you have less education you have less money, so you're more in poverty, I guess. And poverty has always had a, a, a, correlation with filth, I guess."*

The dilemma as to how to support teachers in developing critical orientations about data had been extended and amplified. This pair of teachers had not approached the data critically and, instead, used a deficit notion about low-income families and their homes to justify another deficit interpretation of the dataset about rodent complaints. In other words, an underlying deficit perspective about the cleanliness of the students' homes led to and was simultaneously reinforced by their misinterpretation of the rat complaint data.

As an experienced teacher educator, I am familiar with deficit thinking (Valencia, 1997) about students or families. I have listened to some teachers refer to students by the standardized-test-score labels assigned to them and claim that certain students cannot learn. I have heard some teachers say that Black or poor or Latin@ parents do not care about their children's behavior in school or claim that entire groups of people do not value education. I know several strategies for how to respond to each of these statements. But in that particular moment, over the presentation of the rodent complaint map, I was at a loss. This claim went unmet by any challenges by other teachers. Instead, the subsequent comments followed up with how a lack of cleanliness in the students' homes and a presence of rodents could possibly impact their ability to do mathematics homework! Unlike the other statements reflecting deficit thinking about students and families that, unfortunately, I have learned to anticipate, I had not expected this interpretation of the rodent complaint map.

REFLECTION ON AND RESPONSE TO THE DILEMMA

There was so much to challenge. First, the map did not state that there are more rats in low-income neighborhoods, just more complaints to the city. Second, what is the conceptual link between immigrants who do not have access to education or higher paying jobs and cleanliness? Other geospatial patterns in the city related to poverty might explain discrepancies in rates of rodent complaints, like more vacant lots, more public housing, less frequent municipal street cleaning or garbage pick-up, or more absent landlords. Yet these teachers understood the pattern in the rodent complaint map to mean that there are more rats in their school's neighborhood and their students' homes and attributed those rats to deficit perceptions of low-income people (see Gorski, 2012, for more on stereotypes about low-income people). Beyond a general dilemma of how to guide teachers toward developing critical orientations about data, I was now hit hard with how to respond to the voiced notion that their students' families or homes are dirty and attract rodents. How effective can teachers be if they think this about their students?

I waited a few moments, hoping that another teacher in the group might challenge this train of thought, but challenges did not emerge. I tried to redirect the group. I said, *"So hold on a sec. Let's just offer some other explanations 'cause I don't think rats are going—they're not looking for where there are lots of people, right? I don't think that's how the rodent mind works."* I actively tried to steer them towards other data that were included in the activity, like rates of renter-occupied units, to help them to consider that this data represented the number of complaints and not the number of rodents. Different from my usual stance as facilitator of activities and discussions, here, I assertively continued, *"You can't say, well, because the area is darker on that map it means there are more rats there, right? That's not what they're saying, right? It means they're getting more complaints."* I did not pause, did not make space to allow teachers to consider this idea in discussion, and instead continued, *"The reason I picked that. I think that's a good example of one that connects up to lots of other things for us to start thinking about like, you know, how does it come to be? I mean, is there a rodent problem here, and if there is, why is it that way?"*

I did not give space for teachers to enter into this dialogue with me, and I do not know how they assimilated my remarks. I continued, *"I'm sorry. I didn't mean to take over when you guys were sharing."* But I had taken over while that group was sharing and perhaps silenced them and the rest of the group in that moment. My adrenaline was rushing, and saying so much without giving pause was uncharacteristic of my facilitating style. The session continued with other planned activities, and I never brought this discussion of the rat map back to the group.

CONCLUSIONS: LESSONS LEARNED

Data about local communities can present valuable opportunities for teachers. The increasing quantification of aspects of daily living, along with new and flexible forms of data visualization, have clear connections to a variety of mathematical concepts and representations. Beyond integrating that data into classroom content, data about neighborhoods, people, and processes can offer teachers insights into patterns that can illuminate aspects of their students' lives. Teachers can use what they learn about students and their communities to support student learning in a variety of ways. This case cautions that data-analysis activities need to be framed so that the teachers can gain experience in developing critical orientations about data.

If I were doing this activity again with teachers, I would lead the group in an example of questioning data categories, to open up space for them to interact with data in critical ways, to show how a category that seems like only a deficit could be interpreted as a benefit. For example, although I asked the teachers to use the data to learn about people who live in this neighborhood in terms of benefits *and* challenges that might face residents, the notion that what might seem like only a challenge could be reimagined as a benefit was left too implicit. Questions about what variables get tracked in the census, what categories are used and why, and how might we understand those categories in new ways could be posed as part of the activity itself. Similarly, with reference to a map like the rodent-complaint map, questions such as "What is this a map of? What patterns do you see and how might you explain that pattern? Can you find several alternative explanations for that pattern?" should be posed explicitly.

My tendency that day was to give teachers latitude and intellectual space to think about and express their ideas, and so I refrained from trying to orient them in any particular way. But in an afterschool meeting, there is not time to process something new and complex and at the same time reframe a conventional orientation to data in a new way. More structure to the activity itself would likely be more effective.

As has been shown in this case, data about local communities typically relates directly or indirectly to race or income. If I were doing this activity again, I would precede it, in an earlier meeting, with an activity to orient teachers specifically to stereotypes about poverty. For example, I would have teachers, in collaboration, generate a list of common stereotypes, for instance, about low-income people so that these notions about a "culture of poverty" (see Gorski, 2008; Ullucci & Howard, 2015) are named, shared, and made explicit. That way, there would be language and understanding in place to challenge deficit notions about poverty in interpreting or explaining patterns in data about students and their communities.

ACKNOWLEDGMENTS

This material is based on work supported by the National Science Foundation under Grant No. 0742614. Any opinions, findings, and conclusions or recommendations expressed in this material are those of the author and do not necessarily reflect the views of the National Science Foundation. Thanks to Haiwen Chu for his assistance with this project, to the participating teachers, and to Marta Civil, Matthew Felton-Koestler, and Joi Spencer for their comments of an earlier version of this chapter.

NOTE

1. I follow Gutiérrez (2013) and use the term Latin@ to indicate both an 'a' and an 'o' ending.

REFERENCES

Gorski, P. (2008). The myth of the "culture of poverty." *Educational Leadership, 65*(7), 32–37.

Gorski, P. (2012). Perceiving the problem of poverty and schooling: Deconstructing the class stereotypes that mis-shape education practice and policy. *Equity & Excellence in Education, 45,* 302–319.

Gutiérrez, R. (2012). Embracing nepantla: Rethinking "knowledge" and its use in mathematics teaching. *Journal for Research in Mathematics Education, 43,* 29–56.

Gutiérrez, R. (2013). The sociopolitical turn in mathematics education. *Journal for Research in Mathematics Education, 44,* 37–68.

Rubel, L. (2012). Centering the teaching of mathematics on urban youth: Learning together about our students and their communities. In J. Bay-Williams and R. Speer (Eds.), *Professional collaborations in mathematics teaching and learning: Seeking success for all* (NCTM 2012 yearbook, pp. 49–60). Reston, VA: National Council of Teachers of Mathematics.

Rubel, L. H., & Chu, H. (2012). Reinscribing urban: High school mathematics teaching in low-income communities of color. *Journal of Mathematics Teacher Education 12,* 39–52.

Ullucci, K., & Howard, T. (2015). Pathologizing the poor: Implications for preparing teachers to work in high-poverty schools. *Urban Education, 50,* 170–197.

Valencia, R. R. (1997). Introduction. In R. R. Valencia (Ed.), *The evolution of deficit thinking* (pp. ix–xvii). London, England: Falmer.

THE FROG IN THE PAN: DEVELOPING CRITICAL AWARENESS IN MATHEMATICS TEACHERS

A Commentary on Rubel's Case

Rodrigo Jorge Gutiérrez and Alice Cook
University of Maryland, College Park

OUR POSITIONALITY

We, the authors of this commentary, are a clinical assistant professor in mathematics education and a graduate student in minority and urban education and math education. We both have experiences designing, implementing, and teaching social justice mathematics in public secondary classrooms and are committed to integrating critical pedagogy into our work as teacher educators. To illustrate, we co-taught a graduate course for practicing middle-school mathematics teachers that focused on the teaching and learning of statistics while emphasizing the development of statistical understanding through the investigation of social issues and injustices. We

Cases for Mathematics Teacher Educators, pages 445–450
Copyright © 2016 by Information Age Publishing

provide this background to communicate that we have attempted to do similar work to that of Rubel and have struggled with some of the same tensions and dilemmas. On most occasions, we left class with at least one of us thinking that class conversations had not gone in the desired direction, feeling dismayed about deficit beliefs expressed by the teachers about race, ethnicity, immigration status, socioeconomic status, gender, and so forth, or that participants were demonstrating resistance to social-justice-oriented discussions. Typically, the other offered some perspective on the long-term goals and the small successes we experienced. Our conversations revealed important perspectives on goals and timing of experiences in order to develop critical analyses, as well as the importance of relevance for the participants in doing this work. It is this broader perspective that guided our analysis of this case.

INTERPRETING THE DILEMMA

Several issues arose in this case as Rubel shifted from framing the dilemma broadly ("how to support teachers in developing critical orientations about data") to more specific considerations about in-the-moment pedagogical actions ("how to respond to the voiced notion that their students' families or homes are dirty and attract rats"). In this commentary, we will discuss two main issues: designing experiences such that teachers experience shifts in thinking rather than reinforcement of deficit beliefs and the challenge of timing and planning for this work. To aid with interpreting the dilemma, we would like to bring attention to three particular quotes:

> "This pair of teachers had not approached the data critically and, instead, used a deficit notion about low-income families and their homes to justify another deficit interpretation."
> "This case cautions that data-analysis activities need to be framed so that the teacher can gain experience in developing critical orientations about data."
> "But in an afterschool meeting, there is not time to process something new and complex, and at the same time, reframe a conventional orientation to data in a new way."

These quotes point to a major challenge in social justice mathematics—balancing both critical and dominant mathematical goals and demands (Gutiérrez, 2009; Gutstein, 2006, 2007a, 2007b). And by extension, when rushed for time, how does a teacher educator move teachers forward without simply telling his or her opinion? This is analogous to classroom mathematics instruction where teachers experience the tension of devoting time to exploration and concept development when pressured to cover dozens of mathematical topics. However, social justice mathematics has the added challenge of developing multiple knowledge bases, all informed by previous experiences, cultural norms, and ideologies. To this end, we will discuss how teacher educators can use prior and current teacher experiences with social justice to make the work relevant and meaningful.

RESPONDING TO THE DILEMMA

Teacher Experiences and Orientations

The first dilemma we will address is what Rubel refers to as the importance of "experience" for teachers in order to build on and challenge previous beliefs and understandings. Additionally, Rubel reflects on her role as an instructor, questioning her choices and pedagogical moves.

As social justice educators, we come to this work with the explicit intention of challenging the status quo and deficit perspectives of marginalized communities. We expect some resistance to such discussions and developments, much like society typically avoids or undermines these topics. However, we can counter this resistance by making social justice mathematics relevant to the participants' lives. Rubel presents issues within the school community such as poverty and inequitable distribution of city services. Perhaps data more relevant to the *teachers'* lives and experiences would have allowed for further development of a critical perspective as teachers turned to personal, professional, or social issues that they were interested in, knowledgeable about, and concerned for. For example, data on the achievement gap, math performance, teacher salaries, value-added bonuses, and so forth would prioritize teachers' experiences and funds of knowledge. With such data, teachers could offer nuances and alternative interpretations, pointing to additional data and representations. This could create a shared experience in which teachers critically analyze data and representations where they are the *subjects* of critique. It could also inform their analysis when they turn their focus to their students and communities, demonstrating empathy for those being analyzed.

Reflecting on our work as social justice educators is essential. It is only through self-critique, questioning, and willingness to be vulnerable, as Rubel is in her piece, that we can better our craft. In our statistics class, it became clear that a part of the resistance was due to the seeming irrelevance of the topics we selected for class investigation—homelessness of LGBTQ youth and racial profiling on traffic stops by police. The teachers did not find these topics meaningful to *their* own experiences. Although we as the instructors believe strongly that issues of race and sexual orientation are important for all people, especially teachers working with vulnerable populations, we realized we needed to shift our topics to ones that more closely supported what our teachers chose to study (e.g., gender differences in test scores, teachers' out-of-school work hours, and impacts of religion on marriage). We were reminded of the importance of meeting teachers where they are and facilitating their growth from their foundations.

Time for Critical and Professional Development

Professional-development experiences on social justice mathematics must take a long-term approach with sustained interactions. Otherwise, participants will maintain noncritical positions (Bartell, 2013; Gonzalez, 2008; Rodriguez & Kitchen, 2005) and continue to assimilate new information into their well-established ideologies. This is what the teachers in the case did when presented with data on the school's community. Because these deficit interpretations are precisely where the teacher educator's work needs to focus and because learning to ask critical questions of data is such a complex skill, it is possible that these teachers needed more experiences and more time.

In Rubel's case, she created a "2-year professional-development project about culturally relevant mathematics pedagogy," focusing this case on one specific afterschool meeting in the project's first year. Although Rubel critiques her own responses, we believe it is important to maintain perspective on how this *one* activity fits into the bigger picture. For example, Rubel wonders if this activity should be scaffolded and structured more tightly in order to promote critical analysis. However, with the long-term goal of helping teachers approach data critically, we wonder if these activities would be better implemented during the summer portion of the project. Concentrated time could be dedicated to investigations, allowing for teacher research and facilitator responsiveness. Questions that remain at the end of one session could be addressed the next day with new information provided by the facilitator. Teachers could dedicate themselves to the academic exercise in ways unlikely during afterschool sessions when they are distracted by the realities of daily instruction.

Alternatively, given the demanding goals and objectives of social justice mathematics, this particular investigation could benefit from being extended to more than one afterschool session. Rubel reflects that "other geospatial patterns in the city related to poverty might explain discrepancies," which could have helped teachers identify more "positives" in the data. This reflection points directly to two essential components of statistical investigations: the demands for more data and various interpretations. So rather than solely searching for ways to scaffold this one activity, perhaps

Rubel could note these initial perspectives and reflections in order to develop the *next* activity to help push the teachers' thinking. For example, ask the group what other data they would need to support or challenge their initial ideas, and then bring more data and different representations back to the group. Rather than providing structure to move teachers to the desired interpretation, honor the complexity of the social investigation and provide more data, ask more questions, and produce a variety of analyses.

CONCLUSION

The aim of our commentary was to provide perspective on the role of individual activities and interactions when considering the broader goals of social justice mathematics. We, along with Rubel, are self-critical and constantly reflecting on how to improve every aspect of our craft. However, in order to attend to teachers' long-term development, we must view their responses not as a missed opportunity to correct noncritical analysis, but as the expected initial stages from which most teachers must begin the journey of self-reflection, social critique, and praxis (Freire, 1970). Shifts in teacher beliefs, orientation towards students, and teaching practices require sustained effort and guidance. Thankfully, Rubel is engaged in such work and considers the professional implications of both her successes and struggles.

REFERENCES

Bartell, T. G. (2013). Learning to teach mathematics for social justice: Negotiating social justice and mathematical goals. *Journal for Research in Mathematics Education, 44,* 129–163.

Freire, P. (1970). *Pedagogy of the oppressed* (M. B. Ramos, Trans.). New York, NY: Seabury Press.

Gonzalez, L. (2008). Mathematics teachers as agents of change: Exploring teacher identity and social justice through a community of practice. Retrieved from http://ezproxy.library.arizona.edu/login?url=http://proquest.umi.com/pqdweb?did=1608368721&Fmt=7&clientId=43922&RQT=309&VName=PQD

Gutiérrez, R. (2009). Embracing the inherent tensions in teaching mathematics from an equity stance. *Democracy & Education, 18*(3), 9–16.

Gutstein, E. (2006). *Reading and writing the world with mathematics: Towards a pedagogy for social justice* (1st ed.). New York, NY: Routledge.

Gutstein, E. (2007a). Connecting community, critical, and classical knowledge in teaching mathematics for social justice. *The Montana Mathematics Enthusiast, Monograph 1,* 109–118.

Gutstein, E. (2007b). Possibilities and challenges in teaching mathematics for social justice. *Philosophy of Mathematics Education Newsletter, 22.*

Rodriguez, A. J., & Kitchen, R. S. (Eds.). (2005). *Preparing mathematics and science teachers for diverse classrooms: Promising strategies for transformative pedagogy.* Mahwah, NJ: Erlbaum.

COMMENTARY 2

CAN MATHEMATICS PAVE THE ROAD TO SOCIAL JUSTICE?

A Commentary on Rubel's Case

Robert Klein
Ohio University

MY POSITIONALITY

Rubel presents a snapshot of teachers participating in professional development focused on culturally relevant pedagogy. Rubel describes her effort to engage a group of teachers in critical interrogation of data about neighborhoods surrounding Harwood High School. Her goal is to support them in seeing data as a resource in developing "critical orientations" toward the categories and variables that structure the data. Rubel candidly describes a lesson that, in her estimation, goes awry when teachers "had not approached the data critically." Rubel's adrenaline takes over and she "did not give space for teachers to enter into this dialogue with me...and perhaps silenced them." Her desire to help teachers adopt critical orientations is clear throughout, as is her mindset of a self-reflective educator.

Cases for Mathematics Teacher Educators, pages 451–455
Copyright © 2016 by Information Age Publishing
All rights of reproduction in any form reserved.

My commentary derives from my experience teaching mathematics content methods courses. I work in rural Southeastern Appalachian Ohio. My position, as such, begs the question, "How does this expertise position my commentary on Rubel's (very personal) description of the activity?" The answer is, frankly, that although the goals of social justice efforts seem broad if not universal, actions taken to affect social justice are inherently local, deriving from interactions of context, culture, and location. Hence, it would be inappropriate, if not unethical, to position myself as in any better frame to criticize the case here than helping the reader to consider issues that arise in my mind as I read Rubel's case. This commentary questions the broader contexts of social justice work while avoiding claims of understanding the local challenges, inclinations, and opportunities with which Rubel works.

INTERPRETING THE DILEMMA

Using geospatial data to motivate discussions of social injustice is a promising way to connect mathematics and statistics education to real-world challenges. Tate (2008) pointed out that "to seek an understanding of the strengths and problems in our communities is an important civic function" (p. 408) and highlighted the use of Geographic Information Systems (GIS) to explore the geospatial nature of opportunity in two metropolitan areas. Hogrebe, Kyei-Blankson, and Zou (2006) conducted similar analyses using publicly available datasets. The Rural School and Community Trust's biannual report, *Why Rural Matters* (Johnson, Showalter, Klein, & Lester, 2014) uses large, public datasets and basic statistical analyses to the explicit political ends of highlighting areas of need for rural education policy. In each of these cases, professional researchers with extensive training conduct the analyses.

Social-justice-oriented mathematics activities (e.g., Gutstein & Peterson, 2005) provide teachers with materials to help students develop critical orientations. Case studies detail how culture becomes "relevant" even when cultural relevance is not a primary instructional goal. Lubienski's (1997) attempt to develop students' "critical consumption" of statistical representations backfired:

> lower-SES students equated persuasion with lying, whereas higher-SES students equated persuasion with careful data collection and representation. Although this difference pertained to persuasion in the media, might it also influence or relate to the way students view mathematical arguments in the classroom? Might there be clashes among different views of acceptable mathematical reasoning? (p. 57)

Can We Have Both Social Justice and Mathematics?

Dowling and Burke (2012) analyzed one of Gutstein's social justice mathematics lessons, noting the difficulty of balancing political and mathematical goals. They asked whether it is possible to have lessons whose objectives include both mathematics and social justice. One difficulty occurs when "the results of mathematical manipulation are then pushed back into the political arena, providing a simplistic and misleading response to a complex sociological problem" (pp. 100–101). This seems to have happened with Rubel's third group that offered superficial explanations of the data, then extrapolated to judgments about their students' hygiene. Dowling and Burke concluded:

[A]ddressing social inequalities demands explicit, dissonant strategies, referred to here as interrogation. However...interrogation itself is likely to lead to misinterpretation where the mathematical activity is foregrounded and mathematics is likely to lose out where it is not. Ultimately, we may be left with the choice of whether to do politics or to teach mathematics. (p. 87)

Bartell (2013) similarly found that "teachers acknowledged a tension in negotiating mathematical and social justice goals. For [them], negotiation of the mathematics and social justice goals resulted in lessons that seemed to divide these two foci" (p. 31). Bartell (2011) offered a framework for supporting teachers to develop a "caring with awareness" as one critical mindset.

Responding to the Dilemma

So is this work just difficult and complex, as the work of Lubienski (1997), Bartell (2011), and Gutstein and Peterson (2005) would suggest, or is it not possible, as Dowling and Burke (2012) (and possibly one reading of Bartell, 2013) might suggest? It is impossible to capture what happened from a limited case taken in isolation, yet the case of the rats reveals insights important for determining whether (and if not, then how) social justice and mathematics might coexist in professional development or classroom lessons.

First, this case as presented suggests that very little mathematics was going on. The data were already (re)presented in forms deemed most appropriate by Rubel, including the normalization of rat complaints per 10,000 residents in an area. *Representation* must be an explicit part of interrogating the data. Data never speak for themselves—representation and interpretation must do that (numerically, visually, or narratively). Rubel asks too much of the data, claiming, "Data that represent a school's neighborhood relative to its surrounding context invite these types of questions and support teachers both in developing critical orientations about data and with data and, in so doing, in refining their political knowledge." If mathematics and social justice are to coexist, then mathematics must be more than a set of facts to be applied to situations (what Dowling & Burke, 2012, called a *fetch strategy*) but a means for seeing, representing, and questioning the world. Also, people don't come with ready-made critical orientations—these are skillsets to be learned and refined. Rubel likely understands that this is a developmental process. She reflects that, developmentally, teachers may not have been ready to interrogate categories and question the assumptions that lead to superficial readings of those data. What does a critical orientation look like along the continuum from "no critical orientation" to "mature (or useful?) critical orientation?" Although Rubel observed that "teachers had not approached the data critically," it is unclear what criteria she was using to determine what mature critical orientations look like. What constitutes evidence of growth or professional development in this area?

Rubel was upset at the teachers' interpretations and also for "shutting them out" of the conversation. This reveals yet another dimension of this work—the teacher or facilitator of professional development almost always occupies a different discursive space than participants in the professional development. Often it is a space of privileged knowledge or position (teacher, professor, etc.) that makes it harder for someone with caring intentions like Rubel to unknowingly exploit that privilege as she did. Having done this more times than I can count, I know that, like Rubel, I constantly reflect on how to guide and facilitate growth without imposing a privileged perspective or outsider's advantage. When a primary goal is something as personal as critically interrogating data about those we care about (Harwood's community broadly), it's all the harder.

NEGOTIATING THE TENSION

Rubel's case evidences the underlying tensions of working as educators to eliminate injustice and to promote care and justice. Whereas Dowling and Burke (2012) decided that one of mathematics or social justice has to take a back seat in any lesson that includes both, I choose to remain a naïve optimist. Technologies used in representing and sharing data and computing statistics continue to improve, applied mathematics and especially mathematical modeling are finding greater acceptance in K–12 curricula, and our world is becoming more transparent. As such, our tools and a broader policy mandate may afford more opportunities to make explicit connections between good mathematics and a more just world. Mathematics offers a powerful way of seeing the world that is and worlds that can be. Yet, de Abreu, Cline, and Shamsi (2002) argued that "understanding of how particular social groups learn, use and transmit knowledge requires consideration of the link between knowledge and values" (p. 124). Classrooms may be one of the last remaining public spaces (certainly among the last with compulsory attendance). Thus it is there that we will have to deal explicitly with issues of social justice, negotiating the world that can be. We may not have the luxury of asking Dowling and Burke's (2012) question if both social justice and mathematics are possible. Since what we have now is social injustice and mathematics together, how in good conscience could we not try, as Rubel does, to have both politics and mathematics in the classroom?

REFERENCES

Bartell, T. G. (2011). Caring, race, culture, and power: A research synthesis toward supporting mathematics teachers in caring with awareness. *Journal of Urban Mathematics Education, 4*(1), 50–74.

Bartell, T. G. (2013). Learning to teach mathematics for social justice: Negotiating social justice and mathematical goals. *Journal for Research in Teaching Mathematics, 44,* 129–163.

de Abreu, G., Cline, T., & Shamsi, T. (2002). Exploring ways parents participate in their children's school mathematical learning: Case studies in multiethnic primary schools. In G. de Abreu, A. Bishop, & N. Presmeg (Eds.), *Transitions between contexts of mathematical practices* (pp. 123–148). Dordrecht, The Netherlands: Kluwer Academic.

Dowling, P., & Burke, J. (2012). "Shall we do politics or learn some maths today?" In H. Forgasz & F. Rivera (Eds.), *Towards equity in mathematics education: Gender, culture, and diversity* (pp. 87–103). New York, NY: Springer.

Gutstein, R., & Peterson, B. (2005). *Rethinking mathematics: Teaching social justice by the numbers.* Madison, WI: Rethinking Schools.

Hogrebe, M. C., Kyei-Blankson, L., & Zou, L. (2006). *Science attainment by content strands in St. Louis area school districts: Technical Report.* St. Louis, MO: Washington University. Center for Inquiry in Science Teaching and Learning. Retreived from http://alhs.wustl.edu/scienceandtechnology/Tech%20Report.pdf

Johnson, J., Showalter, D., Klein, R., & Lester, C. (2014). *Why rural matters 2013– 2014: The condition of education in the 50 states.* Washington, DC: Rural School and Community Trust.

Lubienski, S. T. (1997). Class matters: A preliminary excursion. In J. Trentacosta & M. Kenney (Eds.), *Multicultural and gender equity in the mathematics classroom: The gift of diversity* (pp. 46–59). Reston, VA: National Council of Teachers of Mathematics.

Tate, W. F. (2008). "Geography of opportunity": Poverty, place, and educational outcomes. *Educational Researcher, 37,* 397–411.

COMMENTARY 3

BEING STUDENTS AND TEACHERS OF MATH AND SOCIAL JUSTICE

A Commentary on Rubel's Case

Cynthia Nicol
University of British Columbia

MY POSITIONALITY

Most of the teacher candidates I teach and the practicing teachers I work with have not experienced homelessness or poverty, yet most will during their careers teach children who have. Many who eventually teach in urban areas will not live near the communities and schools in which they teach. And many will teach in diverse classrooms unlike the classrooms where they were once learners. As a mathematics teacher educator and researcher in a large urban area on Canada's Pacific west coast, I strive to provide opportunities for teachers I work with to be prepared to teach children unlike themselves.

For the past 10 years I've worked with indigenous communities to explore the nature, challenges, and possibilities of teaching and learning a

Cases for Mathematics Teacher Educators, pages 457–461
Copyright © 2016 by Information Age Publishing
All rights of reproduction in any form reserved.

mathematics curriculum that is responsive, relational, and relevant to the community, its people, and land. This means working with teachers who have had few opportunities to learn about indigenous ways of knowing, or about colonizing histories and practices, or to be with people of Aboriginal ancestry. I work alongside Aboriginal and non-Aboriginal teachers to better understand positions of privilege as members of a dominant group and to unearth the associated normative practices that continue to provide power to some while marginalizing others. Such work involves a critical response to colonial actions by making space for indigenous ways of knowing and being and thereby working to decolonize both teaching and research practices. I'm particularly interested in mathematics teacher education *with* indigenous education. This means paying attention to opportunities to challenge privileged stances, better understand race and racism, and confront assumptions that can lead to deficit thinking. I don't see myself as an expert but am open to listening to the voices of teachers, students, and their communities. By doing so mathematics education provides an excellent context through which to explore these issues.

INTERPRETING THE DILEMMA

Rubel's case is a good example of mathematics teacher education meeting social justice education. Working with practicing teachers, Rubel provides opportunities for teachers to learn more about their school communities by studying community data. Rubel's intent is for teachers to examine visual-spatial representations of data through maps to critically and "mathematically investigate issues of fairness and equity." Better understanding the communities of their students and their families provides teachers with informed knowledge (vs. opinion) and opportunities to critically analyze assumptions they may have about students and their families. Rubel describes her dilemma as "how to support teachers in developing critical orientations about data" (p. 6). A group of teachers interprets a spatial map of rodent complaints to mean an increased number of rats exists that is further rationalized as acceptable in this low-income area. Teachers make connections among number of complaints about rats, number of rats, low-income households, cleanliness, and ability to focus on completing mathematics homework.

My own interpretation of the dilemma aligns with Rubel's. Teachers, given data and opportunities to question assumptions underlying how these data are collected and possibly interpreted, made conclusions without considering alternatives. The task, instead of opening up discussion to question data, actually appears to reinforce teachers' assumptions of low-income families. Being troubled by such an unexpected interpretation leads to Rubel's dilemma: respectfully responding as a teacher educator in ways that acknowledge the teachers' response while at the same time push their thinking to consider alternatives.

Exploring mathematics through social justice issues provides opportunities for teacher educators to support the teachers we work with in problematizing privilege, oppression and power, racism, and social positionality. Although it may seem for some that a discussion of such issues has little to do with mathematics pedagogy, Giroux (2011), following Freire (1970/2000), argued that

> pedagogy is a political and moral practice that provides the knowledge, skills, and social relations that enable students to explore the possibilities of what it means to be critical citizens while expanding and deepening their participation in the promise of substantive democracy. (p. 155)

Thus the study of mathematics can be a place to examine interpretations of data to support teachers and their students engaging as critical citizens. Bell (2010) referred to these interpretations as stories and argued that developing a counter-storytelling community means critically analyzing practices that "might reify and repeat stock stories developed by the dominant group (p. 18). Rubel finds herself in a position of deciding how to respond to the deficit stories provided by the teachers in their interpretation of the map data. Moving from these stories to Bell's counter-telling stories is a complex process that involves creating a community that values and engages in critical social justice practices (Sensoy & DiAngelo, 2012) that makes visible the relationships among knowledge, authority, and power (Freire, 1970/2000).

RESPONDING TO THE DILEMMA

It is apparent in Rubel's analysis of her dilemma and what she has learned that she does not consider the teachers to be at fault. She notices that they draw upon a cultural-deficit model to explain the data but appears to recognize that these patterns are likely invisible to teachers resulting from their own socialization. Although teachers are not to blame for these patterns of response, they do need to be responsible for interrupting and disrupting them (Dowling & Burke, 2012; Gutstein, 2006; Sensoy & DiAngelo, 2012). As Rubel notes, and I concur through my own efforts to decolonize teaching and research practices, this work requires dedicated time and support. Indigenous scholar Susan Dion (2009) introduced the concept of a perfect stranger where nonindigenous people consider themselves outside the issues or relationships involving Aboriginal people and refer to themselves as strangers. Similarly the teachers working with Rubel may consider themselves as strangers to the people and community of the data they are analyzing. Without time to critically analyze the stories they have heard about socioeconomic status, access to health and education, power, and relationships, their assumptions of those unlike themselves remain uninterrupted. Supporting teachers to examine the stories they have, the assumptions they make, what they know, and how they came to know it is crucial toward constructing their counter-telling stories.

For Rubel's context this could mean prompting teachers to explicitly examine the assumptions made toward reaching the conclusions stated. In supporting teachers to develop such a critical stance, we could ask teachers to engage in understanding the data and posing questions about the data before committing to a conclusion about the data. This could lead teachers to more informed conclusions as well as to critically reflect on the interpretations shared.

Mathematics education can be a context for teachers to recognize how their own social positionality informs how they understand and interpret data, which data they choose to accept uncritically, which data they are more critical of, and how such data and mathematical interpretations can be used to reify some social injustices and interrupt other injustices. However, as Rubel experienced, activities provided to teachers such as analyzing community data through spatial maps may only provide the teacher educator with glimpses of some of the layers of assumptions brought to the learning context. The dilemma for teacher educators is what to do next: stay within the context of our obligations to teach mathematics or move to our social-justice obligations to interrupt injustices.

Dowling and Burke (2012) argued that in order to engage in learning that doesn't reify established stereotypes or create alternative stereotypes, teachers would need to move out of the activity of teaching math toward the activity of teaching critical social justice. They suggest "we can be both mathematics educators and political activists, just not at the same time" (p. 101). Perhaps so, but if we and the teachers we work with were better prepared to engage in activities that take a critical social justice approach, we may be able to bring the study of mathematics and social justice closer together. Without opportunities to practice and study critical social-justice perspectives beyond what is experienced through everyday living or through mathematics learning, "dominant society does not prepare us to think critically about or develop the language and skills to discuss these issues" (Sensoy & DiAngleo, 2012, p. 165). As mathematics teachers and teacher educators we need more opportunities to participate in a counter-storytelling community, one where we have the courage and strength to examine our own social positionality so that we can work together within and outside the study of mathematics for a more socially just society.

REFERENCES

Bell, L. A. (2010). *Storytelling for social justice. Connecting narrative and the arts in anti-racist teaching.* New York, NY: Routledge.

Dion, S. (2009). *Braiding histories: Learning from Aboriginal people's experiences and perspectives.* Vancouver, Canada: UBC Press.

Dowling, P., & Burke, J. (2012). Shall we do politics or learn some maths today? In H. Forgasz and F. Rivera (Eds.), *Towards equity in mathematics education: Gender, culture and diversity* (pp. 87–103). New York, NY: Springer.

Freire, P. (2000). *Pedagogy of the oppressed* (M. Ramos, Trans.). New York, NY: Bloomsbury Academic. (Originally published 1970)

Giroux, H. (2011). *On critical pedagogy.* New York, NY: Bloomsbury Academic.

Gutstein, E. (2006). *Reading and writing the world with mathematics: toward a pedagogy for social justice.* New York, NY: Routledge.

Sensoy, O., & DiAngelo, R. (2012). *Is everybody equal? An introduction to key concepts in social justice education.* New York, NY: Teachers College Press.

CHAPTER 20

"LET ME BE YOUR CULTURAL RESOURCE"

Facilitating Safe Spaces in Professional Development

Anita A. Wager
University of Wisconsin-Madison

In this case I explore how, as a facilitator of a professional-development program focused on equitable mathematics pedagogy, I failed to recognize opportunities to take up the offer of one teacher to be "a cultural resource" and provide a safe space for others to engage in work that was new to them.

CONTEXT

Several years ago I designed and facilitated a semester-long professional-development (PD) seminar focused on equitable mathematics pedagogy (Hand, 2012; Wager, 2010, 2014). The PD was one of several that had been designed and facilitated by University faculty and graduate students with

Cases for Mathematics Teacher Educators, pages 463–468
Copyright © 2016 by Information Age Publishing

the goal of supporting local teachers' implementation of more equitable instruction in their mathematics classrooms. This PD was a 3-credit graduate course that met 10 times over the semester for 3 hours each session. Seventeen volunteer third- to fifth-grade teachers participated in the PD and the study. The teachers brought with them a broad range of experiences related to equity and mathematics. For example, some had grown up in privileged White middle-class areas with limited exposure to diversity whereas others encountered moral outrage against racism at an early age. With respect to diversity among the teachers, one self-identified as African American, one as Native American who grew up in an adoptive White family, and 15 self-identified as White. All of the teachers taught in ethnically, economically, and linguistically diverse elementary schools.

A central goal of the PD was to create a community for teachers to develop equitable mathematics pedagogy by identifying connections between mathematics and the three constructs of culturally relevant pedagogy articulated by Ladson-Billings (1995)—academic achievement, cultural competence, and social justice. In order to support teachers to identify these connections in their mathematics teaching, I elaborated on the indicators for each (as described by Ladson-Billings, 2001, in *Crossing Over to Canaan*) by mapping them onto research in mathematics education. This mapping was intended to provide a framework for teachers to implement equitable mathematics pedagogy. The seminar was organized into three 3-week blocks, one for each of the constructs of culturally relevant pedagogy defined through Ladson-Billings's work. During each of these blocks the teachers read articles and engaged with an activity designed to help them explore how the indicators of each construct could be incorporated into their mathematics practice. Examples of some of the indicators and connections to mathematics are set forth in Table 20.1.

THE DILEMMA

I was a graduate student facilitating the PD and studying it for my dissertation. From my first interaction with the teachers, it was clear that the work would be challenging and that I might not have been prepared for some of the issues that arose. Although I was aware that in a PD on equitable mathematics pedagogy issues of inequity (be they classism, racism, sexism, etc.) would surface, I was not really prepared to respond to these surfacings. How could I provide a safe space for teachers with deep and personal experiences with inequities in schools as well as those who were just beginning to consider these issues? I am a White woman who grew up with a deep awareness of social issues and commitment to improving equity in the mathematics education, but I was a novice in this context.

TABLE 20.1 Selected Mapping of Indicators of Culturally Relevant Pedagogy to Mathematics Education

Constructs	Indicators of Culturally Relevant Pedagogy	Indicators of Equitable Mathematics Pedagogy
Academic Achievement	The teacher knows the content, the learner, and how to teach content to the learner.	The teacher attends to and builds on student thinking by providing opportunity to explain and justify thinking.
	The teacher supports critical consciousness toward the curriculum.	The teacher allows for or encourages students to use more than one strategy.
	The teacher encourages academic achievement as a complex conception not amenable to a single, static measurement.	The teacher focuses on mathematical proficiency in a variety of strands (conceptual understanding, procedural fluency, strategic competence, adaptive reasoning, and productive disposition).
Cultural Competency	The teacher uses student culture as a basis for learning.	The teacher uses mathematics from the students' cultures in designing lessons.
	The teacher promotes a flexible use of students' local and global cultures.	The teacher uses contexts from the students' cultures in math problems and activities.
Social Justice	The teacher plans and implements academic experiences that connect students to the larger social context.	The teacher incorporates social justice into lessons or discussion.

On the first night of the PD, one of the teachers, Grace, who is African American, told me in front of the rest of the class that she knew most of what we would be talking about in the PD and stated that she would be the "cultural resource" for the class. Grace did not really explain what she meant by cultural resource, but it made me nervous that she would undermine my supposed expertise. I worried she was essentially saying that she was more qualified to facilitate the course than I was. Multiple tensions arose for me and the teachers from Grace's voluntary role as cultural resource, particularly because some of the teachers were coming to the PD to work on issues of equity in mathematics for the first time and did not know how to take up Grace's offer or her later recommendations for changing their practice.

The first assignment was to prepare a mathematical and multicultural autobiography for which I asked teachers to write about their own experiences as a learner and teacher of mathematics and their interactions with equity. I began the second session of the PD by having teachers share an excerpt of their choosing from their autobiography. The ways that the teachers initially positioned themselves in terms of equitable mathematics pedagogy was evidenced by what they shared—10 shared reflections on how their experiences with (in)equity had changed their worldview, 5 reported a history of their interactions with equity but did not reflect on how those interactions had an impact on their practice or perspectives, while 2 shared their experiences around mathematics. The differences in what the teachers shared later proved to be gauges of their comfort level in talking about equity. Grace was the last person to share and read a deeply poignant and beautiful poem about how anger over racist and other oppressive experiences had dominated her life but that at this point the anger had been replaced with disappointment that after so many years of being angry, nothing had changed. She prefaced her reading by "acknowledging all of you for your commitment to doing this kind of work" and shared that though her poem pointed fingers it was not meant to be personal and that she respected everyone present. Grace's poem set the tone for the balance of the session and, in some ways, the rest of the PD.

Later that session, in a conversation about ability grouping, Annie (one of the teachers who shared about her experiences with equity but had not yet seen the connection to teaching mathematics) described a geometry activity she had done with her fifth-grade students. In the activity students used shoeboxes to represent houses and measured various aspects of the boxes. In describing the ways in which the activity was accessible mathematically to all students, she said "any kid can draw a window" and other kids were "measuring angles." Just as Annie finished explaining the lesson, Grace asked in a tone of voice that came across as more of a challenge than a question, "How many of your students actually live in a house?" She then explained why she did not think contexts such as houses were appropriate when many students lived in apartment buildings. There was a rather heated debate between Annie and Grace, and ultimately other teachers interjected to change the direction of the conversation. I let the exchange proceed without commenting—something I regret now.

As a biracial lesbian woman who had been called overweight, Grace brought experiences of oppression in multiple targeted groups (see Allies for Change, n.d.; Stinson & Spencer, 2013). A lifetime of oppressive acts and comments (both direct and indirect) toward her had left Grace angry but also passionate about teaching others how to avoid reifying her schooling experiences with the students they taught. Her efforts to support and teach others were sometimes well received and appreciated, but sometimes she raised her voice and accused others of failing to recognize the cultural boundedness of their practice. It was when her anger came through as she endeavored to support others that I felt conflicted. As the facilitator of the PD, I was unsure how to provide a safe space for both Grace and Annie (and all the other Graces and Annies I would encounter in future PDs) to learn and develop their practice. And, quite honestly, as a doctoral student who was studying the PD for my dissertation, I was worried how it would impact my research. I was left wondering—when should I, as facilitator, intervene? How should equal status be supported in the PD?

REFLECTION ON THE DILEMMA

As a mathematics teacher educator committed to equity, I wanted to support Annie to interrogate her practice to recognize how her choice of context was not connected to many of her students' life experiences. And, further, that by selecting that context she was (un)intentionally confirming that mathematics was for some students but not others (Gutiérrez, 2012). To do this, I thought that Annie needed scaffolding and prompting so she could discover for herself what her choices implied for students' access to learning and opportunities to see mathematics reflected in their lives. On the other hand, I wanted to validate Grace's immediate recognition that the activity was culturally biased and also acknowledge that her frustration with Annie's choice was justified. I too was stunned that Annie did not see houses as a context that was problematic but wanted to ease Annie into that awareness.

Then: My Response to the Situation

My response at the time was avoidance. As facilitator, I was attempting to let the teachers guide the discussion, and in doing so I validated neither Grace's nor Annie's perspectives. Even worse than my in-the-moment lack of response to the geometry house project was my inability to take Grace up on her offer to be our culture resource in an authentic way. I wasn't sure how to do it, and I wasn't sure how it would impact my study.

Now: What I Wish I Had Done

As the only African American woman, Grace offered a wonderful gift— herself as a cultural resource to the rest of the teachers (and me) in nego- tiating new territory of examining how mathematical practices could be equitable. She brought to the group a perspective none of us had—that as both a student and teacher of mathematics in a racist and inequitable system. I wish I had taken her up on the offer by collaborating with her to plan the discussions for the PD and providing a space for her to provide feedback to others. I could have made space for Grace to share her experi- ences with inequity as a student and how she approached equitable teach- ing. I also wish I had talked to her about how to scaffold the teachers who were engaging around the idea of equitable mathematics pedagogy for the first time. This would not have been hard to do. I had observed in Grace's classroom and saw how she scaffolded children's learning—she could certainly do the same with the teachers in our group.

I also wish that I had thought more purposefully about how to differentiate experiences for the teachers in the PD. I had access to interviews that teachers conducted with each other around the three constructs of culturally relevant pedagogy and the connection to mathematics, and multicultural autobiographies to use as data to plan differentiated instruction based on where teachers were. By using that information, I could have provided safe spaces for those who were relatively new to discussions of equity to unpack their understandings without fear of saying the wrong thing or being accused of being inequitable. In fact, I could have had Grace facilitate the group of teachers who were more experienced in equitable mathematics, and I could have facilitated the other. Although I regret not having taken this action at the time, I learned from the experience and have used autobiographies to purposefully engage practicing teachers in PD and prospective teachers in my methods courses. As always, I learn as much if not more from my students as they from me.

REFERENCES

Allies for Change. (n.d.) Glossary of terms. Retrieved from http://www.alliesfor-change.org/documents/Glossary.pdf

Gutiérrez, R. (2012). Context matters: How should we conceptualize equity in mathematics education? In B. Herbel-Eisenmann, J. Choppin, D. Wagner, & D. Pimm (Eds.), *Equity in discourse for mathematics education: Theories, practices, and policies* (pp. 28–50). New York, NY: Springer.

Hand, V. (2012). Seeing culture and power in mathematical learning: Toward a model of equitable instruction. *Educational Studies in Mathematics, 80,* 233–247.

Ladson-Billings, G. (1995). But that's just good teaching! The case for culturally relevant pedagogy. *Theory Into Practice, 34,* 159–165.

Ladson-Billings, G. (2001). *Crossing over to Canaan: The journey of new teachers in diverse classrooms.* San Francisco, CA: Jossey-Bass.

Stinson, D. W., & Spencer, J. A. (Eds.). (2013). Privilege and oppression in the mathematics preparation of teacher educators [Special issue]. *Journal of Urban Mathematics Education, 6*(1).

Wager, A. A. (2010). Teacher positioning and equitable mathematics pedagogy. In M. Q. Foote (Ed.), *Mathematics teaching & learning in K–12: Equity and professional development* (pp. 77–92). New York, NY: Palgrave.

Wager, A. A. (2014). Noticing children's participation: Insights into teacher positionality toward equitable mathematics pedagogy. *Journal for Research in Mathematics Education, 45,* 312–350.

COMMENTARY 1

OPENING SPACES
IN MATHEMATICS TEACHER
EDUCATION

A Commentary on Wager's Case

Corey Drake
Michigan State University

MY POSITIONALITY

I am a mathematics teacher educator who has taught elementary mathematics methods to prospective teachers for many years. Over the past several years, I have increasingly tried to integrate topics related to equity and children's home- and community-based funds of knowledge into my mathematics methods courses (see, e.g., Turner, Drake, Roth McDuffie, Aguirre, Bartell, & Foote, 2012). I have taught these methods courses at several different large state universities with groups in which 90% or more of the students identify as female and White, as do I. My students also tend to be in their early 20s, born and raised in small towns or suburbs in the Midwestern United States, neither of which apply to me. In many other ways, large and small, visible and invisible, my students are both similar to and different

Cases for Mathematics Teacher Educators, pages 469–473
Copyright © 2016 by Information Age Publishing
All rights of reproduction in any form reserved.

from me and our experiences with mathematics, with schooling, and with teaching and learning that have unfolded in a country and education system that privileges certain identities and oppresses others. Because of our similarities and differences and because we each bring important perspectives, I try to facilitate classrooms that are open spaces for sharing, questioning, and learning. This begins with the first day of class, when students write their "mathematics stories"—their stories of themselves as mathematics learners and teachers—stories they return to at the end of the semester (LoPresto & Drake, 2005).

Given my commitment to creating these open spaces—and also given that I still have a lot to learn in this area, particularly when it comes to opening spaces for topics related to race, class, equity, and social justice—I read Anita's case with great interest. And, although our contexts of working with prospective and practicing teachers are somewhat different, I appreciated the dilemmas raised by this case—and also appreciated Anita's willingness to share and reflect on those dilemmas.

INTERPRETING THE DILEMMA

In my reading of the case, there are two interrelated dilemmas or tensions at work, both related to the preparation of and knowledge base for mathematic teacher educators to create and sustain these open spaces. One dilemma concerns the knowledge mathematics teacher educators need in order to anticipate and respond to the *content* of the contributions of adult participants in contexts in which race and class and mathematics learning and teaching are all the focus of discussion. The second dilemma is related to the interpersonal preparation and knowledge needed to *interact* productively across race, power, and potentially other divides.

In terms of the first dilemma, I was repeatedly reminded while reading the case of how important it is to understand—for any topic—where learners (in this case, teachers) are in their understandings of the topic, as well as the multiple pathways they might take to develop more advanced understandings of that topic. "Facilitating safe spaces" requires, in part, an understanding of the journeys participants will take within those spaces and what the boundaries of those spaces could be. Anita notes that using Grace's contributions to scaffold the learning of the other teachers in the group was something she would have liked to do but was difficult to do in the moment. I argue that knowing something more about teacher development related to mathematics and social justice can reduce the challenge of this dilemma for teacher educators.

The field of mathematics education knows a great deal about some kinds of pathways—for example, in the cognitive realm, a great deal of research has focused on identifying children's entry points into understanding whole-number operations and the ways in which they develop more efficient strategies and deeper understandings (e.g., Carpenter, Fennema, Franke, Levi, & Empson, 1999; Fosnot & Dolk, 2001). A parallel body of research has investigated how teachers learn about the development of children's mathematical thinking, how teachers use that knowledge to support children's development, and even how teacher educators can support teachers in acquiring this knowledge.

We know less—and need to know more—about how teachers, prospective and practicing, access and engage with ideas related to equity, culture, funds of knowledge, or social justice, though we are beginning to develop a research base related to their various and multiple entry points into these ideas (e.g., Taylor, 2012; Turner & Drake, 2016; Turner et al., 2012; Wager, 2012). We know still less about how teacher educators learn to facilitate teachers' learning in these areas—a gap that this current volume is designed to help address. I acknowledge here that understanding the pathways along which teachers learn about and experience the intersections of identity and mathematics is complex, as the racialized and gendered mathematical experiences of individuals are not only cognitive, but also social, emotional, personal, and located in contexts that are laden with inequities in privilege and power. Nonetheless, I argue that the dilemma that Anita has shared calls attention to the need for mathematics teacher educators to *know* more—and have access to more information—about the range of ways in which teachers may enter into and engage in learning about equity and mathematics education.

In Anita's case, I would be much more likely to engage with Grace's offer if I knew something about how to connect the shared experiences with the overall goals of the course and how Grace's contributions could be used to support the learning of participants at various points in their pathway toward greater understanding.

At the same time, as Anita also emphasizes, supporting adult learners in learning to teach mathematics in ways that attend to culture, race, and class is only about the content of this learning but also about the emotional, interactive, and identity aspects of this learning. This dilemma is further complicated by differences in status based on roles, positions, and life experiences—and, in Anita's case—even further complicated by concerns about shifts in status that might result from particular interactions. To the extent that we as a field can develop understandings of the multiple pathways of teacher learning that attend to both content and identity—and highlight the ways in which teacher learning happens in interaction with colleagues, with students, and with teacher educators—the better prepared we as mathematics teacher educators will be to respond in productive ways to participants' contributions.

RESPONDING TO THE DILEMMA

So, how would I have responded in this case? I am not sure, but I imagine that in the moment I might have responded in a very similar way. Although I have learned over the years to anticipate many of the things that teachers and students will say, unanticipated comments about difficult topics still leave me unsure how to respond and, most often, silent. Moreover, I worry not only about the contribution to which I have replied with silence, but also about the comments, questions, and suggestions that I have unintentionally silenced, that remain unspoken because the classroom space was not open enough to support those contributions.

How would I like to have responded in this case? Similar to Anita, I would hope to be able to respond in ways that build on the multiple strengths of each participant—drawing on an understanding of the shared pathways of teacher development and the uniquely individual experiences of each participant. I would like to be able to anticipate the kinds of resources and experiences that teachers bring to these contexts and to therefore anticipate how to respond. Fundamentally, I would like to be able to take a strengths-based and funds-of-knowledge approach to working with adult learners as well as children.

What would I need to be able to do to respond in this way? Responding gracefully, directly, and in ways that are productive for the learning of the individual and the group (including the mathematics teacher educator) requires practice, self-awareness, and a willingness to engage—important qualities that we do not typically develop in the regular course of graduate school or faculty life. Similar to the ways in which the field is moving toward preparing teachers in ways that involve repeated practice, simulations (Dieker, Rodriguez, Lignugaris/Kraft, Hynes, & Hughes, 2014), rehearsals (Lampert et al., 2013), and role-plays (Gutiérrez, 2013) to address the content, pedagogical content, and political knowledge (Gutiérrez, 2013) needed for teaching mathematics, new (and experienced) mathematics teacher educators need opportunities to practice, rehearse, and role-play the practices and interactions needed for teaching mathematics teachers. Only by engaging in and practicing this work—and engaging with those who are both similar to and different from ourselves in many ways—can we learn the interactional skills and practices necessary for opening spaces that value the contributions and support the learning of all participants and recognize that learners, particularly adult learners, often have much to teach. This work is critically important, and I appreciate the contributions of this case—and of this book more generally—in providing tools for advancing the work.

REFERENCES

Carpenter, T. P., Fennema, E., Franke, M. L., Levi, L., & Empson, S. B. (1999). *Children's mathematics: Cognitively guided instruction*. Portsmouth, NH: Heinemann.

Dieker, L. A., Rodriguez, J. A., Lignugaris/Kraft, B., Hynes, M. C., & Hughes, C. E. (2014). The potential of simulated environments in teacher education: Current and future possibilities. *Teacher Education and Special Education 37*, 21–33.

Fosnot, C. T., & Dolk, M. (2001). *Young mathematicians at work: Constructing multiplication and division*. Portsmouth, NH: Heinemann.

Gutiérrez, R. (2013). Why (urban) mathematics teachers need political knowledge. *Journal of Urban Mathematics Education, 6*(2), 7–19.

Lampert, M., Franke, M. L., Kazemi, E., Ghousseini, H., Turrou, A. C., Beasley, H., Cunard, A., & Crowe, K. (2013). Keeping it complex: Using rehearsals to support novice teacher learning of ambitious teaching. *Journal of Teacher Education, 64*, 226–243.

LoPresto, K., & Drake, C. (2005). What's your (mathematics) story? *Teaching Children Mathematics, 11*, 266–271.

Taylor, E. V. (2012). Supporting children's mathematical understanding: Professional development focused on out-of-school practices. *Journal of Mathematics Teacher Education, 15*, 271–291.

Turner, E. E., & Drake, C. (2016). A review of research on prospective teachers' learning about children's mathematical thinking and cultural funds of knowledge. *Journal of Teacher Education, 67*, 32–46.

Turner, E. E., Drake, C., Roth McDuffie, A., Aguirre, J. M., Bartell, T. G., & Foote, M. Q. (2012). Promoting equity in mathematics teacher preparation: A framework for advancing teacher learning of children's multiple mathematics knowledge bases. *Journal of Mathematics Teacher Education, 15*, 67–82.

Wager, A. A. (2012). Incorporating out-of-school mathematics: From cultural context to embedded practice. *Journal of Mathematics Teacher Education, 15*, 9–23.

NOSOTRAS SPACES: COBUILDING TRANSFORMATIONAL BRIDGES

A Commentary on Wager's Case

Carlos A. LópezLeiva
University of New Mexico

MY POSITIONALITY

The dilemma presented by Wager regarding the lack of agreement on issues of equity during intercultural comunication is a familiar topic to me in several ways. First, personally, I grew up as a member of mainstream society in Guatemala, and I did not have a racialized experience of overt discrimination as many indigenous people there do. Through direct observation and experiences with my indigenous friends and students, I became aware of discrimination, but it was not until I came to the U.S. that cultural and racial differences became my own lived experience. Through this transition, not only did I become an "alien" and a Hispanic man but also

---

Cases for Mathematics Teacher Educators, pages 475–478
Copyright © 2016 by Information Age Publishing
475

by looking as I do, some people have considered me and made me feel "different"—though we all are—and, at times, through glances, reactions, and/ or avoidances, have also marked me as a "dangerous" person, such as a mugger or a thief. This picture becomes even more complicated, as I speak my second language, English, with an accent. (LópezLeiva, 2012, p. 143)

Second, in my work with preservice teachers (PSTs) and graduate students, I have engaged with them in reflective processes on issues of language, race, and culture. These students' responses have ranged from seriously embracing to seriously resisting and mocking these issues. For example, in conversations on readings about racism and cultural marginalization, some students have commented: "I was discriminated against by being brown and ugly. As a teacher, I want to interrupt similar situations with my students." "I don't get why some people are always so angry and can't think beyond race. I'm not a racist!" Such comments were made in the same group; obviously our familiarity with issues of equity seems related to our life experiences. A transformative process into committing to support equity issues resides at the intersection of knowing, caring, and acting (Banks, 2013). As a Latino instructor, my work is often perceived as pursuing a personal agenda and the result of anger through which I want to "take revenge" on students from privileged backgrounds. Having become aware of such tensions, I strive to focus on supporting dialectical classroom interactions and emphasizing our common cultural and evolving nature as human beings as well as the generative power of knowing, caring, and acting.

INTERPRETING THE DILEMMA

I believe that in the dilemma described by Wager, the tension arose by asking teachers, especially Annie, to shift into thinking about and learning from the community—those who have experienced oppression, those whose voices have been silenced or ignored—in order to implement equitable pedagogic approaches in mathematics. The inclusion of this knowledge, a transformative intellectual-knowledge approach, in the curriculum is a complex and challenging process for teachers (Sleeter, 2005). It represents a shift in the resources that teachers access and make use of in order to plan for and teach mathematics. It is like starting anew. In Annie's case, she proposed the past successful idea of using shoeboxes to represent houses and learn about geometry. Within the context of the PD on the topic of culturally responsive mathematics education, however, the idea also needed to be relevant to the particular learners' experiences and knowledge. Grace— as an African American woman and insider of a target community—also presented her knowledge built on her familiarity with the community. In my limited knowledge of the circumstances, I believe that the dilemma emerged when differences between two perspectives were explicitly marked and placed in conflict rather than serving as the valuable foundation for a dialogue that could lead to increased mutual knowledge. As Allan Johnson (2006) has argued, the trouble is not in the differences themselves, but in how people use "difference to include or exclude, reward or punish, credit or discredit, elevate or oppress, value or devalue, leave alone or harass" (p. 16). Grace's and Annie's perspectives on using learners' housing experiences to teach geometry became a problem not because they represented different perspectives but because these perspectives were antagonistic and were not negotiated and processed as sources for learning from each other.

RESPONDING TO THE DILEMMA

These circumstances make me think about the need for developing a third space between Grace and Annie, that is, a dialectical communication that would bridge these teachers' perspectives to create mutual understanding. "The pact of interpretation is never simply an act of communication between the I and the You designated in a statement," as Bhabha (1994) argued; rather, "the production of meaning requires that these two places be mobilized in the passage through a Third Space" (p. 53). In other words, these teachers' positions became a dilemma of "us" versus "them" or *nos/otras*, to speak with Anzaldúa's terms (Anzaldúa & Keating, 2000). Instead, Anzaldúa asserted that on issues of diversity we need to come to a "we" or unified *nosotras* stance: "The future belongs to those who cultivate sensitivities to differences and who use these abilities to forge a hybrid consciousness that transcends the 'us' versus 'them' mentality and will carry us into a nosotras position bridging the extremes of our cultural realities" (p. 254).

From this perspective, I believe that Annie and Grace have much to learn from each other, just as we are learning from them through Wager's account. Various projects promote bridge building across cultures and help teachers engage in transformative intellectual learning. For example, through shared storytelling and reflection, Schmidt (1999) and Wlazlinski and Cummins (2013) promoted communication among people from different cultures to learn from and understand each other better. In my math methods class, I am learning with the PSTs that by conversing with their young students, PSTs learn about student interests, life experiences, and preferences, so they design math tasks that address such knowledges. Further, the ABCs communication model (Finkbeiner, 2006) has been also a useful process to promote bridge building among the students that I work with.

From a transformative intellectual-knowledge perspective, I understand the relevance of Grace's argument for building on learners' knowledge and experiences. Furthermore, from a dialectical learning perspective, I think that teachers are cultural resources to learn from one another, and the community itself could be the main source of knowledge. Regarding mathematics, I think that having students investigate not only the geometry but also the social-justice issues related to the types of housing (single-family houses, apartment buildings, housing projects, etc.) that people have access to might also be a useful context to use mathematics to read, become aware of, and stand against such inequities in our worlds (Gutstein, 2003).

As a result, when I think about creating safe spaces to promote teacher development, I believe that safe spaces encourage *nosotras* stances toward transformative intellectual knowledge in mathematics curricula. Nevertheless, the critical nature of these spaces is inherently tensional. In these learning spaces tension and discomfort are expected as part of promoting awareness, reflection, personal growth, and social change. Throughout this process we are to learn about ourselves and others, for "the 'other' can be a part of oneself in an inner dialogue or it can be another person one is having a dialogue with" (Finkbeiner, 2006, p. 35). I believe that the questions that Wager asks in her case study reflect this tensional, evolving process of working towards equity in mathematics education.

REFERENCES

Anzaldúa, G., & Keating, A. (2000). *Interviews.* London, England: Psychology Press.
Banks, J. A. (2013). *An introduction to multicultural education* (5th ed). Boston, MA: Pearson.
Bhabha, H. (1994). *The location of culture.* London, England: Routledge.
Finkbeiner, C. (2006). Construction of third space: The principles of reciprocity and cooperation. In P. Ruggiano Schmidt & C. Finkbeiner (Eds.), *The ABC's of cultural understanding and communication: National and international applications* (pp. 19–42). Greenwich, CT: Information Age.
Gutstein, E. (2003). Teaching and learning mathematics for social justice in an urban, Latino school. *Journal for Research in Mathematics Education, 34,* 37–73.
Johnson, A. G. (2006). *Privilege, power, and difference.* New York, NY: McGraw Hill.
LópezLeiva, C. A. (2012). Realizations through my own skin. In N. S. Maldonado & L. L. DiBello (Eds.), *Hispanic/Latino American families in the United States: An introduction for educators and care providers* (pp. 83, 142–143). Olney, MD: Association for Childhood Education International.
Schmidt, P. R. (1999). Know thyself and understand others. *Language Arts, 76,* 332–340.
Sleeter, C. (2005). *Un-standardizing curriculum: Multicultural teaching in the standards-based classroom.* New York, NY: Teachers College Press.
Wlazlinski, M. L., & Cummins, J. (2013). Using family stories to foster parent and pre-service teacher relationships. In E. M. Olivos, O. Jimenez-Castellanos, & A. M. Ochoa (Eds.), *Bicultural parent engagement* (pp. 39–57). New York, NY: Teachers College Press.

SEEING THE PROBLEM BEFORE ATTEMPTING TO SOLVE IT: THE ROLE OF NOTICING SOCIOPOLITICAL NARRATIVES IN EQUITY-FOCUSED WORK

A Commentary on Wager's Case

Jennifer M. Langer-Osuna
Stanford University

Wager describes her experience facilitating professional-development work meant to engage teachers in discussion about teaching mathematics for equity. Her case focuses primarily on the complex experiences and tense interactions of two teachers, Grace, a queer woman of color, and Annie, a White woman. Wager frames the event as a missed opportunity for supporting productive dialogue about equitable mathematics teaching. I am

Cases for Mathematics Teacher Educators, pages 479–483
Copyright © 2016 by Information Age Publishing

grateful for her willingness to share this experience and for the opportunity to reflect on this case as a mathematics teacher educator with my own share of missed opportunities while attempting to do similar work.

MY POSITIONALITY

I was born and raised in Miami, Florida, as the daughter of parents who emigrated from Cuba during the communist revolution. Because of the particular role Miami played in that history, I grew up in a city where my identity, bilingualism, and culture were reflected back to me on a regular basis. This sense of belonging was particularly salient while attending an all-girl high school that was almost completely made up of Latinas. All social cliques— the cool kids, the nerdy kids, the rebel kids– reflected my cultural and gender identity back to me. As one of the nerdy girls, I experienced myself in mathematics classes as powerful, engaged, and unremarkable for being so. I graduated and went away to college in Pennsylvania with a strong mathematics identity and plans to double major in mathematics and psychology.

My undergraduate experience was a shock to my identity. Claims related to belonging, to being "the girl next door," were simply disallowed. While in the social sphere of friendships and romances these interactions were blatant, in the academic sphere, they took on particular patterns, especially in the mathematics department. I entered my undergraduate mathematics classes with a positive orientation to the discipline. I participated, enthusiastically and competently, as if I belonged in that space and would thrive in it. However, those around me took up my engagement as odd, at best. I could see clearly that other students in my courses, mostly Anglo-American men, were navigating the space as if it were perfectly normal that they were there. I recognized the experience because it had long been my own. And I recognized as obvious that I did not have a way to make those same claims and have them recognized in this community. Increasingly I sensed my alienation from mathematics. Though it continued to be an area of deep intellectual curiosity and one I academically excelled in, I ultimately left the mathematics department. It was painful to be there.

I focused instead on my increasing role in a psychology research lab where students were afforded the intellectual autonomy to explore and offer ideas. The professor fostered a creative intellectual space where those who entered had a fundamental right to contribute. I ultimately followed in his footsteps and became a professor with a learning lab of my own. However, along the way to my own career in research, the theories of learning I first encountered seemed inadequate to explain how learning took place in spaces like my undergraduate mathematics classes. I eventually turned to theories that framed learning as a process of becoming.

Theoretically, I've become interested in the construct of *belonging* and in locating belonging in classroom interactions. Increasingly, my work points to belonging as linked to relationships of authority; that is, we belong in spaces where we have the right to contribute and where our contributions are taken up by others in ways that position us with competence. We experience *un*-belonging in spaces where we struggle to gain and maintain the floor long enough to contribute or where our contributions are repeatedly ignored, discounted, or rejected. Furthermore, these local interactions draw on broader sociopolitical narratives. It wasn't coincidence that I had particular experiences in Miami and others in Pittsburgh or in an all-girl school versus a tech-oriented and male-dominated university. The classrooms I experienced as a learner drew from particular narratives about gender, ethnicity, and language that framed assumptions about mathematics and me.

INTERPRETING THE DILEMMA

In order to engage productively in discussions about equity, teachers and teacher educators must first recognize the sociopolitical narratives that shape their experiences in the world and their locations within those narratives. Those of us who navigate society unmarked may be blind, often willfully so, to the experiences of others. Teachers from culturally dominant backgrounds who are committed to equity-focused work need to start by making these invisible narratives visible because what emerges into view *is* the focus of equity work.

In the focal case, Grace offered to be a "cultural resource" by sharing her knowledge about teaching students from marginalized communities. In doing so, Grace positioned herself with intellectual and social authority, which made others in the room uncomfortable. Grace and Annie both expressed discomfort in ways that reflected their histories and positional identities. Grace expressed deep frustration. She did so not only in her poem, which discussed her relationship with anger over time, but also in interactions with other PD participants through a raised voice and a willingness to accuse others of failing to recognize privilege. Grace's response was reasonable. Her experiences with marginalization were being marginalized in a space intentionally designed to confront issues related to inequality. Annie's discomfort arose from her struggle to notice how her pedagogical choices were related to her privilege. She was not yet in a place to recognize what insights Grace's experiences could offer to Annie as a classroom teacher.

RESPONDING TO THE DILEMMA

Wager wrote that, of the 17 participating teachers, 10 claimed that experiences with inequality shifted their worldviews. Annie had not yet made this claim. She was among the 5 teachers who were still struggling to make connections between their experiences with inequality and their perspectives or teaching practice. I agree with Wager when she reflects that different groups might have been formed to structure different goals among teachers. In my own work, I have explored activities to help make invisible sociopolitical narratives visible to educators. Recently, I used a satirical blogpost entitled, "What if we talked about monolingual White children the way we talk about low-income children of color?" (Flores, 2015) to highlight the absurdity of how we frame bilingual and bidialectical children of color from deficit perspectives. Experiencing the interpretation of dominant communities from a deficit perspective, even in satire, exposes the power and bias of the existing narratives that are otherwise difficult to notice. Once teachers establish common ground on the dominant narratives that shape the lives of students, they can more productively discuss the implications for their own classrooms.

This awareness must emerge first because it clarifies what social justice work is aiming toward and what aspects of that work are in teachers' locus of control. For instance, an awareness of the dominant narratives that shape the mathematical lives of bilingual Latinos can lead teachers to intentionally establish local classroom norms that center diverse linguistic competencies. Turner, Dominguez, Maldonado, and Empson (2013) described an after-school mathematics club that enabled students of differing language backgrounds to share and debate mathematical ideas. This counters dominant narratives that position the Spanish language as marginal to the intellectual work of the mathematics classroom such that English learners have little opportunity, or perceived rights, to make mathematical contributions. Further, monolingual English-speaking students rarely get the opportunity to consider the ideas of their Spanish-speaking peers. The afterschool math club utilized the teacher and the other bilingual students in the club as resources for establishing a dynamically bilingual space accessible to their entire sense-making community.

In my own work with teachers, we have watched videos of classrooms in order to notice how students engage with one another around problems. We watch video with the sound off to first notice spatial markers of central and marginalized engagement. Engaged students lean in toward and gaze at one another, have access to the activity space, and are generally oriented toward one another and the work. Marginalized students are spatially blocked from the activity space, are less able to capture their peers' gaze, and are generally oriented away from the work. We then watch the video again with sound on to notice what talk was occurring during those different spatial arrangements. We discuss the ways students and teachers use talk to establish relationships of authority and how these interactions served to enable or constrain possibilities for engagement.

Culturally relevant pedagogy is fundamentally about allowing children to feel belonging in mathematics. That sense of belonging does not solely exist in the task's story context; it also exists in how students' voices are reflected back to them. A relevant context, such as families living in apartments, makes the mathematical problem more accessible because a student can more easily visualize the space, imagine it in her mind's eye. But as she engages in the problem with others around her, she also needs to hear her ideas taken seriously as if they mattered to the work being accomplished in the classroom.

REFERENCES

Flores, N. (2015, July 6). *What if we talked about monolingual White children the way we talk about low-income children of color?* [Weblog post]. Retrieved from https://educationallinguist.wordpress.com/2015/07/06/what-if-we-talked-about-monolingual-white-children-the-way-we-talk-about-low-income-children-of-color/

Turner, E., Dominguez, H., Maldonado, L., & Empson, S. (2013). English learners' participation in mathematical discussion: Shifting positionings and dynamic identities. *Journal for Research in Mathematics Education, 44*, 199–234.

ABOUT THE EDITORS

Dorothy Y. White is an Associate Professor of Mathematics Education in the College of Education at the University of Georgia. Her research focuses on equity and culture in mathematics education by examining ways to: prepare mathematics teachers of diverse student populations, build and support teacher learning communities, and develop models of collaborative mathematics planning. She serves on several committees for national and local organizations including the National Council of Teachers of Mathematics, the Association of Mathematics Teacher Education, and the Clarke County School District's Parent Advisory Board. She teaches undergraduate and graduate mathematics methods courses and provides professional development in mathematics for classroom teachers.

Sandra Crespo is a Professor of Mathematics Education and Teacher Education at Michigan State University. She researches learning and teaching practices that disrupt and redistribute power dynamics in the mathematics classroom. She uses design experiments at different grain sizes to introduce students, prospective teachers, and practicing teachers to empowering learning and teaching practices. Her goal is to learn with and from them what it takes to relearn mathematics and teach mathematics in ways that affirm and encourage traditionally marginalized students to experience mathematics as a purposeful and meaningful human activity. She teaches undergraduate and graduate mathematics education courses and provides professional development to teacher learning communities.

Cases for Mathematics Teacher Educators, pages 485–486
Copyright © 2016 by Information Age Publishing

Marta Civil is a Professor of Mathematics Education and the Roy F. Graesser Chair in the Department of Mathematics at The University of Arizona. Her research looks at cultural, social, and language aspects in the teaching and learning of mathematics; connections between in-school and out-of-school mathematics; and parental engagement in mathematics. She has led several funded projects working with children, parents, and teachers, with a focus on developing culturally responsive learning environments, particularly with Latina/o communities. She usually teaches mathematics courses for prospective and practicing elementary teachers, as well as courses in mathematics education research for graduate students.

Imani Masters Goffney is an Assistant Professor of Mathematics Education at the University of Maryland–College Park. Her research focuses on mathematics instruction and on interventions designed to improve its quality and effectiveness, especially for students not traditionally served well by our educational system. In particular, she studies the ways in which teachers use mathematical knowledge for teaching in equitable ways. She takes up this research by serving as principal investigator for projects funded by the National Science Foundation and the Greater Texas Foundation. She teaches undergraduate and graduate mathematics education courses.

Mathew D. Felton-Koestler is an Assistant Professor of Mathematics Education in the Department of Teacher Education at Ohio University. His primary research focus is on prospective and practicing teachers' beliefs about connecting mathematics to real-world contexts, and in particular to social and political issues. He also studies mathematical modeling, especially as a means of supporting teachers in connecting mathematics to students' out-of-school interests and to social justice issues. He has taught content courses for prospective elementary teachers, frequently works with teachers in professional development contexts, and teaches methods courses for prospective elementary and middle grades teachers.

Joi A. Spencer is an Associate Professor of Mathematics Education at the University of San Diego. Her research has examined the mathematics learning opportunities of students in some the poorest schools in the country. Joi was a member of the National Science Foundation's Diversity in Mathematics Education Center for Learning and Teaching and a recipient of an American Education Research Association (AERA) Dissertation Year Fellowship for her dissertation, Balancing the Equation: African American Students' Opportunities to Learn Mathematics with Understanding in Two Central City Schools. She teaches mathematics methods courses to prospective secondary mathematics teachers and leads teams of mathematics and STEM teachers in improving their practice through Lesson Study.